Lecture Notes in Artificial Intelligence

Edited by J. G. Carbonell and J. Siekmann

Subseries of Lecture Notes in Computer Science

Marcos Faundez-Zanuy   Léonard Janer
Anna Esposito   Antonio Satue-Villar
Josep Roure   Virginia Espinosa-Duro (Eds.)

Nonlinear Analyses and Algorithms for Speech Processing

International Conference
on Non-Linear Speech Processing, NOLISP 2005
Barcelona, Spain, April 19-22, 2005
Revised Selected Papers

 Springer

Series Editors

Jaime G. Carbonell, Carnegie Mellon University, Pittsburgh, PA, USA
Jörg Siekmann, University of Saarland, Saarbrücken, Germany

Volume Editors

Marcos Faundez-Zanuy
Léonard Janer
Antonio Satue-Villar
Josep Roure
Virginia Espinosa-Duro
Escola Universitària Politècnica de Mataró, Spain
E-mail: {faundez,leonard,satue,roure,espinosa}@eupmt.es

Anna Esposito
Second University of Naples, Department of Psychology
Caserta, Italy
and
International Institute for Advanced Scientific Studies
Vietri sul Mare, Italy
E-mail: iiass.annaesp@tin.it

Library of Congress Control Number: 2005938710

CR Subject Classification (1998): I.2.7, J.5, C.3

ISSN   0302-9743
ISBN-10   3-540-31257-9 Springer Berlin Heidelberg New York
ISBN-13   978-3-540-31257-4 Springer Berlin Heidelberg New York

Springer is a part of Springer Science+Business Media

springer.com

© Springer-Verlag Berlin Heidelberg 2005
Printed in Germany

Typesetting: Camera-ready by author, data conversion by Scientific Publishing Services, Chennai, India
Printed on acid-free paper   SPIN: 11613107   06/3142   5 4 3 2 1 0

Preface

We present in this volume the collection of finally accepted papers of NOLISP 2005 conference. It has been the third event in a series of events related to Nonlinear speech processing, in the framework of the European COST action 277 "Nonlinear speech processing".

Many specifics of the speech signal are not well addressed by conventional models currently used in the field of speech processing. The purpose of NOLISP is to present and discuss novel ideas, work and results related to alternative techniques for speech processing, which depart from mainstream approaches.

With this intention in mind, we provide an open forum for discussion. Alternate approaches are appreciated, although the results achieved at present may not clearly surpass results based on state-of-the-art methods.

The call for papers was launched at the beginning of 2005, addressing the following domains:

1. Non-Linear Approximation and Estimation
2. Non-Linear Oscillators and Predictors
3. Higher-Order Statistics
4. Independent Component Analysis
5. Nearest Neighbors
6. Neural Networks
7. Decision Trees
8. Non-Parametric Models
9. Dynamics of Non-Linear Systems
10. Fractal Methods
11. Chaos Modeling
12. Non-Linear Differential Equations
13. Others

All the main fields of speech processing are targeted by the workshop, namely:

1. Speech Coding: The bit rate available for speech signals must be strictly limited in order to accommodate the constraints of the channel resource. For example, new low-rate speech coding algorithms are needed for interactive multimedia services on packet-switched networks such as the evolving mobile radio networks or the Internet, and nonlinear speech processing offers a good alternative to conventional techniques. Voice transmission will have to compete with other services such as data/image/video transmission for the limited bandwidth resources allocated to an ever growing, mobile network user base, and very low bit rate coding at consumer quality will see increasing demand in future systems.

2. Speech Synthesis: New telecommunication services should include the capability to produce speech in a "natural way"; to this end, much research is required for improving the voice quality of text-to-speech and concept-to-speech systems. Enriched output signals of self-excited nonlinear feedback oscillators are expected to permit matching synthetic voices better to human voices. In this area, the COST Action has build on results obtained in signal generation by COST Action 258 "The Naturalness of Synthetic Speech".

3. Speaker Identification and Verification: Security in transactions, information access, etc. is another important question to be addressed in the future, and speaker identification/verification is perhaps one of the most important biometric systems, because of its feasibility for remote (telephonic) recognition without additional hardware requirements. This line of work has built on results from COST Action 250 "Speaker Recognition in Telephony".

4. Speech Recognition: Speech recognition plays an increasingly important role in modern society. Nonlinear techniques allow us to merge feature extraction and classification problem and to include the dynamics of the speech signal in the model. This is likely to lead to significant improvements over current methods which are inherently static.

In addition, other topics have been discussed in detail, such as Voice Analysis, where the quality of the human voice is analyzed (including clinical phonetics applications) and where techniques for the manipulation of the voice character of an utterance are developed, and Speech Enhancement, for the improvement of signal quality prior to further transmission and/or processing by man or machine.

After a careful review process, 33 papers were accepted for publication, including the contribution of two invited speakers. A total of 15 sessions containing 43 papers were accepted for presentation, covering specific aspects like speaker recognition, speech analysis, voice pathologies, speech recognition, speech enhancement and applications.

NOLISP 2005 was organized by the Escola Universitària Politècnica de Mataró, for the European COST action 277, "nonlinear speech processing". Sponsorship was obtained from the Spanish Ministerio de Educación y Ciencia (TEC2004-20959-E), and Fundació Catalana per a la Recerca i la Innovació. We also want to acknowledge the support of ISCA, EURASIP and IEEE.

We would like to express our gratitude to the members of the NOLISP organizing committee, and to all the people who participated in the event (delegates, invited speakers, scientific committee). The editors would like to address a special mention to the people who helped in the review process as special or additional reviewers.

Finally, we would like to thank Springer, and specially Alfred Hofmann, for publishing this post-conference proceedings.

September 2005

Marcos Faundez-Zanuy, EUPMt
Léonard Janer, EUPMt
Anna Esposito, IASS
Antonio Satue-Villar, EUPMt
Josep Roure-Alcobé, EUPMt
Virginia Espinosa-Duró, EUPMt

Organization

Scientific Committee:

Frédéric Bimbot	IRISA, Rennes, France
Gérard Chollet	ENST, Paris, France
Marcos Faundez-Zanuy	EUPMt, Barcelona, Spain
Hynek Hermansky	OGI, Portland, USA
Eric Keller	University of Lausanne, Switzerland
Bastiaan Kleijn	KTH, Stockholm, Sweden
Gernot Kubin	TUG, Graz, Austria
Jorge Lucero	University of Brasilia, Brasil
Petros Maragos	Nat. Tech. Univ. of Athens, Greece
Stephen Mclaughlin	University of Edimburgh, UK
Kuldip Paliwal	University of Brisbane, Australia
Bojan Petek	University of Ljubljana, Slovenia
Jean Schoentgen	Univ. Libre Bruxelles, Belgium
Jordi Sole-Casals	Universitat de VIC, Barcelona, Spain
Isabel Trancoso	INESC, Portugal

Organising Committee:

Marcos Faundez-Zanuy	EUPMt
Jordi Sole-Casals	UVIC
Léonard Janer	EUPMt
Josep Roure	EUPMt
Antonio Satue-Villar	EUPMt
Virginia Espinosa-Duró	EUPMt

Scientific Reviewers:

Amir Hussain
Andrew Morris
Anna Esposito
Atonio Satue-Villar
Bastiaan Kleijn
Bojan Petek
Dijana Petrovska
Enric Monte-Moreno
Erhard Rank

Fernando Diaz-de-María
Gernot Kubin
Jean Rouat
Jean Schoentgen
Jordi Solé-Casals
Léonard Janer
Marcos Faundez-Zanuy
Miguel-Ángel Ferrer-Ballester
Mohamed Chetouani
Najim Dehak
Pedro Gómez-Vilda
Peter Murphy
Robert van Kommer
Stephen McLaughlin
Yannis Stylianou

The sponsorship and support of:

- ESCOLA UNIVERSITÀRIA POLITÈCNICA DE MATARÓ (EUPMt), Spain
- EUROPEAN COMMISSION COST ACTION 277: Nonlinear Speech Processing
- MINISTERIO DE CIENCIA Y TECNOLOGÍA (TEC2004-20959-E), Spain
- FUNDACIÓ CATALANA PER A LA RECERCA I LA INNOVACIÓ
- INTERNATIONAL SPEECH COMMUNICATION ASSOCIATION (ISCA)
- EUROPEAN ASSOCIATION FOR SIGNAL, SPEECH AND IMAGE PROCESSING (EURASIP)
- IEEE INDUSTRIAL APPLICACIONS SOCIETY

are gratefully acknowledged

Table of Contents

Voice Pathologies

Speech Recognition

Speech Enhancement

Applications

The COST-277 European Action: An Overview

Marcos Faundez-Zanuy[1], Unto Laine, Gernot Kubin,
Stephen McLaughlin, Bastiaan Kleijn, Gerard Chollet, Bojan Petek,
and Amir Hussain

[1] Escola Universitaria Politècnica de Mataró, Spain
faundez@eupmt.es

Abstract. This paper summarizes the rationale for proposing the COST-277 "nonlinear speech processing" action, and the work done during these last four years. In addition, future perspectives are described.

1 Introduction

COST-277 is an innovative approach: so far, cost actions where focused on a single application field: Speaker Recognition in Telephony (COST-250), Naturalness of synthetic speech (COST-258) [1], Spoken Language interaction in telecommunication (COST-278), etc. However, there are strong arguments for a global approach, which considers speech processing from a general point of view, rather than focussing on a single topic. Section 2 summarizes the rationale for this general approach and the goals of COST-277. Section 3 summarizes the work done inside the framework of COST-277 and section 4 is devoted to results and future lines.

2 Rationale for a Speech Processing COST Action

The four classical areas of speech processing:

1. Speech Recognition (Speech-to-Text, StT)
2. Speech Synthesis (Text-to-Speech, TtS and Code-to-Speech, CtS)
3. Speech Coding (Speech-to-Code, StC with CtS) and
4. Speaker Identification &Verification (SV)

have all developed their own methodology almost independently from the neighboring areas. (See the white arrows in the Figure 1.)

This has led to a plurality of tools and methods that are hard to integrate. Some of the ideas of COST action were to study the following fundamental questions:

- Are there any parametric, discrete models or representations for speech useful for most or even all of the tasks mentioned?
- What properties should these representations have?
- How can the parameters of these models be estimated automatically from continuous speech?

M. Faundez-Zanuy et al. (Eds.): NOLISP 2005, LNAI 3817, pp. 1–9, 2005.

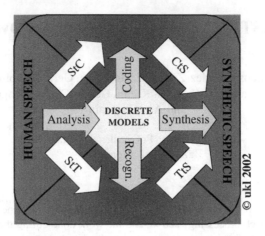

Fig. 1. Classical, separate speech processing areas (white arrows) and an advanced, multifunctional, integrated platform utilizing common discrete models and representations for speech signals

In Fig. 1 natural, human speech is on the left and its synthetic counterpart on the right. Two main methods to compress the speech information are depicted in the middle. The "written speech" refers to standard orthography of the language or to phonetic writing. The "coded speech" refers to a plurality of different coding methods and parametric representations.

The coded speech is less compressed and may have high quality whereas the written speech is strongly compressed and without any side information it has lost, e.g., the identity and the emotional state of the speaker.

The simplest codes, like PCM, can be called *one-quality-layer* codes or representations (see Fig. 2). The code is directly related to one and only one quality, attribute or dimensionality, e.g., signal amplitude or sound pressure. These simplest coding methods do not apply models. Model free methods lead to universal coding where the waveform may represent any type of time varying signal: temperature, music, speech etc.

Two-quality-layer codes and representations make the primary separation between source (excitation) and filter (vocal tract and lip radiation). They apply source-filter theory and related models. The filter is typically assumed to be all-pole. All of the possible zeroes of the signal together with the temporal fine structures are modeled by the source (e.g., CELP). These methods may take into consideration the non-uniform frequency resolution of the human auditory system by applying auditory frequency scales (PLP, WLP).

The modeling can be developed further, e.g., by including aspects of articulation and/or related spectral dynamics. These codes can be called *three-quality-layer* codes. The corresponding methods and models are here called "discrete models". Further, when more complicated structures are found and coded we approach phonetic qualities, descriptions, and codes (IPA). Finally, linguistic qualities and structures are modeled and coded (speech understanding).

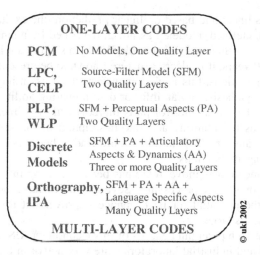

Fig. 2. Different levels of coding and representations of speech signals

The "discrete models" area in the middle of Fig. 1 denotes methods that are scalable and able to produce variable quality, depending on the purpose and capacity of the target system.

Advanced models and methods may be linear, non-linear or combinations of both.

The models and methods should not only help to integrate different speech processing areas, but in addition they should -if possible- posses the following features and reflect:

- Properties of human perception (auditory aspect)
- Properties related to articulatory movements (motoric aspect)
- Inherent features of phonemes or subsegmentals
- Allow mappings between speakers (normalization)
- Robustness: Insensitivity to ambient noise and channel (wireless and packet-based ones) distortions
- Underlying dynamics of the speech signal.

The parametric models capable of reflecting aspects of speech production could help to understand the "hidden structures" of the speech signal. They could provide tools for more detailed analysis and study on the (acoustical) coding principles of phones or diphones in different contexts of continuous speech.

All this should be of help in understanding the mechanisms of the coarticulation, too. This problem and weakness in current speech recognition schemes should be transformed into power and strength useful in other speech processing areas as well. Phonetic research -especially related to articulatory phonetics and subsegmentals-could benefit of these new tools and methods.

Deeper understanding and efficient modeling the reflections of the speech production mechanism in continuous speech signal and in its phones are in the focus of COST-277.

Source-filter models are almost always part of speech processing applications such as speech coding, synthesis, speech recognition, and speaker recognition technology.

Usually, the filter is linear and based on linear prediction; the excitation for the linear filter is either left undefined, modeled as noise, described by a simple pulse train, or described by an entry from a large codebook. While this approach has led to great advances in the last 30 years, it neglects structure known to be present in the speech signal. In practical applications, this neglect manifests itself as an increase in bit rate, a less natural speech synthesis, and an inferior discriminating ability in speech sounds. The replacement of the linear filter (or parts thereof) with non-linear operators (models) should enable us to obtain an accurate description of the speech signal with a lower number of parameters. This in turn should lead to better performance of practical speech processing applications.

For the reasons mentioned above, there has been a growing interest in the usage of non-linear models in speech processing. Several studies have been published that clearly show that the potential for performance improvement through the usage of non-linear techniques is large.

Motivated by the high potential benefits of this technology, US researchers at well-known universities and industrial laboratories are very active in this field. In Europe, the field has also attracted a significant number of researchers. The COST-277 (Non-linear speech processing) project is the first step towards the creation of a scientific community and the possibility of European collaborative efforts. The initial COST-277 working plan is published in [2].

3 Overview of COST-277 Research Activities

COST-277 started on June 2001 and will officially finish in June 2005. Section 4.2 contains a list of participating countries. During these last four years, several meetings, workshops and training schools have been organized and articulated the research activities. Four main working groups have been established and worked close together:

1. WG1: Speech Coding.
2. WG2: Speech Synthesis.
3. WG3: Speech and speaker recognition.
4. WG4: Voice Analysis and enhancement.

The main scientific events inside COST-277 action are described in the next subsections.

3.1 Management Committee Meetings

The administrative and scientific matters have been discussed in several meetings, whose minutes can be found on the website. The three initial MCM (0, 1, 2) have been organized for the start up of the action, and the remaining ones have included other activities summarized in the next sections.

MCM-0 (pre-inaugural meeting): September, 1999, Budapest (Hungary)
MCM-1 (Inaugural meeting): June, 2001, Brussels (Belgium)
MCM-2 (Unofficial EUROSPEECH meeting): September, 2001, Aalborg (DK)
MCM-3 (Vietri Sul Mare meeting): 6th/7th December, 2001, Vietri Sul Mare (Italy)
MCM-4 (Graz meeting): 11th/12th April, 2002, Graz (Austria)

MCM-5 (Unofficial EUSIPCO'02 meeting): September, 2002, Toulouse (France)
MCM-6 (Edinburgh meeting): $2^{nd}/3^{rd}$ December 2002, Edinburgh (UK)
MCM-7 (Le Croisic meeting): 20th to 23th May, 2003, Le Croisic (France)
MCM-8 (Laussane meeting): $5^{th}/6^{th}$ September, 2003, Laussane (Switzerland)
MCM-9 (Limerick meeting) : $15^{th}/16^{th}$ April, 2004, Limerick (Ireland)
MCM-10 (2nd Vietri sul Mare meeting): 13th-17th September'04
MCM-11 (Barcelona meeting): 19th-22th April'05
MCM-12 (Creete meeting): 19th-23th September'05.

3.2 Special Sessions Organized in Signal Processing Conferences

Five special session have been organized in well-established conferences:

1. International Workshop on Artificial Neural Networks (IWANN'01): Held in June 2001 in Granada (Spain). 3 technical presentations.
2. European Speech Processing Conference: Held in September 2002 in Toulouse (France). 5 technical presentations.
3. International Workshop on Artificial Neural Networks (IWANN'03): Held in June 2003 in Menorca Island (Spain). 4 technical presentations.
4. European Speech Processing Conference: Held in September 2004 in Vienna (Austria). 5 technical presentations.
5. International Conference on Artificial Neural Networks (ICANN'05): To be held in Poland in September 2005.

3.3 Workshops

Two main workshops, named NOLISP, have been organized:

1. NOLISP'2003: Held in May 2003 in Le Croisic (France). 32 technical presentations.
2. NOLISP'2005: Held in April 2005 in Barcelona (Spain). 42 technical presentations.

3.4 Training Schools

Two training schools have been organized:

1. Vietri Sul Mare (Italy): Held in September 2004. 36 technical presentations [4].
2. Crete (Greece): To be announced.

3.5 Short Term Scientific Missions

Two short term missions have been organized:

1. STM-1: Held during 19 June 2003 to 21 June 2003 from Belgium (Brussels) to Laussane (Switzerland): Synthesis of disordered speech and the insertion of voice quality cues into text-to-speech systems.
2. STM-2: from Graz (Austria) to Canada: research on auditory modeling (summer '04).

3.6 EU Framework Programm 6 (FP6) Initiatives

Two expressions of interest were submitted to the European Commission within the FP6 programme:

1. Advanced Methods for Speech Processing (AMSP):
   http://eoi.cordis.lu/dsp_details.cfm?ID=38399
2. Human Language Technology Portability (HLTport):
   htt p://eoi.cordis.lu/dsp_details.cfm?ID=32189.

4 Results and Future Lines

One of the great successes of COST-277 has been the increase of contributions between different countries and other COST actions. This has let to deal and study new research topics, summarized in section 4.3.

4.1 Collaboration with Other COST Actions

COST-219ter: Accessibility for All to Services and Terminals for Next Generation Networks
COST-219ter has showed a strong interest on the work "Enhancement of Disordered Male Voices" done by a COST-277 team. Possible future interactions between both COST actions are being studied.

COST-275: Biometric-Based Recognition of People over the Internet
Several COST-277 members have attended regular workshops of COST-275 and presented results related to Speaker recognition. In addition, COST-277 has produced the COST-277 database for speaker recognition, which is suitable for the study of new techniques such as:

- Speech signal watermarking for improving the security on remote biometric applications.
- Speech signal bandwidth extension for improving the naturalness of encoded speech.

This database will be available in 2006 [3]. In addition, a joint brochure was disseminated at major conferences, at the beginning of the action.

COST-276: Information and Knowledge Management for Integrated Media Communication Systems
Contacts have been established with COST-276 WG-4 due to the interest on Speech watermarking. However, although COST-276 has interest on speech watermarking, they are more focus on audio (music) watermarking. Thus, COST-277 has a more mature technology for speech, which will be transferred to COST-276.

COST-278: Spoken Language Interaction in Telecommunication
COST-278 members will attend the NOLISP'05 workshop in Barcelona and will present some topics and problems that could be addressed with NL speech processing.

On the other hand, NL speech feature extraction for speech recognition will be presented to COST-278. In addition, a joint brochure was disseminated at major conferences, at the beginning of the action.

4.2 Collaboration Between Different Countries

In order to summarize the different collaborations between institutions, we have just representted in a matrix the collaborations between different countries, made possible thanks to COST-277 action. Next diagram summarizes the inter-country collaborations.

	A	B	CAN	CH	CZ	D	E	F	UK	GR	I	IRL	LT	P	S	SI	SK
A	▓	▓				▓									▓		
B	▓		▓					▓	▓		▓	▓			▓		
CAN	▓	▓						▓									
CH		▓			▓	▓						▓					
CZ				▓							▓						
D				▓			▓	▓									
E	▓		▓			▓		▓							▓	▓	
F		▓	▓			▓	▓		▓						▓		
UK		▓						▓				▓					
GR																	
I		▓			▓												
IRL		▓		▓					▓								
LT														▓			
P													▓				
S	▓						▓	▓								▓	
SI	▓						▓								▓		
SK																	▓

A shadowed cell means contribution between respective file and row countries (joint publications and/or Short Term Missions).

Country codes

A	Austria
B	Belgium
CAN	Canada
CH	Switzerland
CZ	Czech Republic
D	Germany
E	Spain
F	France
UK	United Kingdom
GR	Greece
I	Italy
IRL	Ireland
LT	Lithuania
P	Portugal
S	Sweden
SI	Slovenia
SK	Slovakia

4.3 Scientific Results

A detailed explanation of scientific achievements is beyond the goal of this paper, and can be found in our website www.nolisp2005.org thus, we restrict this section to an enumeration of research activities:

- Analysis and synthesis of the phonatory excitation signal by means of a polynomial waveshaper.
- Modulation frequency and modulation level owing to vocal micro-tremor.
- Decomposition of the vocal cycle length perturbations into vocal jitter and vocal microtremor and comparison of their size in normophonic speakers.
- Acoustic primitives of phonatory patterns.
- Multivariate Statistical Analysis of Flat Vowel Spectra with a View to Characterizing Disordered Voices.
- Relevance of bandwidth extension for speaker identification and verification.
- Waveform speech coding using non-linear vectorial predictors.
- SVM-Based Lost Packets Concealment for ASR Applications Over IP.
- Space–time representation.
- Nonlinear masking of a time–space representation of speech.
- Adaptive nonlinear filtering and recognition with neurons.
- Nonlinear masking and networks of oscillatory neurons.
- Speech structure and masking.
- Speech analysis.
- What can predictive speech coders learn from speaker recognizers?
- Nonlinear features for speaker recognition.
- Speaker recognition improvement using blind inversion of distortions.
- Isolating vocal noise in running speech via bi-directional double linear prediction analysis.
- On the bandwidth of a shaping function model of the phonatory excitation signal.
- Speech signal watermarking: a way to improve the vulnerability of biometric systems.

4.4 Future Lines

COST-277 will officially finish in June 2005. However, a final event will be held in the last semester of 2005 in the form of a training school. Afterwards, the Nonlinear speech processing community should survive without the European Science Foundation founding. In NOLISP'2003 it was stated the interest for keep on working on this topics, and to stay close to the speech processing community, rather than nonlinear processing groups (image, communications, etc.). Probably, a good idea would be the establishment of an ISCA Special Interest Group (SIG) on Nonlinear speech processing.

You can keep informed by looking at the website!

Acknowledgement

This work has been supported by FEDER and the Spanish grant MCYT TIC2003-08382-C05-02, and European COST action 277 "Non-linear speech processing".

B. Petek acknowledges grant 3311-04-837052 from Ministry of Education, Science and Sport of Republic of Slovenia.

References

1. Keller, E. et al (Eds.) Improvements in Speech Synthesis. John Wiley 2002.
2. Chollet G., Faundez-Zanuy M., Kleijn B., Kubin G., McLaughlin S., and Petek B. "A description of the cost-277 nonlinear speech processing action working-plan". Vol. III pp.525-528 EUSIPCO'2002, Toulouse.
3. Faundez-Zanuy M., Hagmüller M., Kubin G., Nilsson M., Kleijn W. B., "The COST-277 speech database". ISCA Workshop NOLISP'2005 LNCS, Springer Verlag 2005.
4. G. Chollet, A. Esposito, M. Faundez-Zanuy, M. Marinaro Eds., "Advances in nonlinear speech processing and applications". LNCS 3435, Springer Verlag 2005.

Neuro-fuzzy Logic in Signal Processing
for Communications: From Bits to Protocols

Ana Pérez-Neira[1], Miguel A. Lagunas[2], Antoni Morell[1], and Joan Bas[2]

[1] Universitat Politècnica de Catalunya, Campus Nord UPC, Building D5,
Dept. of Signal Theory and Communications, c/ Jordi Girona, 1-3,
08034 Barcelona, Spain
anuska@gps.tsc.upc.edu
[2] Parc Mediterrani de la Tecnologia (PMT), Av. Canal Olímpic S/N,
08860 – Castelldefels, Barcelona, Spain
{m.a.lagunas, joan.bas}@cttc.es

Abstract. The present work shows how communication systems benefit from fuzzy logic. From signal processing applications, which process bits at the physical layer in order to face complicate problems of non-Gaussian noise, to practical and robust implementations of these systems and up to higher layers in the communication chain, which are engaged in the protocol design. The ability for modeling uncertainty with a reasonable trade-off between complexity and model accuracy, makes fuzzy logic a promising tool.

1 Introduction

Since the introduction of fuzzy logic in the engineering field, this discipline has been very successful in automatic control [1] with applications such as autonomous robot navigation, auto focus cameras, image analysis and diagnosis systems. A proof of this success can be found in the number of companies developing hardware and software for developing fuzzy systems (Accel Infotech Pte, Ltd., Adaptive Informations Systems, American NeuraLogix, Fujitsu, Oki Electronic, OMRON Corporation, Toshiba, SGS-Thomson, Siemens, etc.).

The present work shows how communication systems benefit also from fuzzy logic systems. From signal processing applications that process bits at the physical layer in order to face complicate problems of non-Gaussian noise, to practical and robust implementations of these systems and up to higher layers in the communication chain, which are engaged in the protocol design. The ability for modeling uncertainty with a reasonable trade-off between complexity and model accuracy makes fuzzy logic a promising tool.

In the 90's, Bart Kosko [2-3] and Jerry Mendel [4-5] began to study the application of fuzzy logic and set theory to the signal processing field. Since then various works have appeared focused on fuzzy logic under the intelligent signal processing framework [6-7]. Fuzzy systems are able to build up mathematical models from linguistic knowledge and do not require statistical knowledge, although they can incorporate it, offering an scalable design with the available information. Communication systems can

M. Faundez-Zanuy et al. (Eds.): NOLISP 2005, LNAI 3817, pp. 10–36, 2005.

benefit from these features to gain in robustness and in fast acquisition and tracking, as the present work shows in the application of interference canceling in CDMA (Code Division Multiplex Systems). Another important feature is that fuzzy systems offer also physical interpretability, this helps adjust the parameters of the system in an easy and friendly way. The second application presented in this work at bit level takes advantage of this feature in order to design a robust beamformer for communication signals, resulting in an easy implementable and tunable beamformer when compared with other existing techniques in the literature. These applications together with others developed by the authors in [8-9] are based on the capability of fuzzy logic to implement model-free function approximations. All these works aim to achieve the intelligent filtering that Lofti Zadeh in 1954 stated in his work "General filters for separation of signal and noise" [10]. We could summarize it saying that intelligent and robust filtering consist in decision making. Decision making that should operate not with an statistical model but with the available data: *"Since the extent to which a signal can be separated from noise is ultimately limited by the statistical data available to the designer, a more rational approach would be to start with all the available data about the signal and noise an try to design a filter that would use it fully and most effectively,"[10].*

However, we should talk about the benefits of fuzzy logic and systems with caution. If expert or linguistic knowledge is not available to carry out the decision making in an "intelligent" way, fuzzy logic systems can be used as filters or classifiers that generalize the ones designed on a statistical basis (e.g. minimum mean square error, maximum likelihood, maximum a posteriori probability); thus, offering much more flexibility and possibilities than the classical statistical systems, but presenting a greater complexity that results difficult to cope with. Concerning this aspect, the present work compares different fuzzy and non-fuzzy classifiers. In spite of the greater flexibility that fuzzy systems offer due to the fuzzy instead of crisp classification thresholds, the conclusion is that the fuzzy systems only stand out when expert knowledge and not only data is available in the design.

At the protocol level in the communication stack, known research applications are queuing and buffer management, distributed access control, hand-off control, load management, routing, call acceptance, policing congestion mitigation, bandwidth allocation, channel assignment, network management, etc [11-17]. All of them take advantage of: the flexibility of fuzzy logic, its ability to cope with different types of inputs and its decision making structure. Protocols are in fact controllers that have to make decisions based on many different variables; thus, the appropriateness of fuzzy logic. Part of the present chapter is dedicated to applications regarding hand-off algorithms [18-19], combining distance measurements with received signal strength to decide hand-off while keeping quality of Service. Although the work focuses on horizontal hand-off in WLAN (Wireless Local Area Network), we point out that in the emerging multimedia systems, hand-off is also considered vertically, as a switching between different systems covering the same area and service. As an example, consider switching between GPRS, UMTS and a satellite segment, as in [20].

Next some of the mentioned applications are presented: section 2 describes the interference canceller for CDMA, section 3 is dedicated to the robust beamformer, section 4 is devoted to fuzzy classification and finally, section 5 discusses on the use of fuzzy logic for hand-off control. Finally conclusions come. For a review of fuzzy logic and systems in signal processing we refer to the tutorial in [5].

2 Fuzzy Logic Canceller for CDMA

The new communication standards require high capacity to support the increasing demand of multimedia services. In order to achieve the high capacity, the standards propose to reduce the cell site and reutilize frequency or codes. However, this strategy ask for more sophisticated signal processing techniques that are able to cope with the increases level of interference. This section focus on a CDMA system, where K spread spectrum users are received at each single-user terminal. The signal model for the sample k of the received signal is

$$r_k = A_k s_k + i_k + n_k \tag{1}$$

where s_k represents the binary information $(\in \pm 1)$ of the desired user. The interference i_k can be either analog or digital and n_k models both the thermal noise (AWGN or additive white Gaussian noise) and the multiple access interference or MAI as (2) formulates

$$n_k = \sum_{l=1}^{K} \sqrt{P_l}\, s_k^l + \sqrt{P_N}\, w_k \tag{2}$$

where s_k^l is the binary sequence of the undesired user j, and w_k is the thermal noise. P_j and P_N represent the corresponding powers. Next section is devoted to the design of the fuzzy canceller.

2.1 Formulation of the Fuzzy Interference Canceller

The canceller subtracts the interference signal from the received one in (1). Therefore, it is necessary a non-linear filter able to estimate i_k in a non-Gaussian noise environment. When the interference is analog, the conditional mean estimator of (3) is the optimal one

$$\hat{i} = E\{i \,|\, r\} = \int_i i p(i \,|\, r)\, di \tag{3}$$

where p(i/r) represents the a posteriori probability of the interfering signal. Applying the Bayes theorem and the signal model in (1), p(i/r) can be equated as

$$p(i \,|\, r) = \frac{p(i) \sum_{p=1}^{Q} \lambda_p \exp\!\left(\dfrac{-z_p^T z_p}{2\sigma_w^2} \right)}{\int_i p(i) \sum_{p=1}^{Q} \lambda_p \exp\!\left(\dfrac{-z_p^T z_p}{2\sigma_w^2} \right)} \tag{4}$$

where $z_p = \left[r_p - m_p - i \right]^T$, r_p embraces all the received samples and m_p consists of all the possible noise and MAI states. Finally, p(i) is the a priori probability of the interference and λ_p is the a priori probability of the p noise state.

Combining (3) and (4) we get the following equation for the conditional mean of the interference

$$
\hat{i} = \int_i i p(i \mid r) di \approx \sum_{l=1}^{L} \sum_{j=1}^{Q} i_l \frac{w_{j,l} \exp\left(\frac{-z_{j,l}^T z_{j,l}}{2\sigma_w^2}\right)}{\sum_{l=1}^{L} \sum_{o=0}^{K^N} w_{o,l} \exp\left(\frac{-z_{o,l}^T z_{o,l}}{2\sigma_w^2}\right)} \tag{5}
$$

where $w_{j,l}$ is the product of $p(j)$ and λ_l . Note that (5) can be seen as a fuzzy system with LxQ linguistic rules, exponential membership functions, Sugeno inference and centroid defuzzification. In other words, i_l represents the output centroid for the l-th rule, which is weighted by w_{jl}. Therefore, (5) can be reformulated as (6)

$$
i_{fuzzy} = \sum_{j=1}^{M} i_j \Phi_j \tag{6}
$$

where Φ_j is the fuzzy basis function of rule j. Under high Signal to Noise ratio conditions, the fuzzy interference estimator of (6) approximates the maximum a posteriori estimator as equated in (7)

$$
i_{fuzzy} = \sum_{j=1}^{M} i_j \Phi_j \approx i_m \Phi_m \tag{6}
$$

where Φ_m is the fuzzy basis function that presents a maximum value, which is close to one.

2.2 Expert Knowledge in the Fuzzy Interference Canceller

The fuzzy system is designed based on 4 variables: $\hat{i}_{k-3}, \hat{i}_{k-2}, r_{k-1}, r_k$. In order to reduce the fuzzy rule base, the variable r_{k-1} has been taken as reference of the input universe of discourse. Therefore, the input vector $\mathbf{x}$ is $\mathbf{x} = \left[\hat{i}_{k-3} - r_{k-1} \ \hat{i}_{k-2} - r_{k-1} \ r_k - r_{k-1}\right]^T = \left[x_1 \ x_2 \ x_3\right]^T$. The fuzzy rule base has the following structure

If x_1 is A_1^j and x_2 is A_2^j and x_3 is A_3^j THEN y is B_j with weight w_j

where A_i^j, B_j are the input and output fuzzy sets respectively for rule j. The input and output variables have been modeled with F=7 fuzzy sets. The mean of the fuzzy sets takes the values $\{-3, -2, -1, 0, 1, 2, 3\}$ and the variance equal to 0.5. The fuzzy rule base is initialized in order to model the slow evolution of the narrow band interference when compared with the sampling time [21]. Finally, the weights are tuned by the Least Mean Square (LMS) algorithm of (7).

$$\theta_{j,k} = \theta_{j,k-1} + e_k \cdot \phi_{j,k-1} \cdot \left[w_{j,k-1} - \hat{i}_{fuzz_k} \right] / \theta_{j,k-1} \tag{7}$$

where the error e_k is

$$e_k = \left(r_k - \hat{i}_k \right) - sign \left(r_k - \hat{i}_k \right) \tag{8}$$

2.3 Simulations

Next the fuzzy system has been compared with the two-sided linear filter of [22-23] with 10 coefficients. This coefficients and the the centroids of the fuzzy filter have been adapted with a decision directed LMS. Fig. 1 shows the Signal to Noise and Interference Ratio improvement when the interference is modeled as an autoregressive process. The SNR is equal to 20 dB. 10 Monte Carlo runs have been conducted with 660 samples for training each and 8000 for evaluation. Note the better performance of the fuzzy filter. In Fig. 2 the Bit Error Rate (BER) has been evaluated for SNR=20 dB and a multiple access interference of K=3 users and spreading factor of SF=11. As the weighted fuzzy filter takes into account the states of the MAI, it outperforms both the linear and the non-weighted filter.

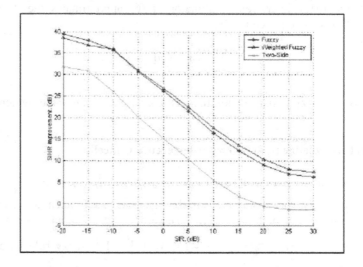

Fig. 1. Suppression of an AR interference. SNR= 20 dB.

Fig. 3 evaluates the performance of the fuzzy filter depending on the Signal to Noise ratio (SNR) for SIR equal to –20 dB. The only interference in the scenario has been modeled as an autoregressive process. The fuzzy filter has been compared with the optimum filter (DDK) and with the minimum mean square error one (MMSE). Note that for low power of Gaussian noise (i.e. in an interference limited scenario), the fuzzy filter outperforms the other ones. Finally, Fig. 4 shows the fast acquisition time of the fuzzy interference canceller due to the initial expert knowledge incorporated in the rule base.

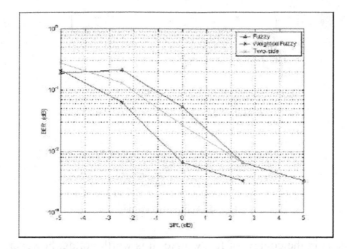

Fig. 2. Suppression of a digital interference. SNR= 20 dB and MAI of 3 users.

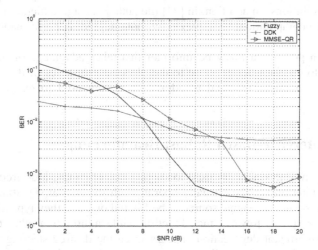

Fig. 3. Suppression of a digital interference. SIR= -20 dB and autoregressive interference.

2.4 Conclusions

The designed fuzzy filter is able to cope with both analog and digital interference even in the presence of MAI. On the other hand, due to the difficulty of statistical modeling, classical filters, which relay just on statistics, are not able to cope with this complex situation. The initial fuzzy rule base is built up from expert knowledge and can be trained with data whenever available; thus, approaching to the optimum MAP interference estimator. Therefore, the system is scalable with the available information and if only a short training is possible, the expert knowledge incorporated in the rule base guarantees a better performance than existing interference cancellers. In fuzzy systems optimality is pursued by emulating an expert operator. This is the best

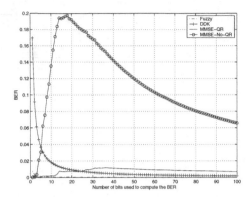

Fig. 4. Acquisition time

that can be done in the context of fuzzy logic. It is worth noting, however, that in all cases of queuing control where a mathematically optimal solution is known, as it is the case of interference estimation in CDMA, the fuzzy system yields precisely the same optimal solutions. In addition, in [24] the authors design a hierarchical rule base that reduces the computational complexity without degrading performance in most of the cases.

Next section presents a fuzzy beamformer that takes advantage of the fuzzy rule base in order to obtain a design close to the physical properties of the scenario and, therefore, easy to implement and to adjust when compared to existing non-fuzzy beamformers.

3   Fuzzy-Inference-Based Robust Beamforming

Adaptive array beamformers are used to spatially discriminate a desired signal from the unwanted interferers. They usually operate with the spatial signature or steering vector associated with the signal source to be filtered out, and they typically produce a constant array response towards that direction while minimizing other contributions (see Fig. 5). Significant degradation appears when the desired steering vector is not known exactly [25]. It is specially noticeable when the number of snapshots is low (i.e. the so called sample support problem) and gets worse for high Signal to Noise plus Interference Ratio (SNIR). The phenomenon is that desired information is treated as interference and consequently nulled.

Numerous methods have been proposed. We classify them depending on the knowledge they require from the uncertain desired Direction of Arrival or DOA. Traditional approaches require a nominal DOA and its corresponding uncertainty range [26-28]. The approach in these techniques is to gain robustness related to DOA errors at the expenses of decreasing interference and noise suppression and in general the problem of interference within the uncertainty range for the DOA is not addressed. A different approach consists in computing a DOA estimate and proceed as if the DOA were already known [29], yielding the so-called Direction-based techniques. Finally, the third group of techniques in complexity order contains subspace techniques, as in

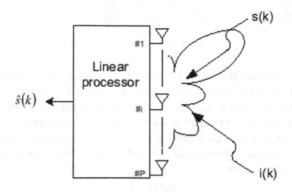

Fig. 5. An example of spatial linear processing to filter out a signal s(k) in presence of an interference i(k)

[30] or references therein, where the signal plus interference subspace is estimated in order to reduce mismatch. However, they also suffer from significant performance degradation when the available data doesn't provide good estimates. The above mentioned techniques resort to different robust signal processing schemes, as for instance: regularization, minimax/worst-case design or Bayesian approaches.

This paper uses fuzzy logic as another tool worth considering when imprecise a priori knowledge of input characteristics makes the sensitivity to deviations from assumed conditions an important factor in the design [31]. We derive a direction-finding based beamformer that describes DOA imprecision by means of fuzzy sets, which does not make statistical assumptions on interference and noise. An important issue in robust beamforming is the design or adjustment of parameters, whose values trade-off between robustness and array gain. In that sense, physical interpretability as in the proposed fuzzy techniques is always desirable.

This part is organized as follows. Section 3.1 states the problem. The fuzzy inference based beamformers are developed in Section 3.2, Section 3.3 parameter design, Section 3.4 presents performance examples and a summary is given in Section 3.5.

3.1 Problem Statement

The narrowband beamformer is a linear processor and consists of a set of P complex weights that combine the signals received at P sensors with the objective of filtering out a signal s(k) (see (9)) that impinges the array from a specific spatial direction (see Fig. 5)

$$\hat{s}(k) = \boldsymbol{w}^H \boldsymbol{x}(k) \tag{9}$$

where k is time index, $\mathbf{x}(k) = \left[x_1(k) \cdots x_p(k)\right]^T$ is the complex vector of array observations, $\mathbf{w}(k) = \left[w_1(k) \cdots w_p(k)\right]^T$ is the complex vector of beamformer weights, P is the number of array sensors. The base band observation data $\mathbf{x}(k)$ is

$$x(k) = s(k) + i(k) + n(k)$$
$$= s(k) \cdot a_d + i(k) + n(k) \tag{10}$$

where s(k) represents the desired signal contribution, i(k) the interference and n(k) is noise. Note that we have decomposed s(k) into desired signal waveform s(k) and desired signal steering vector a_d, which contains the spatial information. It is easily modeled resorting to wave propagation theory and array geometry.

The weights are chosen accordingly to some optimum criterion, such as maximum SNIR, minimum mean square error or minimum power distortionless response (MPDR). All of them equate

$$w_{opt} = \mu R_x^{-1} a_d \tag{11}$$

where μ is a scale factor and $R_x = E\{xx^H\}$ is the data covariance matrix. A beamformer having this form is often referred to as the "optimum beamformer". Its performance strongly depends on both a_d and R_x in those common practical applications where they are obtained through estimation. In this paper we consider no knowledge of the desired DOA and that K snapshots are available for the estimation of both the covariance matrix and the DOA. The array is assumed to be calibrated, so errors in the spatial signature come form the DOA estimate.

The Sample Matrix Inversion (SMI) method is used for adaptive beamforming. The weights are updated every K snapshots using the K-sample covariance matrix of (12), as well as the DOA estimate, which is obtained with the Capon estimator; thus, obtaining the so-called Capon beamformer. Other estimates are possible, however, without loss of generality for the proposed techniques

$$\hat{R}_x = \frac{1}{K} \sum_{k=1}^{K} xx^H \tag{12}$$

Direction-finding based beamformers suffer from significant performance degradation when the DOA estimates are not reliable, because of low number of snapshots or low SNR. Next section develops an adaptive beamformer which balances the user of observed data and approximate DOA knowledge.

3.2 Fuzzy Inference Based Beamformer

Assuming partial knowledge about the desired DOA (i.e. a nominal DOA and uncertainty region), we aim to use the capability of fuzzy systems to approximate any continuous function on a compact (closed and bounded) domain to obtain a reliable estimate of s(k). Fuzzy theory states that it is always possible to find a finite number of rule patches (which describe the system) to cover the graph of f while keeping the distance $|f(x) - F(x)|$ as small as we please, being F(x) the fuzzy approximation function.

We consider one input variable to the system, the DOA candidate u_d that we extract from input data (Capon estimator). In order to take into account DOA errors, we define an interval of possible DOA values, the prior and L fuzzy sets $\{A_i, i = 1...L\}$ are placed equispaced along it. They describe the imprecise DOA estimate. As far as we are also concerned about practicality and implementation issues, we choose input fuzzy sets to be triangular. Their membership degree over the variable u can be expressed as

$$
\mu_{A_i} = \begin{cases} 1 - \frac{|u - u_i|}{amp \cdot d_p} & if \ |u - u_i| \leq amp \cdot d_p \\ 0 & if \ |u - u_i| > amp \cdot d_p \end{cases} \tag{13}
$$

with $d_p = \frac{u_{d_{min}} - u_{d_{max}}}{L}$, where $u_{dmin} - u_{dmax}$ stands for the prior width. Widths are set using amp and $u_i = \frac{d}{2}(2(i-1) - (L-1))$.

The choice for the output fuzzy sets follows from the conditional mean beamfomer w_{CM} in (14)

$$
\begin{aligned}
\hat{s}(k) &= E\{\hat{s}(k)/X\} \\
&= \sum_{i=1}^{L} p(u_i/X) E\{\hat{s}(k)/X, u_i\} \\
&= \sum_{i=1}^{L} p(u_i/X) w_{opt}^{H}(u_i)x = w_{CM}^{H} x
\end{aligned} \tag{14}
$$

where X represents the available data set and $p(u_i/X)$ is the a posteriori probability density function (pdf). The global estimation problem is divided into L smaller problems that assume fixed input parameters u_i and that are gain controlled by the probability of each possible signal incoming direction u_i given the data set.

In light of (14), it is clear that the conditional mean beamformer is optimal as far as L is big enough and the a posteriori pdf is assumed perfect. In that asymptotic case, the conditional mean beamformer is the optimum beamfomer in the minimum mean square error sense. When a finite (and low) number of optimum beamformers are available, a fuzzy approximator that does not rely on statistics but on expert knowledge can be derived. The output fuzzy sets B_i are designed accordingly as singletons placed at $\mathbf{w}_{opt}^{H}(u_i)\mathbf{x}$, which, in other words, is the output of the "optimum beamfomer" pointing at u_i. Due to philosophy similarities, we use the Bayesinan beamformer [28] as a benchmark for comparison. Indeed, it is derived from (14) and uses a parametric model for the pdf.

To completely describe the system, assuming we implement additive fuzzy systems with singleton fuzzification, product inference and centroid defuzzification [5], we establish the rules Ri, i=1...L that relate inputs with outputs:

Ri: IF the DOA is Ai THEN the desired signal is Bi

It can be shown that the final expression of the spatial filter is the one given in (15)

$$\boldsymbol{w}_F = \sum_{i=1}^{L} \frac{\mu_{A_i}(\hat{\mu}_d)}{\sum_{j=1}^{L} \mu_{A_i}(\hat{\mu}_d)}\; \boldsymbol{w}_{opt}(u_i) \tag{15}$$

Fig. 6 shows an operation example of the beamformer for the case L=3. Note that although statistics play an important role in computing both the output fuzzy sets and the input to the system, the rules are transparent to statistics and rely to a knowledge based approach. Thus, the system is not so model-dependent and consequently model-limited as for example the Bayesian one. Another important aspect is how we tune parameters to incorporate expert knowledge. This is discussed in the next section.

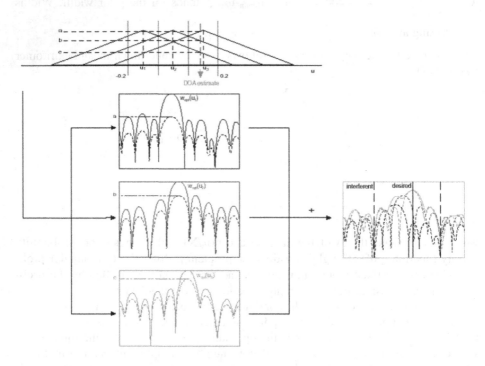

Fig. 6. Beamforming example for the case L=3

3.3 Parameter Design

The fitting quality of the designed fuzzy beamformer will strongly depend on the election of the parameters related to the input fuzzy sets (width, mean and number) and the number of snapshots K. The number of fuzzy sets L is established on complexity criteria. For the means and widths, u_i and *amp* respectively, it is possible to tune them with learning algorithms such as the stochastic gradient descent on the squared error [31]. In that case a training sequence is necessary. For practicality reasons, this work only considers the set up of the fuzzy sets widths and keeps fixed

Table 1. Heuristic rules for tuning the fuzzy sets widths function of SNR and in-prior SIR

SNR	In-pr. SIR	Width
$\geq 0dB$	$\geq 2dB$	$\frac{1}{6}L$
$0dB \geq$ SNR $> -10dB$	$\geq 2dB$	$\frac{1}{2}L$
$\leq -10dB$	$\geq 2dB$	L
$\forall$	$< 2dB$	$1.5\ L$

means. It is done heuristically as shown in Table1 in order to make the beamformer easy tuning and user friendly. Note that the scenario is described by the SNR and the in-prior Signal to Interference ratio (SIR).

The DOA description by means of fuzzy sets is less critical; actually, good empirical results are achieved when a low number of fuzzy sets (L=6) is used.

If information reaches the sensors spatially spread or distributed over some known angular region, it is possible to incorporate this information into the fuzzy system thanks to non-singleton fuzzification.

Finally, the last parameter to fix is the number of snapshots K. It is generally chosen as large as possible to get a good estimate of the data correlation matrix and DOA but small enough so that temporal fluctuations may be tracked. Randomness in the covariance matrix causes the cancellation of the desired signal in the $\mathbf{w}_{opt}$ beamformer of (15). One way of diminishing these problems is using diagonal loading [26] at the expenses of less interference suppression.

Next section presents some significative results.

3.4 Simulations

We assume a uniform linear array with 10 omnidirectional sensors spaced half a wavelength apart. The uncertainty in the DOA of the desired signal is over the region [-0.2, 0.2] and it has been equally divided into L=6 intervals. We are mostly interested in evaluating the performance of the proposed fuzzy beamformer from the point of view of array gain. This figure of merit is defined as

$$G = \frac{SNIR_{out}}{SNIR_{in}} = \frac{|\mathbf{w}^H \mathbf{a}_d|^2}{\mathbf{w}^H \rho_n \mathbf{w}} \tag{16}$$

where $\mathbf{w}$ stands for the weights of the beamfomer, $\mathbf{a}_d$ for the desired signal steering vector and ρ_n represents noise plus interferers normalized covariance matrix.

Just to make the understanding of the proposed beamformer easier, the results of the fuzzy beamformer formulated in (15) are presented together with those obtained by the Minimum Power Distortionless Response (MPDR) beamformer of (11) and the Bayesian beamformer of [28]. Note that the fuzzy inference based beamformer combines both philosophies: it is a direction-finding based beamformer, such as the MPDR one, and is able to cope with DOA uncertainties such as the Bayesian beamformer. This fact motivates the election of these two techniques, although they imply less computational complexity. We note in addition that there are many other approaches, each one having its own advantages, problems and applications, where neither is absolutely better than the other. In [28], the Bayesian beamformer is extensively compared with the linearly constrained minimum variance beamformer using directional constraints and a subspace beamformer. Thus, for the sake of clarity we do not include them in this paper and refer to the conclusions in [28]. The main goal of this simulation section is to show that the fuzzy inference beamformer is an alternative technique easy to implement and worth considering in scenarios with specific features such as different interference conditions.

Next we make a comparative study between the Bayesian and fuzzy beamformers. The objective is to see how easy is to adjust parameters in both beamformers. In the Bayesian technique, we have to adjust the γ parameter, that establishes the confidence given to the calculated a posteriori probabilities. The basic parameter for the fuzzy beamformer is the width of the fuzzy sets. In this study, we consider all other variables without error, i.e. no DOA error, perfect covariance matrix, etc. . . , although the presence of interference inside the prior interval plays an important role for deciding the signal to be focused. Bayesian beamformer computes itself an estimate of the pdf of the DOA from the data, while fuzzy beamformer departs from a given estimate (therefore we study it both focusing desired or interferent fonts). The different choices for the parameters establish a trade-off between robustness and performance.

Of key importance is how sensitive the performance of the beamformer is to the setting of its design parameters. Thus, high sensitivity implies low practicality of the beamformer. Figures 7 and 8 depict the array gain variations along with the fuzzy sets widths (fuzzy) and γ parameter (Bayesian), respectively. Three different SNR are considered, and an in-prior interference is simulated. Note that there is in general a trade-off between array gain at high SNR and acceptable performance at low SNR (for the fuzzy beamformer we assume that at SIR < 0dB, the DOA estimate points towards the interference). The reader can appreciate the smoother evolution that fuzzy beamformer provides. Finally, Fig. 9 shows the robustness or less sensitivity of the fuzzy system for different values of DOA misadjustment when compared with the Bayesian system. The presence of interference within the uncertainty range of the desired signal is not taken into account in the statistical model of the Bayesian beamformer; thus, its worst performance.

3.5 Conclusions

This work makes use of fuzzy logic systems as universal model-free function approximators and proposes a fuzzy inference based beamformer. The obtained beamformer is a direction-finding based technique that offers a robust approximation of the conditional mean estimate of the desired signal. The term robust is quite wide and this work focuses on the problem of DOA uncertainty in scenarios where interference signals are present. We note that no constraints have been imposed on the nature of the sources (i.e.point or spread). As expected, due to the soft DOA quantization, the fuzzy approach presents a "graceful" degradation when the working conditions are different from those expected. Additionally, the robustness of the presented beamformer applies also when adjusting its design parameters (fuzzy sets widths and number of beams). Because of the fuzzy systems interpretability, the parameters are easy to set once the scenario is known, thus demonstrating its practicality.

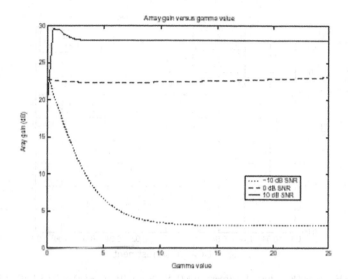

Fig. 7. Bayesian beamformer array gain versus γ value at different SNR, u_d=0.14, interferent directions u_{int}= [-0.5, 0.6, -0.07] and INR=[20, 20, 0] dB

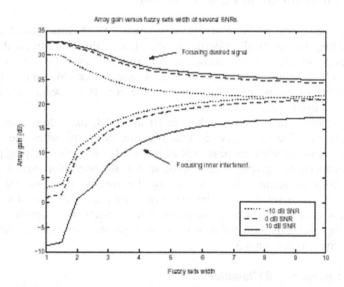

Fig. 8. Fuzzy beamformer array gain versus fuzzy sets width at different SNR focusing either desired or in-prior interference signal. Same scenario as Fig. 7.

Next section 4 is devoted to fuzzy classification for signal separation in 2-Dimensional spaces. It shows the greater flexibility that fuzzy systems offer in front of classical classifiers. Better results are then obtained in most of the cases, however, in order to take advantage of the great potential of fuzzy systems, expert knowledge would be needed.

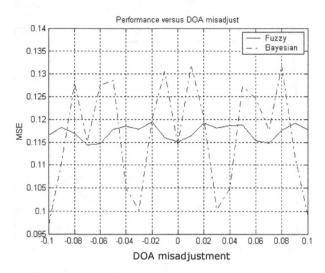

Fig. 9. MMSE versus DOA misadjustment, $u_d=0$, SNR = 0dB, $u_{int}=\{-0.5, 0.6\}$, INR=$\{20, 20\}$ dB

4 Fuzzy Logic for Signal Classification

This work addresses the problem of signal separation for 2-Dimensional spaces. Instead of resorting to statistical properties of the signals, this work treats the problem as one of image segmentation. Variants of known fuzzy classifiers are studied and compared with existing techniques, as the Unsupervised Maximum Likelihood (MLU) classifier or the watershed technique. The goal is the separation of seismic waves collected from experimental data.

Whenever there is uncertainty in the statistical model, fuzzy logic can be useful. This is maybe the case of supervised classification problems when the number of training data is low, or when there is lack of knowledge in the underlying parametric model as it is the case of geophysical signals, such as the ones addressed in this paper. This work aims at seismic wave separation by means of signal classification. Section 4.1 reviews a selection of the existing techniques [1,2] and studies how to treat fuzziness in order to better manage uncertainty. Section 4.2 applies these techniques to image segmentation for seismic wave separation or identification. Finally conclusions come in Section 4.3.

4.1 Fuzzy Unsupervised Classifiers

We focus on unsupervised classifiers that are going to be applied to 2-Dimensional signal separation, also called image segmentation. The algorithms like clustering or fuzzy C-means (FCM) [32-33], unsupervised Maximum Likelihood (MLU) [34] and watershed (W) [35] are the ones to be studied. From the simulations that we have carried out, we can conclude that a variant of the FCM, the so-called FACM (proposed by Gustafson and Kessel [32-33]), is the one that on average gives a better performance.

4.1.1 Fuzzy a C-Means

This algorithm minimizes the following distance of samples $\mathbf{x}_k$ to the cluster centers $\mathbf{v}_i$

$$J_m = \sum_{i=1}^{C}\sum_{k=1}^{N}(\mu_{ik})^m (\mathbf{x}_k - \mathbf{v}_i)^T \mathbf{A}_i (\mathbf{x}_k - \mathbf{v}_i) \tag{17}$$

where

$$\underset{1\leq i\leq C}{\forall} \mathbf{A}_i = [\rho_i \det C_{Fi}]^{\frac{1}{D}} C_{Fi}^{-1} \qquad \rho_i = \det(\mathbf{A}_i) \tag{18}$$

where D is the dimension of the feature space (e.g. D=3) and

$$C_{Fi} = \left[\sum_{k=1}^{N}(\mu_{ik})^m (\mathbf{x}_k - \mathbf{v}_i)(\mathbf{x}_k - \mathbf{v}_i)^T\right]\left[\sum_{k=1}^{N}(\mu_{ik})^m\right]^{-1} \tag{19}$$

where the fuzziness is controlled by factor "m".

Along the study that we have carried out we have observed a better performance when matrix Ai (the Mahalanobis distance) is introduced to the Fuzzy c-means. In this way the Mahalanobis distance defines an ellipsoid with an specific volume centered at each cluster (that we can refer to an image). Also better performance is obtained if the membership function is designed as

$$\underset{\substack{1\leq i\leq C \\ 1\leq k\leq N}}{\forall} \mu_{ik} = \left(\frac{1}{d_{ik}^{md}}\right)^{\frac{1}{m-1}}\left[\sum_{j=1}^{C}\left(\frac{1}{d_{jk}^{md}}\right)^{\frac{2}{m-1}}\right] \tag{20}$$

where

$$d_{ik} = (\mathbf{x}_k - \mathbf{v}_i)^T \mathbf{A}_i (\mathbf{x}_k - \mathbf{v}_i) \tag{21}$$

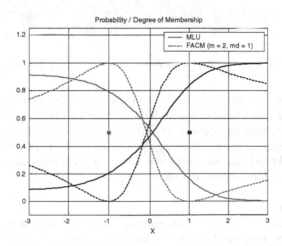

Fig. 10. Comparison of membership functions between MLU and FACM (classes with different variance show different borders)

Note that with respect to the Fuzzy c-means we have incorporated the parameter "md", which helps to better tune the membership function. For md>1, the membership degree of those points close to the cluster center are emphasized with respect to those more far apart.

A more closed mathematical analysis [36] reveals that FACM is like a version of the MLU, that uses a more generic kernel function as the Gaussian one used by MLU (we can see an example in Fig. 10). Therefore, FACM has more degrees of freedom to adapt to the data; thus, offering better results if it is properly tuned.

4.1.2 Fuzzy Watershed

The conventional morphological segmentation technique is the watershed transform [35]. The idea of watershed is drawn from a topographic analogy. Consider the gray-level intensity as a topographic relief. Find the minima and "pierce" them. Immerse the whole relief into water and let the water flood the areas adjacent to the piercing points. As the relief goes down some of the flooded areas will tend to merge; prevent this happening by raising infinitely tall dams along the watershed lines. When finished, the resulting network of dams defines the watershed of the image. Each of the lakes that have been formed are called catchment basins and correspond to the resulting classes of the classifier. Fig. 11 shows an example of watershed in a section of topographic surface.

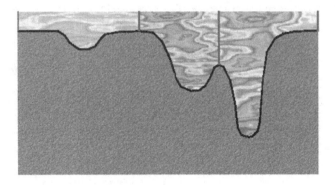

Fig. 11. Example of watershed in a section of topographic surface

Note that in the watershed segmentation there is no intersection between regions. As there is a great deal of ambiguity in the segmentation process, we studied the possibility of fuzzy membership degrees to allow the different catchment basins to intersect (in an algorithm that we call fuzzy Watershed or FW). However, the lack of a clear feature or knowledge to design the fuzzy membership functions, makes this extra degree of freedom in general useless as we show next.

4.1.3   Simulations

We have generated 12 different 2-Dimensional data of 100,000 samples each in order to evaluate the performance of the studied classifiers. Table 2 shows the probability of misclassification and Table 3 the mean error when reconstructing the image from the

Table 2. Misclassification error

Classi-fier	Gaussian	NonGaus-sian
CM	0.2404	0.3505
FCM	0.2436	0.3536
ACM	0.1226	0.1693
FACM	0.1175	0.1446
MLU	0.1071	0.1455

Table 3. Reconstruction error

Classi-fier	Gaussian	NonGaus-sian
CM	$7.447 \; 10^{-8}$	$3.056 \; 10^{-8}$
FCM	$3.256 \; 10^{-8}$	$1.755 \; 10^{-8}$
ACM	$2.839 \; 10^{-8}$	$2.240 \; 10^{-8}$
FACM	$6.707 \; 10^{-9}$	$9.397 \; 10^{-9}$
MLU	$1.131 \; 10^{-11}$	$1.305 \; 10^{-8}$

classified data. Although the FACM, the ACM and the UML are very similar in performance, note the better behavior of the FACM in front of the ACM, that is the c-means with the Mahalanobis distance. Note also the similar performance of the FACM and the unsupervised ML, although the UML presents a worst behavior in front of non-Gaussian shapes.

Next section considers in addition the watershed technique and fuzzy variants for the seismic image segmentation.

4.2   Separation of Seismic Signals

This application departs from a series of temporal signals measured in geological prospecting. The aim is to separate the different component waves.

4.2.1   Introduction to Seismic Prospecting

Seismic prospecting allows to know the structure of the earth underneath. A small explosive detonates on the surface and an array of sensors measures the generated waves in the subsoil. There are as many seismic waves as layers (between 6 and 10 Km of depth), see Fig.12. As the transmission speed of the waves in the different materials is known, the subsoil composition can be studied by analyzing the amplitude variations of each wave if the terrain of the surface is known.

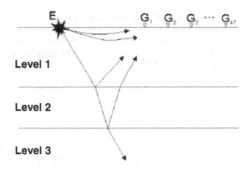

Fig. 12. Seismic profile

The generated waves belong to three different classes: i) waves P (primary), ii) waves S (secondary), and iii) waves L (long). Waves P are internal and longitudinal and fastest than waves S, which are internal but transversal. Waves L are superficial and of big amplitude and they cause the damages during the earthquakes. The explosions during seismic prospection cause mainly P and S waves.

4.2.2 Experimental Data
The experimental data that has been used in this work consists of a sequence of 47 temporal signals of 512 samples each, which, after an explosion, have been captured by each of the 47 seismic sensors. Fig. 13 shows the data, where we can see 4 different waves that separate as they propagate along the array because of the different propagation speeds.

4.2.3 Wave Separation
Before initiating the separation process, the data are pre-processed by means of the wavelet transform, which extracts the most relevant features in order to help the classifier in the separation process.

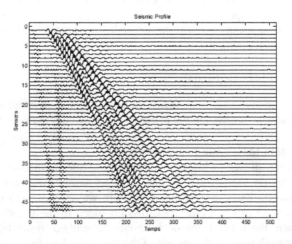

Fig. 13. Experimental seismic data: 47 sensors and 512 temporal samples at each sensor

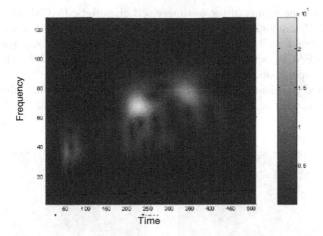

Fig. 14. Modulus of the time-frequency representation of sensor 47

The wavelet transform [6] obtains a representation in time and frequency for each of the signals that are measured at each sensor. Fig. 14 shows the modulus (scalogram) of the wavelet transform for sensor number 47. Note that 4 energy centers can be observed, which correspond to the 4 different temporal waves that propagate along the sensors. The separation is carried out by considering this energy distribution: each energy concentration is considered a different class.

Once the scalogram has been properly divided into segments by the appropriate classifier, the inverse wavelet transform is applied in order to obtain the separated signal in the time domain.

Before working with the experimental data, test or synthetic data has been used in order to evaluate the different methods described in section 4.1. 15 test signals have been generated by mixing different waveforms: sinusoids, wavelet kernels as Morlet type, Mexican hat, and Gaussian.

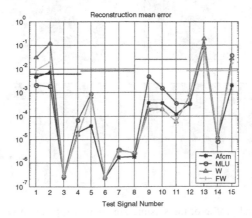

Fig. 15. Mean error when recovering each of the 15 temporal signals

Fig. 15 shows the results of the unsupervised classifiers: FACM, ML, W and FW for each of the 15 temporal mixtures. Note that although the FACM behaves well, it is not the best option for all the signals. In general, for low level of superposition in the scalogram, the FACM is the best, for medium level, the ML is to be chosen and, finally, for high superposition level, the W presents the best results.

When applied to seismic data, the FACM presents an additional advantage when compared to the other techniques. FACM does not need to look for the scalogram

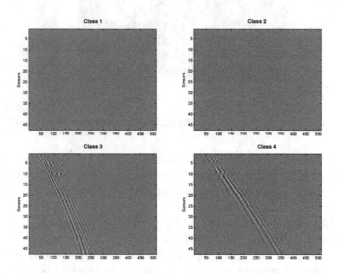

Fig. 16. Spatial-temporal profile of the seismic signal after separation via watershed

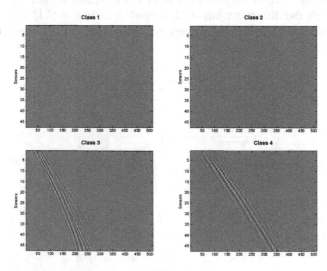

Fig. 17. Spatial-temporal profile of the seismic signal after separation via FACM (m=3, md=4 and m=2 for background class)

maxima image by image in a "manual" way. FACM can be initialized (i.e. cluster centers and matrix norm) with an image of well-separated clusters, as for instance the image obtained from the last sensor 47, and use the final parameters of one classification for initializing the classification of the next image. Thus taking advantage of the smooth evolution of the signal from sensor to sensor. We can also add one extra class used for background separation, leaving apart all the data points that doesn't bring any energy to the classes.

After extensive simulations, we can conclude that there are not substantial differences among the methods, although the FACM behaves in general better than the others. As the watershed is the most used technique for image segmentation, in Fig. 16 we compare it against FACM in Fig. 17.

Note in Fig. 16 that from sensor 15, the classes become too close together for the watershed to separate them properly. In Fig. 17 these problems disappear because the FACM is able to follow the classes thanks to the initialization with signal from sensor number 47. Finally, Fig. 18 shows the 4 temporal waves in this last sensor after separation with FACM.

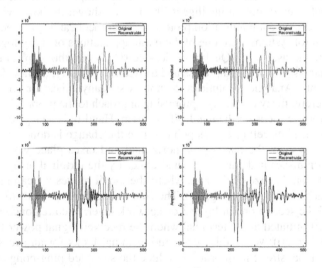

Fig. 18. Four waves after separation with FACM, comparison with the temporal signal of sensor 47

4.3 Conclusions

In this work fuzzy classifiers appear as a good alternative for image segmentation with the aim of seismic wave separation. With a proper tuning of their parameters their flexibility, when compared to other classifiers, allows them to adapt to many classification problems. Specifically, for the seismic wave classification problem the devised FACM techniques results in a good trade off between performance and complexity. This paper does not take into account expert knowledge of the seismic signal when designing the fuzzy system; thus, leaving open this point, which, to the authors believe can give promising results.

In the communication field, there is now a growing interest in reconfigurable receivers and cognitive radio. The study carried out in this work can be directed to blind recognition of the communication system in use (i.e. 3G or 3G systems); more specifically, to blind recognition or classification of the spectrum: channel bandwidth and shape, in order to reconfigure the entire architecture of the terminal with the appropriate software.

Up to now, the present work has applied fuzzy logic for signal filtering or separation at the bit or sample level. However the decision making of the rule base is also very useful for upper communication layers in the protocol stack that take into account input variables of different nature: congestion state, available load and total interference among others.

5 Fuzzy Logic at the Protocol Level: Horizontal Hand-Off

We have discussed on the application of fuzzy systems at the bit level so far. However, one of the main features of fuzzy systems is their explicit decision making. This feature is useful to carry out an intelligent filtering, as shown in the previous sections, but also to help design communication protocols. A protocol can be viewed as a control system, and control systems were one of the main applications of fuzzy logic. Protocols have to provide the users with Quality of Service (QoS) and this implies to cope with subjective variables, which are imprecise and difficult to quantify, as they depend on the user requirements. Analytical solutions do not exist many times for communications protocols; therefore, fuzzy control is a promising approach to the problem.

In this section we focus on the hand-off problem. Hand-off takes place when a movil terminal changes its cell or access point. When the change is done within the same communication system, the hand-off is horizontal and takes place when the QoS of the terminal diminishes and can be initiated either by the terminal or by the base station. Depending on the size of the area where the movil moves we can talk of micromobility or macromobility (see Fig. 19). The first one requires a hand-off at layer 2 or link layer, and the second one at layer 3 or network layer. We are concerned with the layer 2 hand-off initiated at the terminal when the received signal power falls below a threshold (see Fig. 20). When this occur, the terminal looks for an access point that offers it more signal strength. In order to reduce the so-called ping-pong effect among cells, the threshold has a hystereris. The hysteresis margin introduces a tolerance level above the received signal level from the current BS. In addition, there is a delay in the hand-off due to hysteresis. Thus a major obstacle facing the traditional hand-off with its hysteresis margin is the speed with which a hand-off can be made. It is for this reason that a better solution to hand-off is required, one that provides a fast yet stable performance. We propose to use fuzzy logic to design the thresholds so as to reduce the delay that the hand-off introduces in the signal transmission.

The thresholds are going to be design depending on the terminal profile, as for instance, its speed. Fig. 21 shows the proposed fuzzy controller for the hand-off threshold. Three fuzzy sets describe the universe of discourse for the speed and previous threshold is described with 5 fuzzy sets. Triangular fuzzy sets, min-max inference and centroide defuzzification are chosen for simplicity. The aim of the rule base is to optimize the threshold depending on the terminal speed. For high speed we would like to reduce the number of hand-offs because the terminal may go through many cells in a short time, thus the threshold should increase. On the other hand, for a slow terminal movement we would decrease the threshold level.

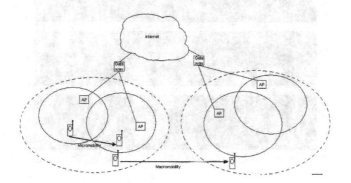

Fig. 19. Micromobility vs. macromobility

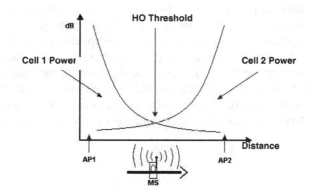

Fig. 20. Hand-off threshold at layer 2

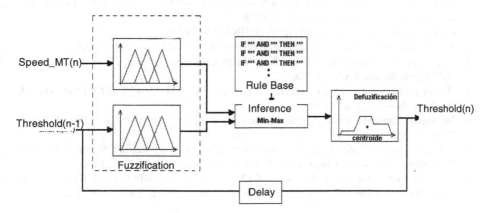

Fig. 21. Fuzzy controller for the hand-off threshold

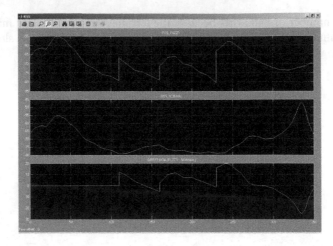

Fig. 22. Final received strength for the fuzzy controlled hand-off and for the conventional one

Fig. 22 plots the final power received by the access point for each of the strategies. In average, the fuzzy system offers more power, thus, better quality. However, depending on the application, voice for instance, the abrupt changes in the power might not be desirable. In case of file transfer, they are irrelevant. This fact motivates to incorporate the service type into the design of the final system. Future work is to incorporate more QoS variables in order to take the hand-off decision. Another aspect is the so-called soft hand-off, where two base stations or access points are received simultaneously during the hand-off. Fuzzy logic can then been used as an access point fusion technique.

References

1. S. Ghosh, Q. Razouqi, H.J. Schumacher and A. Celmins, "A Survey of Recent Advances in Fuzzy Logic in Telecomunnications Networks and New Challanges", IEEE Transactions on Fuzzy Systems, Vol. 6, No. 3, Aug. 1998
2. Fuzzy Engineering, Prentice Hall, 1996 (ISBN 0-13-124991-6).
3. Neural Networks and Fuzzy Systems, Prentice-Hall, 1991 (ISBN 0-13-611435-0).
4. *Uncertain Rule-Based Fuzzy Logic Systems: Introduction and New Directions*, Prentice-Hall, Upper Saddle River, NJ, 2001
5. "Fuzzy Logic Systems for Engineering: a Tutorial," Jerry M. Mendel, *IEEE Proc.*, vol. 83, no. 2, pp. 345-377, March 1995.
6. Intelligent Signal Processing, with S. Haykin (co-editors), IEEE Press , 2001 (ISBN 0-7803-6010-9).
7. "Special Issue on Fuzzy Logic in Signal Processing," Eurasip Journal, Signal Processing, ISSN 0165-1684, vol. 80, no.6, Junio 2000 Guest Editor: A. Pérez-Neira Processing
8. "Fuzzy Adaptive Signal Predistorter for OFDM Systems," EUSIPCO'00, J. Bas, A. Pérez-Neira, Vol. IV, ISBN 952-15-0447-1, Tampere, Septiembre 4-8, 2000 (Finlandia).

9. A.Pérez-Neira, M.A. Lagunas, A. Jové, A. Artés ,"A fuzzy logic filter for coherent detection in mobile communication receivers," IX European Signal Processing Conference (EUSIPCO), Proceedings ref.: 960-7620-06-2, pp. 1601-1604, 8-11 Septiembre 1998, Rhodes (Grecia).

10. L. Zadeh, "General filters for separation of signal and noise", Proc. Symposium on information networks, pp.31-49, 1954.

11. R.-J. Li and E.S. Lee, "Analysis of fuzzy queues", Comput. Math. Applicat., Vol. 17, No. 7, pp. 1143-1147, 1989.

12. H.M. Prade, "An outline of fuzzy or possibilistic models for queuing systems", Proc. Symp. Policy Anal. Inform. Syst., Durham, NC, pp. 147-153, 1980.

13. A.R. Bonde and S. Ghosh, "A comparative study of fuzzy versus "fixed" thresholds for robust queue management in cell-switching networks", IEEE Trans. Networking, Vol. 2, pp. 337-344, Aug. 1994.

14. S. Pithani and A.S. Sethi, "A fuzzy set delay representation for computer network routing algorithms", Proc. 2nd. Int. Symp. Uncertainty Modeling Anal., College Park, MD, pp. 286-293, Apr. 1993.

15. Y. Tanaka and S. Hosaka, "Fuzzy control of telecommunications networks using learning technique", Electron. Commun. Japan, Vol. 76, Pt. 1, No. 12, pp. 41-51, Dec. 1993.

16. R.-G. Cheng and C.-J. Chang, "Design of a fuzzy traffic controller for ATM networks", IEEE Trans. Networking, Vol. 4, pp. 460-469, Jun. 1996.

17. K.-R. Lo and C.-J. Chang, "A Neural Fuzzy Resource Manager for Hierarchical Cellular Systems Supporting Multimedia Services", IEEE Trans. Veh. Tech., Vol. 52, No. 5, pp. 1196-1206, Sep. 2003.

18. G. Edwards and R. Sankar, "Hand-off using fuzzy logic", Proc. IEEE GLOBECOM, Singapore, Vol. 1, pp. 524-528, Nov. 1995.

19. S.S.-F. Lau, K.-F. Cheung and J-C.I. Chuang, "Fuzzy logic adaptive handoff algorithm", Proc. IEEE GLOBECOM, Singapore, Vol. 1, pp. 509-513, Nov. 1995.

20. P.M.L. Chan, R.E. Sheriff, Y.F. Hu, P. Conforto and C. Tocci, "Mobility Management Incorporating Fuzzy Logic for a Heterogeneous IP Environment", IEEE Comm. Magazine, pp. 42-51, Dec. 2001.

21. Joan Bas, Ana I. Pérez-Neira, Differential Fuzzy Filtering for Adaptive Line Enhancement in Spread Spectrum Communications, accepted in Signal Processing de Eurasip.

22. W,Wu,"New Nonlinear Algorithms for Estimating and Suppressing Narrowband Interference in DS Spread Spectrum Systems", in IEEE Trans on Commun, vol.44,nº4, pp. 508-515 April 1996.

23. K.Wang, Y.Yao, "New Nonlinear Algorithms for Narrowband Interference Suppression in CDMA Spread-Spectrum Systems," in IEEE J. Selected Areas in Commun., vol.17,nº12,pp.2148-2153, Dec.1999.

24. J. Bas, Ana I. Pérez-Neira, An Scalable Fuzzy Interference Canceller for DS-CDMA Systems, accepted at Intelligent System (John Wiley & Sons), Special Issue on Soft computing for Modelling, Simulation and Control on Non-linear Dynamical Systems, November, 2003.

25. L.C. Godara. Error analysis of the optimal antenna array processors. In IEEE Trans. Aerosp. Electron. Syst., vol. AES-22, July 1986, pp. 395-409.

26. B.D. Carlson. Covariance matrix estimation errors and diagonal loading in adaptive arrays. In vol.43, Nov. 1995, pp. 2724-2732.

27. Theory and Application of Covariance Matrix Tapers for Robust Adaptive Beamforming. In IEEE Trans. on Signal Processing, vol. 47, no.4, April 1999.

28. K.L. Bell, Y. Ephraim and H.L. Van Trees. A Bayesian Approach to Robust Adaptive Beamforming. In IEEE Transactions on Signal Processing, vol. 48, n. 2, Feb. 2000.
29. J. Yang and A.L. Swindlehurst. The effects of array calibration errors on DF-based signal copyperformance. In IEEE Trans.Signal Processing, vol.43, Nov. 1995, pp. 2724-2732.
30. C. Lee and J. Lee. Eigenspace-based adaptive array beamforming with robust capabilities. In IEEE trans. on AP., vol. 45, no.12, Dec. 1997.
31. S. A. Kassam and H. Vincent Poor. Robust techniques for Signal Processing: A survey. In IEEE Proceedings, March 1985.
32. J.C. Bezdek, Pattern Recognition with Fuzzy objective function algorithms, 2nd ed. New York: Plenum Press, 1981.
33. Czogala, E.; Leski J., Fuzzy and Neuro-Fuzzy Intelligent Systems, Heidelberg [etc.]: Physica-Verlag, 2000.
34. R.O. Duda et al., Pattern Classification, 2nd edition New York: John Wiley and Sons cop., 2001.
35. S. Beucher et al., "Use of watersheds in contour detection", Proc. International workshop on image processing, real-time edge and motion, Rennes, September, 1979, pp: 1928-1931.
36. J. Serra, A. Pérez-Neira, "Fuzzy systems for seismic wave separation," Proceed. Eusflat 05 conference, Barcelona, 7-9 Septiembre 2005.

Connected Operators for Signal and Image Processing

Philippe Salembier

Universitat Politecnica de Catalunya, Barcelona, Spain
philippe@gps.tsc.upc.edu
http://gps-tsc.upc.es/imatge/

1 Introduction

Data and signal modeling for images and video sequences is experiencing important developments. Part of this evolution is due to the need to support a large
number of new multimedia services. Traditionally, digital images were represented as rectangular arrays of pixels and digital video was seen as a continuous
flow of digital images. New multimedia applications and services imply a representation that is closer to the real world or, at least, that takes into account part
of the process that has created the digital information. Content-based compression and indexing are two typical examples of applications where new modeling
strategies and processing tools are necessary:

- For content-based image or video compression, the representation based on
  an array of pixels is not appropriate if one wants to be able to act on objects,
  to encode differently the areas of interest, or to assign different behaviors to
  the entities represented in the image. In these applications, the notion of
  object is essential. As a consequence, the data modeling has to include, for
  example, regions of arbitrary shapes to represent objects.
- Content-based indexing applications are also facing the same kind of challenges. For instance, the video representation based on a flow of frames is
  inadequate for many video indexing applications. Among the large set of
  functionalities involved in a retrieval application, let us consider browsing.
  The browsing functionality should go far beyond the "fast forward" and
  "fast reverse" allowed by VCRs. One would like to have access to a table of
  contents of the video and to be able to jump from one item to another. This
  kind of functionality implies at least a structuring of the video in terms of
  individual shots and scenes. Of course, indexing and retrieval involve also a
  structuring of the data in terms of objects, regions, semantic notions, etc.

In both examples, the data modeling has to take into account part of the
creation process: an image is created by projection of a visual scene composed
of 3D objects onto a 2D plane. Modeling the image in terms of regions is an
attempt to know the projection of the 3D object boundaries in the 2D plane.
Video shots detection also aims at finding what has been done during the video
editing process and where boundaries between elementary components have been

M. Faundez-Zanuy et al. (Eds.): NOLISP 2005, LNAI 3817, pp. 37–65, 2005.
© Springer-Verlag Berlin Heidelberg 2005

introduced. In both cases, the notion of region turns out to be central in the modeling process. Note that regions may be spatial connected components but also temporal or spatio-temporal connected components in the case of video.

Besides the modeling issue, it has to be recognized that most image processing tools are not suited to region-based representations. For example, the vast majority of low level processing tools such as filters are very closely related to the classical pixel-based representation of signals. Typical examples include linear convolution with an impulse response, median filter, morphological operators based on erosion and dilation with a structuring element, etc. In all cases, the processing strategy consists in modifying the values of individual pixels by a function of the pixels values in a local window.

Early examples of region-based processing can be found in the literature in the field of segmentation. For example, the classical *Split & Merge* algorithm [1] defines first a set of elementary regions (the split process) and then interacts directly on these regions allowing them to merge under certain conditions.

Recently, a set of morphological filtering tools called *Connected Operators* has received much attention. Connected operators are region-based filtering tools because they do not modify individual pixel values but directly act on the connected components of the space where the image is constant, the so-called *flat zones*. Intuitively, connected operators can remove boundaries between flat zones but cannot add new boundaries nor shift existing ones. The related literature rapidly grows and involves theoretical studies [2, 3, 4, 5, 6, 7, 8, 9, 10], algorithm developments [11, 12, 13, 14, 15, 16] and applications [17, 18, 19, 20]. The goal of this paper is 1) to provide an introduction to connected operators for gray level images and video sequences and 2) to discuss the techniques and algorithms that have been up to now the most successful within the framework of practical applications.

The organization of this paper is as follows: The following section introduces the notation and highlights the main drawbacks of classical filtering strategies. Then, the next section presents the basic notions related to connected operators and discuss some early examples of connected operators. In practice, the two most successful strategies to define connected operators are based either on reconstruction processes or on tree representations. Both approaches are discussed in separate sections. Finally, conclusions are given in the last section.

2   Classical Filtering Approaches

In this section, we define the notation to be used in the sequel and review some of the basic properties of interest in this paper [21, 22]. We deal exclusively with discrete images $f[n]$ or video sequences $f_t[n]$ where n denotes the pixel or space coordinate (a vector in the case of 2D images) and t the time instant in the case of a video sequence. In the lattice of grey level functions, an image f is said to be smaller than an image g if and only if:

$$f \leq g \Longleftrightarrow \forall n, f[n] \leq g[n] \tag{1}$$

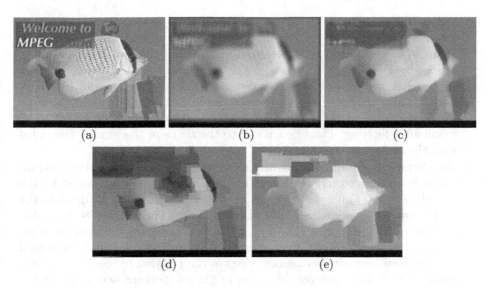

Fig. 1. Example of classical filters: (a) original image, (b) low-pass filter (7x7 average), (c) median (5x5), (d) opening (5x5), (e) closing (5x5)

An operator ψ acting on an input f is said to be:

- *increasing*: $\forall f, g, \quad f \leq g \Longrightarrow \psi(f) \leq \psi(g)$
  (The order is preserved by the filtering)
- *idempotent*: $\forall f, \quad \psi(\psi(f)) = \psi(f)$
  (Iteration of the filtering is not needed)
- *extensive*: $\forall f, \quad f \leq \psi(f)$
  (The output is always greater than the input)
- *anti-extensive*: $\forall f, \quad \psi(f) \leq f$
  (The output is always smaller than the input)
- *a morphological filter*:   if it is increasing and idempotent
- *an opening*:   if it is an anti-extensive morphological filter
- *a closing*:   if it is an extensive morphological filter
- *self-dual*: $\forall f, \quad \psi(f) = -\psi(-f)$
  (Same processing is for bright & dark components)

Almost all filtering techniques commonly used in image processing are defined by a computation rule and a specific signal $h[n]$ that may be called impulse response, window or structuring element. Let us review these classical cases:

– Linear convolution and impulse response: the output of a linear translation-invariant system is given by: $\psi_h(f)[n] = \sum_{k=-\infty}^{\infty} h[k]f[n-k]$. The impulse response, $h[n]$, defines the properties of the filter. An example of linear filtering result is shown in Fig. 1(b). The original image shown in Fig. 1(a). As can be seen, most of the details of the original image are attenuated by the filter (average of size 7x7). However, details are not really removed but simply blurred. The characteristics of the blurring is directly related to the extension and shape of the impulse response.

- Median filter and window: The output of a median filter with window W is defined by: $\psi_W(f)[n] = Median_{k \in W}\{f[n-k]\}$. Here also the basic properties of the filter are defined by its window. An example is shown in Fig. 1(c). Here, small details are actually removed (for example the texture of the fish). The major drawback of the filtering strategy is that every region tends to be round after filtering. This effect is due to the shape of the window combined with the median processing.

- Morphological erosion/dilation and structuring elements: morphological dilation by a structuring element $h[n]$ is defined in a way similar to the convolution: $\delta_h(f)[n] = \bigvee_{k=-\infty}^{\infty}(h[k] + f[n - k])$, where $\bigvee$ denotes the supremum (or maximum in the discrete case). The erosion is given by: $\epsilon_h(f)[n] = \bigwedge_{k=-\infty}^{\infty}(h[k] - f[n + k])$, where $\bigwedge$ denotes the infimum (or minimum in the discrete case). In practice, erosion and dilation are seldom used on their own because they do not preserve the position of contours. For example, the dilation enlarges the size of bright components and decreases the size of dark components by displacing their contours. However, they provide a simplification effect: a dilation (erosion) removes dark (bright) components that do not fit within the structuring element. Based on these two primitives, morphological opening and closing can be constructed.

The opening is given by: $\gamma_h(f) = \delta_h(\epsilon_h(f))$ and the closing by: $\varphi_h(f) = \epsilon_h(\delta_h(f))$. These operators are morphological filters (that is, at the same time, increasing and idempotent). Moreover, the opening is anti-extensive (it removes bright components) whereas the closing is extensive (it removes dark components). The Processing results are shown in Fig.s 1(d) and 1(e). In the case of opening (closing) with a square structuring element of size 5x5, small bright (dark) components have been removed. As can be seen, the contours remain sharp and centered on their original position. However, the shape of the components that have not been removed are not perfectly preserved. In both examples, square shapes are clearly visible in the output image. This is due to the square shape of the structuring element.

Once a processing strategy has been selected (linear convolution, median, morphological operator, etc.), the filter design consists in carefully choosing a specific signal $h[n]$ which may be the impulse response, the window or the structuring element. While most people would say that this is the heart of the filter design, our point here is to highlight that, for image processing, the use of $h[n]$ has some drawbacks. In all examples of Fig. 1, $h[n]$ is not related to the input signal and its shape clearly introduces distortions in the output. The distortion effect depends on the specific filter, but for a large range of applications requiring high precision on contours, none of these filtering strategies is acceptable.

To reduce the distortion, one possible solution is to adapt $h[n]$ to the local structures of the input signal. This solution may improve the results but still remains unacceptable in many circumstances. An attractive solution to this problem is provided by connected operators. Most connected operators used in practice rely on a completely different filtering strategy: the filtering is done without using any specific signal such as an impulse response, a window or a structuring element. In fact, the structures of the input signal are used to act on the signal itself. As a result, no distortion related to a priori selected signals is introduced in the output.

3   Connected Operators

3.1   Definitions and Basic Properties

Gray level connected operators act by merging flat zones. They cannot create new contours and, as a result, they cannot introduce in the output image a structure that is not present in the input image. Furthermore, they cannot modify the position of existing boundaries between regions and, therefore, have very good contour preservation properties.

Gray level connected operators originally defined in [2] rely on the notion of partition of flat zones. A partition is a set of non-overlapping connected components or regions that fills the entire space. We assume that the connectivity is defined on the digital grid by a translation invariant, reflexive and symmetric relation[1]. Typical examples are the 4- and 8-connectivity. Let us denote by $\mathcal{P}$ a partition and by $\mathcal{P}(n)$ the region that contains pixel n. A partial order relationship among partitions can be created: $\mathcal{P}_1$ "is finer than" $\mathcal{P}_2$ (written as $\mathcal{P}_1 \sqsubseteq \mathcal{P}_2$), if $\forall n, \mathcal{P}_1(n) \subseteq \mathcal{P}_2(n)$.

It can be shown that the set of flat zones of an image f is a partition of the space, $\mathcal{P}_f$. Based on these notions, connected operators are defined as:

Definition 1. *(Connected operators) A gray level operator ψ is connected if the partition of flat zones of its input f is always finer than the partition of flat zones of its output, that is:*

$$\mathcal{P}_f \sqsubseteq \mathcal{P}_{\psi(f)}, \forall f$$

This definition clearly highlights the region-based processing of the operator: indeed, regions of the output partition are created by union of regions of the input partition. An alternative (and equivalent) definition of connected operators was introduced in [6]. This definition enhances the role of the boundaries between regions and turns out to be very useful to derive leveling.

Definition 2. *(Connected operators) A gray level operator ψ is connected if $\forall f$ input image and $\forall n, n'$ neighboring pixels,*

$$\psi(f)[n] \neq \psi(f)[n'] \implies f[n] \neq f[n'].$$

This definition simply states that if two neighboring pixels of the output image have two different gray level values, they have also two different gray level values in the input image, in other words, the operator cannot create new boundaries.

---

[1] In the context of connected operators, several studies have been carried out on the definition of less usual connectivities. The reader is referred to [22, 23, 24, 9, 8] for more details on this issue.

New connected operators can be derived from the combination of primitive connected operators. The following properties give a few construction rules:

Proposition 1. *(Properties of connected operators)*

- *If ψ is a connected operator, its dual ψ^* defined by: $\psi^*(f) = -\psi(-f)$, is also connected.*
- *If ψ_1, ψ_2 are connected operators, $\psi_2\psi_1$ is also connected.*
- *If $\{\psi_i\}$ are connected operators, their supremum $\bigvee_i \psi_i$ and infimum $\bigwedge_i \psi_i$ are connected.*

3.2   Early Examples of Connected Operators

The first known connected operator is the *binary opening by reconstruction* [25]. This operator eliminates the connected components that would be totally removed by an erosion with a given structuring element and leaves the other components unchanged. This filtering approach offers the advantage of simplifying the image (some components are removed) as well as preserving the contour information (the components that are not removed are perfectly preserved). It can be shown that the process is increasing, idempotent and anti-extensive, that is an opening. Moreover, it was called "by reconstruction" because of the algorithm used for its implementation. From the algorithmic viewpoint, if X is the original binary image, the first step is to compute an erosion with a structuring element B_k of size k, $\epsilon_{B_k}(X)$. This erosion is used to "mark" the connected components that should be preserved. The final result is obtained by progressively dilating the erosion inside the mask defined by the original image:

1. $Y_0 = \epsilon_{B_k}(X)$
2. $Y_k = \delta_C(Y_{k-1}) \bigcap X$, where C is a binary structuring element defining the connectivity, e.g. square of 3x3 (cross) for the 8-connectivity (4-connectivity).
3. Iterate step 2 until idempotence.

The first gray level connected operator was obtained by a transposition of the previous approach to the lattice of gray level functions [22, 11]. It is known as an *opening by reconstruction of erosions*:

1. $g_0 = \epsilon_{h_k}(f)$, where f is the input and h_k a structuring element of size k.
2. $g_k = \delta_C(g_{k-1}) \bigwedge f$, where C is a flat structuring element defining the connectivity, e.g. square or cross.
3. Iterate step 2 until idempotence.

It was shown in [2] that this operator is connected. Intuitively, the erosion acts as a simplification step by removing small bright components. The reconstruction process restores the contours of the components that have not been completely removed by the erosion.

There are several ways to construct connected operators and many new operators have been recently introduced. From the practical viewpoint, the most successful strategies rely on a reconstruction process or on region-tree pruning. Operators resulting from these two strategies are discussed in the sequel.

4 Connected Operators Based on Reconstruction Processes

4.1 Anti-extensive Reconstruction and Connected Operators

The Anti-extensive Reconstruction Process. The most classical way to construct connected operators is to use an anti-extensive reconstruction process. It is defined as follows:

Definition 3. *(Anti-extensive reconstruction) If f and g are two images (respectively called the "reference" and the "marker" image), the anti-extensive reconstruction $\rho^{\downarrow}(g|f)$ of g* under *f is given by:*

$$
\begin{aligned}
g_k &= \delta_C(g_{k-1}) \bigwedge f \text{ and} \\
\rho^{\downarrow}(g|f) &= \lim_{k \to \infty} g_k
\end{aligned}
\tag{2}
$$

where $g_0 = g$ and δ_C is the dilation with the flat structuring element defining the connectivity (3x3 square or cross).

It can be shown that the series, g_k, always converges and the limit always exists. Of course by duality, an extensive reconstruction may be defined:

Definition 4. *(Extensive reconstruction) If f and g are two images (respectively called the "reference" and the "marker" image), the extensive reconstruction $\rho^{\uparrow}(g|f)$ of g* above *f is given by:*

$$
\begin{aligned}
g_k &= \epsilon_C(g_{k-1}) \bigvee f \text{ and} \\
\rho^{\uparrow}(g|f) &= \lim_{k \to \infty} g_k
\end{aligned}
\tag{3}
$$

where $g_0 = g$ and ϵ_C is the erosion with the flat structuring element defining the connectivity (3x3 square or cross).

Note that Eqs. (2) and (3) define the reconstruction processes but do not provide efficient implementations. Indeed, the number of iterations is generally fairly high. The most efficient reconstruction algorithms rely on the definition of a clever scanning of the image and are implemented by First-in-First-out (FIFO) queues. A review of the most popular reconstruction algorithms can be found in [11]. Here, we describe a simple but efficient one: the basic idea of the algorithm is to start from the regional maxima of the marker image g and to propagate them *under* the original image f. The algorithm works in two steps:

1. The *initialization* consists in putting in the queue the location of pixels that are on the boundary of the regional maxima of the marker image. Regional maxima are the set of connected components where the image has a constant gray level value and such that every pixel in the neighborhood of the regional maxima has strictly a lower value. Algorithms to compute regional maxima can be found in [26].
2. The *propagation* extracts the first pixel, n, from the queue (note that n is a pixel of the marker image g). Then, it assigns to each of its neighbors, n',

that have a strictly lower gray level value than $g[n]$ (that is, if $g[n'] < g[n]$), the minimum between the gray level value of n and the gray level value of the pixel of the original image at the same location than n', that is $g[n'] = g[n] \wedge f[n']$. Finally, the pixel n' is introduced in the queue. This propagation process has to be carried on until the queue is empty. The algorithm is very efficient because the image pixels are processed only once.

In practice, useful connected operators are obtained by considering that the marker image g is a transformation $\phi(f)$ of the input image f. As a result, most connected operators ψ obtained by reconstruction can be written as:

$$\psi(f) = \rho^{\downarrow}(\phi(f)|f) \quad \text{(anti-extensive operator), or}$$
$$\psi(f) = \rho^{\uparrow}(\phi(f)|f) \quad \text{(extensive operator).}$$
(4)

In the following, a few examples are discussed.

Size Filtering. The simplest size-oriented connected operator is obtained by using as marker image, $\phi(f)$, the result of an erosion with a structuring element h_k of size k. It is the opening by reconstruction of erosion[2]:

$$\psi(f) = \rho^{\downarrow}(\epsilon_{h_k}(f)|f)$$
(5)

It can be demonstrated that this operator is an opening. By duality, the closing by reconstruction is given by:

$$\psi^*(f) = \rho^{\uparrow}(\delta_{h_k}(f)|f)$$
(6)

An example of opening by reconstruction of erosion is shown in Fig. 2(a). In this example, the original signal f has 11 maxima. The marker signal g is created by an erosion with a flat structuring element which eliminates the narrowest maxima. Only 5 maxima are preserved after erosion. Finally, the marker is reconstructed. In the reconstruction, only the 5 maxima that were present after erosion are visible and narrow maxima have been eliminated. Moreover, the transitions of the reconstructed signal correspond precisely to the transitions of the original signal.

As can be seen, the simplification effect, that is the elimination of narrow maxima is almost perfectly done. However, the preservation effect may be criticized: although the maxima contours are well preserved, their shape and height are distorted. To reduce this distortion, a new connected operator can be built on top of the first one. Let us construct a new marker image, $m[n]$, indicating the pixels where the reconstruction has been inactive, that is where the final result is equal to the erosion.

$$m[n] = \begin{cases} f[n], & \text{if } \rho^{\downarrow}(\epsilon_{h_k}(f)|f)[n] = \epsilon_{h_k}(f)[n] \\ 0 & \text{otherwise.} \end{cases}$$
(7)

---
[2] Note that it can be demonstrated that the same operator is obtained by changing the erosion, ϵ_{h_k}, by an opening, γ_{h_k}, with the same structuring element, h_k.

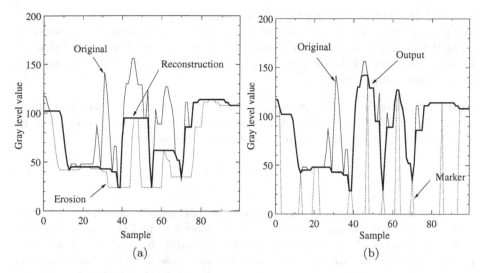

Fig. 2. Size-oriented connected operators: (a) Opening by reconstruction, (b) New marker indicating where the reconstruction has been inactive and second reconstruction

This marker image is illustrated in Fig. 2(b). As can be seen, it is equal to 0 except for the five maxima that are present after erosion and also for the local minima. At that locations, the gray level values of the original image, $f[n]$, are assigned to the marker image. Finally, the second connected operator is created by the reconstruction of the marker, m under f:

$$\psi(f) = \rho^{\downarrow}(m|f) \tag{8}$$

This operator is also an opening by reconstruction. The final result is shown in Fig. 2(b). The five maxima are better preserved than with the first opening by reconstruction whereas the remaining maxima are perfectly removed. The difference between both reconstructions is also clearly visible in the examples of Fig. 3. The first opening by reconstruction removes small bright details of the image: the text in the upper left corner. The fish is a large element and is not

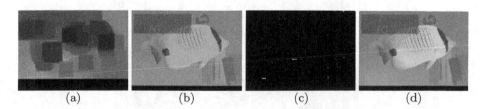

Fig. 3. Size filtering with opening by reconstruction: (a) erosion of the original image of Fig. 1(a) by a flat structuring element of size 10x10, (b) reconstruction of the erosion, (c) marker indicating where the first reconstruction has not been active (Eq. 7) and (d) second reconstruction

removed. It is indeed visible after the first opening by reconstruction (Fig. 3(b)) but its gray level values are not well preserved. This drawback is avoided by using the second reconstruction. Finally, let us mention that by duality closings by reconstruction can be defined. They have the same effect than the openings but on dark components.

Contrast Filtering. The previous section considered size simplification. A contrast simplification can be obtained by substituting the erosion in Eq. 5 by a subtraction of a constant, c, from the original image f:

$$\phi(f) = \rho^{\downarrow}(f - c|f) \tag{9}$$

This operator, known as λ-max operator, is connected, increasing and anti-extensive but not idempotent. Its effect is illustrated in Fig. 4(a). As can be seen, the maxima of small contrast are removed and the contours of the maxima of high contrast are well preserved. However, the height of the remaining maxima are not well preserved. As in the previous section, this drawback can be

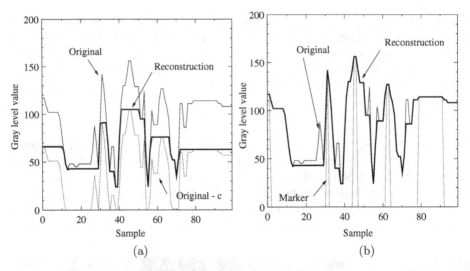

Fig. 4. Contrast-oriented connected operators: (a) Reconstruction of $f - c$, (b) Second reconstruction

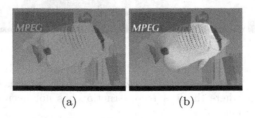

Fig. 5. Contrast filtering: (a) λ-max operator, (b) dynamic opening

removed if a second reconstruction process is used. This second reconstruction process is exactly the same as the previous one defined by Eq. 7 ($m[n] = f[n]$ if $\rho^\downarrow(f - c|f)[n] = f[n] - c$). This second connected operator is an opening. It is called a dynamic opening [27].

The operators effect is illustrated in Fig. 5. Both operators remove maxima of contrast c lower than 100 gray level values. However, the λ-max operator produces an output image of low contrast, even for the preserved maxima. By contrast, the dynamic opening successfully restores the retained maxima.

4.2 Self-dual Reconstruction and Levelings

The connected operators discussed in the previous section were either anti-extensive or extensive. They allow the simplification of either bright or dark image components. For some applications, this behavior is a drawback and one would like to simplify in a symmetrical way all components. From the theoretical viewpoint, this means that the filter has to be self-dual, that is $\psi(f) = -\psi(-f)$.

With the aim of constructing self-dual connected operators, the concept of levelings was proposed in [6] by adding some restrictions in Definition 2:

Definition 5. *(Leveling) The operator ψ is a leveling if $\forall n, n'$ neighboring pixels, $\psi(f)[n] > \psi(f)[n'] \Longrightarrow f[n] \geq \psi(f)[n]$ and $\psi(f)[n'] \geq f[n']$.*

This definition not only states that if a transition exists in the output image, it was already present in the original image (Definition 2) but also that 1) the sense of gray level variation between n and n' has to be preserved and 2) the variation $\|\psi(f)[n] - \psi(f)[n']\|$ is bounded by the original variation $\|f[n] - f[n']\|$.

The theoretical properties of levelings are studied in [6, 7], in particular:

- Any opening or closing by reconstruction is a leveling.
- If ψ_1, ψ_2 are levelings, $\psi_2\psi_1$ is also a leveling.
- If $\{\psi_i\}$ are levelings, their supremum $\bigvee_i \psi_i$, and infimum $\bigwedge_i \psi_i$, are levelings.

The most popular technique to create levelings relies on the following self-dual reconstruction process:

Definition 6. *(Self-dual reconstruction) If f and g are two images (respectively called the "reference" and the "marker"image), the self-dual reconstruction $\rho^\updownarrow(g|f)$ of g with respect to f is given by:*

$$
\begin{aligned}
g_k &= \epsilon_C(g_{k-1}) \bigvee [\delta_C(g_{k-1}) \bigwedge f] \\
&= \delta_C(g_{k-1}) \bigwedge [\epsilon_C(g_{k-1}) \bigvee f] \text{ (equivalent expression) and} \\
\rho^\updownarrow(g|f) &= \lim_{k\to\infty} g_k
\end{aligned}
\tag{10}
$$

where $g_0 = g$ and δ_C and ϵ_C are respectively the dilation and the erosion with the flat structuring element defining the connectivity (3x3 square or cross).

An example of self-dual reconstruction is shown in Fig. 6. In this example, the marker image is constant everywhere except for two points that mark a maximum and a minimum of the reference image. After reconstruction, the output has only

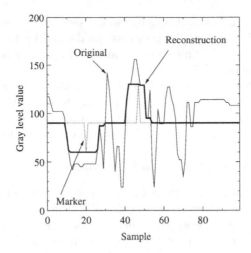

Fig. 6. Example of leveling with self-dual reconstruction

one maximum and one minimum. As can be seen, the self-dual reconstruction is the anti-extensive reconstruction of Eq. 3 for the pixels where $g[n] < f[n]$ and the extensive reconstruction of Eq. 4 for the pixels where $f[n] < g[n]$.

As in the case of anti-extensive reconstruction, Eq. 10 does not define an efficient implementation of the reconstruction process. In fact, an efficient implementation of the self-dual reconstruction can be obtained by combination of the strategies used for anti-extensive and extensive reconstruction processes: the *initialization* step consists in putting in the FIFO queue: 1) the boundary pixels of marker maxima when the marker is smaller than the reference and 2) the boundary pixels of marker minima when the marker is greater than the reference.

The *propagation* step is done in a similar fashion than the one described for anti-extensive reconstruction: the anti-extensive propagation is used when the marker is below the reference and the extensive propagation is used when the marker is above the reference.

In practice, the self-dual reconstruction is used to restore the contour information after a simplification performed by an operator that is neither extensive nor anti-extensive. A typical example is an alternating sequential filter: $g = \varphi_{h_k} \gamma_{h_k} \varphi_{h_{k-1}} \gamma_{h_{k-1}} \cdots \varphi_{h_1} \gamma_{h_1}(f)$, where φ_{h_k} and γ_{h_k} are respectively a closing and an opening with a structuring element h_k. This example is illustrated in Fig.s 7(a) and 7(b). Note the simplification effect which deals with both maxima and minima, and how the contour distortion introduced by the alternating sequential filter is removed by the reconstruction. However, from a theoretical viewpoint, the operator: $\rho^{\uparrow}(\varphi_{h_k} \gamma_{h_k} \cdots \varphi_{h_1} \gamma_{h_1}(f)|f)$ is not self-dual because the alternating sequential filter itself is not self-dual. In order to create a self-dual operator, the creation of the marker has also to be self-dual. Fig.s 7(c) and 7(d) show an example where the marker is created by a median filter (that is self-dual). This kind of results can be extended to any linear filter and the self-dual reconstruction can be considered as a general tool that restores the contour in-

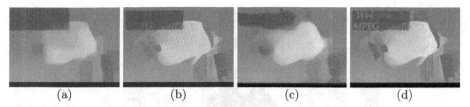

| (a) | (b) | (c) | (d) |

Fig. 7. Size filtering with leveling: (a) Alternating sequential filter of the original image of Fig. 1(a), (b) Self-dual reconstruction of the alternating sequential filter, (c) Median filter and (d) Self-dual reconstruction of the median filter

formation after a filtering process. In other words, the reconstruction allows to create a connected version $\rho^{\updownarrow}(\psi(f)|f)$ of any filter: $\psi(f)$.

5   Connected Operators Based on Region-Tree Pruning

5.1   Tree Representations and Connected Operators

The reconstruction strategies discussed in the previous section can be viewed as tools that work on a pixel-based representation of the image and that provide a way to create connected operators. In this section, we present a different approach: the first step of the filtering process is to construct a region-based representation of the image, then the simplification effect is obtained by direct manipulation of the tree. The approach may be considered as being conceptually more complex than the reconstruction however, it provides more flexibility in the choice of the simplification criterion.

Two region-based representations are discussed in the sequel: the Max-tree / Min-tree [14] and the Binary Partition Tree [15]. The first one leads to anti-extensive connected operators whereas the second one is a basis for self-dual connected operators. Let us discuss first these two region-based representations.

Max-tree and Min-tree. The first representation is called a Max-tree [14]. It enhances the maxima of the signal. Each tree node $\mathcal{N}_k$ represents a connected component of the space that is extracted by the following thresholding process: for a given threshold T, consider the set of pixels X that have a gray level value larger than T and the set of pixels Y that have a gray level value equal to T:

$$X = \{n \text{ , such that } f[n] \geq T\}$$
$$Y = \{n \text{ , such that } f[n] = T\} \tag{11}$$

The tree nodes $\mathcal{N}_k$ represent the connected components of X such that $X \cap Y \neq \emptyset$. A simple example of Max-tree is shown in Fig. 8. The original image is made of 7 flat zones identified by a letter {A,...,G}. The number following each letter defines the gray level value of the flat zones. The binary images, X, resulting from the thresholding with $0 \leq T \leq 2$ are shown in the center of the figure. Finally, the Max-tree is given in the right side. It is composed of 5 nodes that represent the connected components shown in black. The number inside each square

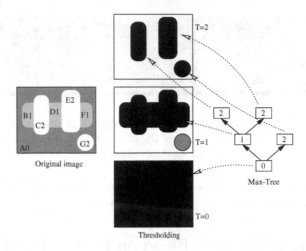

Fig. 8. Max-tree representation of images

represents the threshold value where the component was extracted. Finally, the links in the tree represent the inclusion relationships among the connected components following the threshold values. Note that when the threshold is set to $T = 1$, the circular component does not create a connected component that is represented in the tree because none of its pixels has a gray level value equal to 1. However, the circle itself is obtained when $T = 2$. The three regional maxima are represented by three leaves and the tree root represents the entire support of the image. The computation of Max-tree can be done in an efficient way (see [14] for more details).

Binary Partition Tree (BPT). The second example of region-based representation of images is the BPT [15]. It represents a set of regions obtained from an initial partition that we assume to be the partition of flat zones. The leaves of the tree represent the flat zones of the original signal. The remaining tree nodes represent regions that are obtained by merging the regions represented by the children. The root node represents the entire image support. The tree represents a fairly large set of regions at different scales. Large regions appear close to the root whereas small details can be found at lower levels. This representation should be considered as a compromise between representation accuracy and processing efficiency. Indeed, all possible merging of regions belonging to the initial partition are not represented in the tree. Only the most "likely" or "useful" merging steps are represented in the BPT. The connectivity encoded in the tree structure is binary in the sense that a region is explicitly connected to its sibling (since their union is a connected component represented by the father), but the remaining connections between regions of the original partition are not represented in the tree. Therefore, the tree encodes only part of the neighborhood relationships between the regions of the initial partition. However, as will be seen in the sequel, the main advantage of the tree representation is that it allows the fast implementation of sophisticated processing techniques.

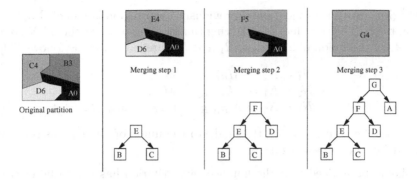

Fig. 9. Example of BPT creation with a region merging algorithm

The BPT should be created in such a way that the most "interesting" or "useful" regions are represented. This issue can be application dependent. However, a possible solution, suitable for a large number of cases, is to create the tree by keeping track of the merging steps performed by a segmentation algorithm based on region merging (see [28, 29] for example). In the following, this information is called the *merging sequence*. Starting from the partition of flat zones, the algorithm merges neighboring regions following a homogeneity criterion until a single region is obtained. An example is shown in Fig. 9. The original partition involves four regions. The regions are indicated by a letter and the number indicates the grey level value of the flat zone. The algorithm merges the four regions in three steps. In the first step, the pair of most similar regions, B and C, are merged to create region E. Then, region E is merged with region D to create region F. Finally, region F is merged with region A and this creates region G corresponding to the region of support of the whole image. In this example, the merging sequence is: $(B, C)|(E, D)|(F, A)$. This merging sequence defines the BPT as shown in Fig. 9.

To completely define the merging algorithm, one has to specify the region merging order and the region model, that is the model used to represent the union of two regions. In order to create the BPTs used to illustrate the processing examples discussed in this paper, we have used a merging algorithm following the color homogeneity criterion described in [29]. Let us define the merging order $O(R_1, R_2)$ and the region model M_R:

– **Merging order:** at each step the algorithm looks for the pair of most similar regions. The similarity between regions R_1 and R_2 is defined by:

$$O(R_1, R_2) = N_1||M_{R_1} - M_{R_1 \cup R_2}||_2 + N_2||M_{R_2} - M_{R_1 \cup R_2}||_2 \qquad (12)$$

where N_1 and N_2 are the numbers of pixels of regions R_1 and R_2 and $||.||_2$ denotes the $\mathcal{L}_2$ norm. M_R represents the model for region R. It consists of three constant values describing the YUV components. The interest of this merging order, compared to other classical criteria, is discussed in [29].

– **Region model:** as mentioned previously, each region is modeled by a constant vector YUV value: M_R. During the merging process, the YUV components of the union of 2 regions, R_1 and R_2, are computed as follows [29]:

$$\text{if } N_1 < N_2 \Rightarrow M_{R_1 \cup R_2} = M_{R_2}$$
$$\text{if } N_2 < N_1 \Rightarrow M_{R_1 \cup R_2} = M_{R_1} \qquad (13)$$
$$\text{if } N_1 = N_2 \Rightarrow M_{R_1 \cup R_2} = (M_{R_1} + M_{R_2})/2$$

As can be seen, if $N_1 \neq N_2$, the model of the union of two regions is equal to the model of the largest region.

It should be noticed that the homogeneity criterion has not to be restricted to color. For example, if the image for which we create the BPT belongs to a sequence of images, motion information can also be used: in a first stage, regions are merged using a color homogeneity criterion, whereas a motion homogeneity criterion is used in the second stage. Fig. 10 shows an example of the *Foreman* sequence. In Fig. 10(a), the BPT has been constructed exclusively with the color criterion described above. In this case, it is not possible to concentrate the information about the foreground object (head and shoulder regions of Foreman) within a single sub-tree. For example, the face mainly appears in the sub-tree hanging from region A, whereas the helmet regions are located below region D. In practice, the nodes close to the root have no clear meaning because they are not homogeneous in color. Fig. 10(b) presents an example of BPT created with color and motion criteria. The nodes appearing as white circles correspond to the color criterion, whereas the dark squares correspond to a motion criterion. The motion criterion is formally the same as the color criterion except that the

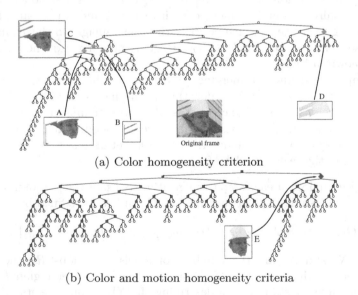

(a) Color homogeneity criterion

(b) Color and motion homogeneity criteria

Fig. 10. Examples of creation of BPT

YUV color distance is replaced by the YUV Displaced Frame Difference. The process starts with the color criterion as in Fig. 10(a) and then, when a given peak signal to noise ratio (PSNR) is reached, it changes to the motion criterion. Using motion information, the face and helmet now appear as a single region E.

As can be seen, the construction of a BPT is fairly more complex than the creation of a Max-tree or a Min-tree. However, BPTs offer more flexibility because one can chose the homogeneity criterion through the proper selection of the region model and the merging order. Furthermore, if the functions defining the region model and the merging order are self-dual, the tree itself is self-dual. The same BPT can be used to represent f and $-f$. The BPT representation is appropriate to derive self-dual connected operators whereas the Max-tree (Min-tree) is adequate for anti-extensive (extensive) connected operators. Note that in all cases, trees are hierarchical region-based representations. They encode a large set of regions and partitions that can be derived for the flat zones partition of the original image without adding new contours.

Filtering Strategy. Once the representation has been created, the filtering strategy consists in *pruning* the tree and in reconstructing an image from the pruned tree. The global processing strategy is illustrated in Fig. 11. The simplification effect of the filter is done by pruning because the idea is to eliminate the image components that are represented by the leaves and branches of the tree. The nature of these components depends on the tree. In the case of Max-trees (Min-trees), the components that may be eliminated are regional maxima (minima) whereas the elements that may be simplified in the case of BPTs are unions of the most similar flat zones. The simplification itself is governed by a criterion which may involve simple notions such as size, contrast or more complex ones such as texture, motion or even semantic criteria.

One of the interests of the tree representations is that the set of possible merging steps is fixed (represented by the tree branches). As a result, sophisticated pruning strategies may be designed. An example of such strategy deals with non-increasing simplification criteria. Mathematically, a criterion $\mathcal{C}$ assessed on a region R is said to be increasing if the following property holds:

$$\forall R_1 \subseteq R_2 \Rightarrow \mathcal{C}(R_1) \leq \mathcal{C}(R_2) \tag{14}$$

Assume that all nodes corresponding to regions where the criterion value is lower than a given threshold should be pruned. If the criterion is increasing, the

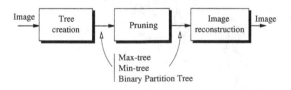

Fig. 11. Connected operators based on Tree representations

pruning strategy is straightforward: merge all nodes that should be removed. It is indeed a pruning strategy since the increasingness of the criterion guarantees that, if a node has to be removed, all its descendants have also to be removed. An example of BPT with increasing decision criterion is shown in Fig. 12. The criterion used to create this example is the size, measured as the number of pixels belonging to the region, which is indeed increasing. Note that this example involves a BPT but the same issue also applies to Max/Min-tree representations.

If the criterion is not increasing, the pruning strategy is not straightforward since the descendants of a node to be removed have not necessarily to be removed. An example of such criterion is the region perimeter. Fig. 13 illustrates this case. If we follow either Path A or Path B in Fig. 13, we see that there are some oscillations of the remove/preserve decisions. In practice, the non-increasingness of the criterion implies a lack of robustness of the operator. For example, similar images may produce quite different results or small modifications of the criterion threshold involve drastic changes on the output.

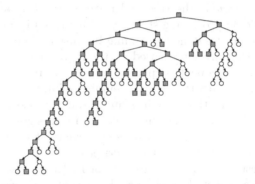

Fig. 12. Example of increasing criterion (size). If a node has to be removed, all its descendants have also to be removed. Gray squares: nodes to be preserved, white circles: nodes to be removed.

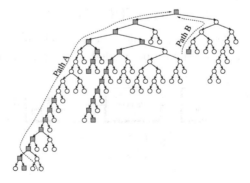

Fig. 13. Example of non-increasing criterion (perimeter). No relation exists between the decisions among descendants (see decisions along path A or path B). Gray squares: nodes to be preserved, white circles: nodes to be removed.

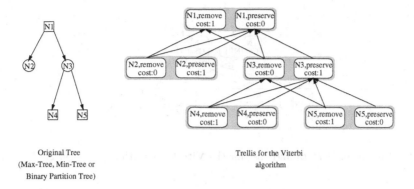

Original Tree
(Max-Tree, Min-Tree or
Binary Partition Tree)

Trellis for the Viterbi
algorithm

Fig. 14. Trellis structure for the Viterbi algorithm. A circular (square) node on the Tree indicates that the criterion value states that the node has to be removed (preserved). The trellis on which the Viterbi algorithm is run duplicates the structure of the Tree and defines a *preserve* state and a *remove* state for each tree node. Paths from *remove* states to child *preserve* states are forbidden so that the decisions are increasing.

A possible solution to the non-increasingness of the criterion consists in applying a transformation on the set of decisions. The transformation should create a set of increasing decisions while preserving as much as possible the decisions defined by the criterion. This problem may be viewed as dynamic programming issue that can be efficiently solved with the Viterbi algorithm.

The dynamic programming algorithm is explained and illustrated in the sequel assuming that the tree is binary. The extension to N-ary trees is straightforward and the example of binary tree is used here only to simplify the notation. An example of trellis on which the Viterbi algorithm [30] is applied is illustrated in Fig. 14. The trellis has the same structure as the tree except that two trellis states, *preserve* $\mathcal{N}_k^P$ and *remove* $\mathcal{N}_k^R$, correspond to each node $\mathcal{N}_k$ of the tree. The two states of each child node are connected to the two states of its parent. However, to avoid non-increasing decisions, the *preserve* state of a child is not connected to the *remove* state of its parent. As a result, the trellis structure guarantees that, if a node has to be removed, its children have also to be removed. The cost associated to each state is used to compute the number of modifications the algorithm has to do to create an increasing set of decisions. If the criterion value states that the node of the tree has to be removed, the cost associated to the *remove* state is equal to zero (no modification) and the cost associated to the *preserve* state is equal to one (one modification). Similarly, if the criterion value states that the node has to be preserved, the cost of the *remove* state is equal to one and the cost of the *preserve* state is equal to zero[3]. The cost values appearing in Fig. 14 assume

---

[3] Although some modifications may be much more severe than others, the cost choice has no strong effect on the final result. This issue of cost selection is similar to the hard versus soft decision of the Viterbi algorithm in the context of digital communications [30].

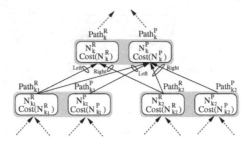

Fig. 15. Definition of *Path* and cost for the Viterbi algorithm (see Eqs. 17, 18 and 19)

that nodes $\mathcal{N}_1$, $\mathcal{N}_4$ and $\mathcal{N}_5$ should be preserved and that $\mathcal{N}_2$ and $\mathcal{N}_3$ should be removed. The goal of the Viterbi algorithm is to define the set of decisions such that:

$$\text{Min} \sum_k \text{Cost}(\mathcal{N}_k) \text{ such that the decisions are increasing.} \tag{15}$$

To find the optimum set of decisions, a set of paths going from all leaf nodes to the root node is created. For each node, the path can go through either the *preserve* or the *remove* state of the trellis. The Viterbi algorithm is used to find the paths that minimize the global cost at the root node. Note that the trellis itself guarantees that this optimum decision is increasing. The optimization is achieved in a bottom-up iterative fashion. For each node, it is possible to define the optimum paths ending at the *preserve* state and at the *remove* state:

– Let us consider a node $\mathcal{N}_k$ and its *preserve* state $\mathcal{N}_k^P$. A path $Path_k$ is a continuous set of transitions between nodes $(\mathcal{N}_\alpha \to \mathcal{N}_\beta)$ defined in the trellis:

$$Path_k = (\mathcal{N}_\alpha \to \mathcal{N}_\beta) \cup (\mathcal{N}_\beta \to \mathcal{N}_\gamma) \cup ... \cup (\mathcal{N}_\psi \to \mathcal{N}_k) \tag{16}$$

The path $Path_k^P$ starting from a leaf node and ending at that state is composed of <u>two</u> sub-paths[4]: the first one, $Path_k^{P,Left}$, comes from the left child and the second one, $Path_k^{P,Right}$, from the right child (see Fig. 15). In both cases, the path can emerge either from the *preserve* or from the *remove* state of the child nodes. If $\mathcal{N}_{k_1}$ and $\mathcal{N}_{k_2}$ are respectively the left and the right child nodes of $\mathcal{N}_k$, we have:

$$\begin{aligned}
Path_k^{P,Left} &= Path_{k_1}^R \bigcup (\mathcal{N}_{k_1}^R \to \mathcal{N}_k^P) \quad \text{or } Path_{k_1}^P \bigcup (\mathcal{N}_{k_1}^P \to \mathcal{N}_k^P) \\
Path_k^{P,Right} &= Path_{k_2}^R \bigcup (\mathcal{N}_{k_2}^R \to \mathcal{N}_k^P) \quad \text{or } Path_{k_2}^P \bigcup (\mathcal{N}_{k_2}^P \to \mathcal{N}_k^P) \, (17) \\
Path_k^P &= Path_k^{P,Left} \bigcup Path_k^{P,Right}
\end{aligned}$$

---

[4] In the general case of an N-ary tree, the number of incoming paths may be arbitrary.

The cost of a path is equal to the sum of the costs of its individual state transitions. Therefore, the optimum (lower cost) path can be easily selected:

If $Cost(Path_{k_1}^R) < Cost(Path_{k_1}^P)$

then { $\quad Path_k^{P,Left} \quad\quad = Path_{k_1}^R \cup (\mathcal{N}_{k_1}^R \to \mathcal{N}_k^P);$

$\quad\quad\quad Cost(Path_k^{P,Left}) = Cost(Path_{k_1}^R); \}$

else { $\quad Path_k^{P,Left} \quad\quad = Path_{k_1}^P \cup (\mathcal{N}_{k_1}^P \to \mathcal{N}_k^P);$

$\quad\quad\quad Cost(Path_k^{P,Left}) = Cost(Path_{k_1}^P); \}$

If $Cost(Path_{k_2}^R) < Cost(Path_{k_2}^P)$ (18)

then { $\quad Path_k^{P,Right} \quad\quad = Path_{k_2}^R \cup (\mathcal{N}_{k_2}^R \to \mathcal{N}_k^P);$

$\quad\quad\quad Cost(Path_k^{P,Right}) = Cost(Path_{k_2}^R); \}$

else { $\quad Path_k^{P,Right} \quad\quad = Path_{k_2}^P \cup (\mathcal{N}_{k_2}^P \to \mathcal{N}_k^P);$

$\quad\quad\quad Cost(Path_k^{P,Right}) = Cost(Path_{k_2}^P); \}$

$Cost(Path_k^P) \quad = Cost(Path_k^{P,Left}) + Cost(Path_k^{P,Right}) + Cost(\mathcal{N}_k^P);$

– In the case of the *remove* state, $\mathcal{N}_k^R$, the two sub-paths can only come from the *remove* states of the children. So, no selection has to be done. The path and its cost are constructed as follows:

$$Path_k^{R,Left} \quad = Path_{k_1}^R \cup (\mathcal{N}_{k_1}^R \to \mathcal{N}_k^R);$$
$$Path_k^{R,Right} = Path_{k_2}^R \cup (\mathcal{N}_{k_2}^R \to \mathcal{N}_k^R);$$
$$Path_k^R \quad\quad = Path_k^{R,Left} \cup Path_k^{R,Right};$$
$$Cost(Path_k^R) = Cost(Path_{k_1}^R) + Cost(Path_{k_2}^R) + Cost(\mathcal{N}_k^R);$$

(19)

This procedure is iterated bottom-up until the root node is reached. One path of minimum cost ends at the *preserve* state of the root node and another path ends at the *remove* state of the root node. Among these two paths, the one of minimum cost is selected. This path connects the root node to all leaves and the states it goes through define the final decisions. By construction, these decisions are increasing and are as close as possible to the original decisions.

A complete optimization example is shown in Fig. 16. The original tree involves 5 nodes. The *preserve* decisions are shown by a square whereas the *remove* decisions are indicated by a circle. The original tree does not correspond to a set of increasing decisions because $\mathcal{N}_3$ should be removed but $\mathcal{N}_4$ and $\mathcal{N}_5$ should be preserved. The algorithm is initialized by creating the trellis and populating its states by their respective cost (see Fig. 14). Then, the first step of the algorithm consists in selecting the paths that go from states $\mathcal{N}_4^R$, $\mathcal{N}_4^P$, $\mathcal{N}_5^R$, $\mathcal{N}_5^P$ to states $\mathcal{N}_3^R$, $\mathcal{N}_3^P$. The corresponding trellis is shown in the upper part of Fig. 16 together with the corresponding costs of the four surviving paths. The second step iterates the procedure between states $\mathcal{N}_2^R$, $\mathcal{N}_2^P$, $\mathcal{N}_3^R$, $\mathcal{N}_3^P$ and states $\mathcal{N}_1^R$, $\mathcal{N}_1^P$. Here again, only four paths survive. They are indicated in the central diagram of Fig. 16. Finally, the last step consists in selecting the path of lowest cost that terminates at the root states. In the example of Fig. 16, the path ending at the

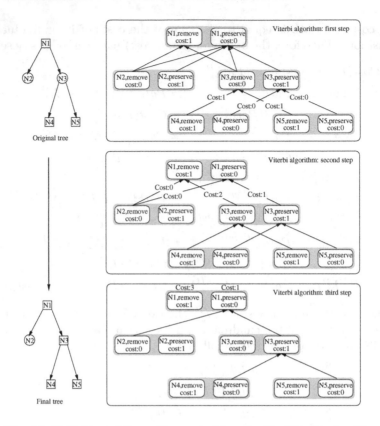

Fig. 16. Definition of the optimum decisions by the Viterbi algorithm

remove state of the root node $(\mathcal{N}_1^R)$ has a cost of 3, whereas the path ending at the *preserve* state $(\mathcal{N}_1^P)$ has a cost of 1. This last path is taken since it corresponds to an increasing set of decisions and involves just one modification of the original decisions. To find the optimum increasing decisions, one has to track back the selected path from the root to all leaves. In our example, we see that the paths hit the following states: $\mathcal{N}_1^P$, $\mathcal{N}_2^R$, $\mathcal{N}_3^P$, $\mathcal{N}_4^P$ and $\mathcal{N}_5^P$. The diagram at the bottom of Fig. 16 shows the final path together with the modified tree. As can be seen, the only modification has been to change the decision of node $\mathcal{N}_3$ and the resulting set of decisions is increasing. A complete example of is shown in Fig. 17. The original tree corresponds to the one shown in Fig. 13. The Viterbi algorithm has to modify 5 decisions along path A and one decision along path B (see Fig. 13) to get the optimum set of increasing decisions.

To summarize, let us say that any pruning strategy can be applied directly on the tree if the decision criterion is increasing. In the case of a non-increasing criterion, the Viterbi algorithm can be used to modify the smallest number of decisions so that increasingness is obtained. These modifications define a pruning strategy. Once the pruning has been performed, it defines an output partition and each region is filled with a constant value. In the case of a Max-tree (Min-

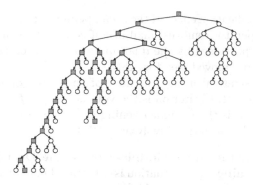

Fig. 17. Set of increasing decisions resulting from the Viterbi algorithm used on the original tree of Fig. 13. Five decisions along path A and one decision along path B have been modified. Gray squares: preserve, white circles: remove.

tree), the constant value is equal to the minimum (maximum) gray level value of the original pixels belonging to the region. As a result, the operator is anti-extensive (extensive). In the case of a BPT, the goal is to define a self-dual operator. Therefore, each region of the output partition has to be filled by a self-dual model, such as the mean or the median of the original pixels belonging to the region.

5.2   Example of Connected Operators Based on Tree Representations

Increasing Criterion ⇒ Direct Pruning. The first example deals with situations where the criterion is increasing. In this case, the comparison of the criterion value with a threshold directly defines a pruning strategy. A typical example is the area opening [12]. One possible implementation of the area opening consists in creating a Max-tree and in measuring the area (the number of pixels) $\mathcal{A}_k$ contained in each node $\mathcal{N}_k$. If the area $\mathcal{A}_k$ is smaller than a threshold, $\mathcal{T}_A$, the node is removed. The area criterion is increasing and the Viterbi algorithm does not have to be used. It can be shown that the area opening is equal to the supremum of all possible openings by a connected structuring element involving

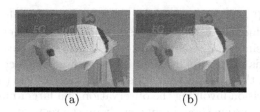

(a)   (b)

Fig. 18. Area filtering: (a) area opening, γ^{area}, (b) area opening followed by area closing, $\varphi^{area}\gamma^{area}$

$\mathcal{T}_A$ pixels. The simplification effect of the area opening is illustrated in Fig. 18(a). As expected, the operator removes small bright components of the image. If this simplified image is processed by the dual operator, the area closing, small dark components are also removed (see Fig. 18(b)).

Using the same strategy, a large number of connected operators can be obtained. For example, if the criterion is the volume: $\sum_{n \in R} f[n]$ (also increasing), the resulting operator is the volumic opening [31]. The reader is referred to [15] to see examples of this situation involving a BPT.

Non-increasing Criterion ⇒ Modification of the Decision (Viterbi algorithm) and Pruning. This situation is illustrated here by a motion-oriented connected operator [14]. Denote by $f_t[n]$ an image sequence where n represents the pixel coordinates and t the time instant. The goal of the connected operator is to eliminate the image components that do not undergo a given motion. The first step is therefore to define the motion model giving for example the displacement field at each position $\Delta[n]$. The field can be constant Δ if one wants to extract all objects following a translation, but in general the displacement depends on the spatial position n to deal with more complex motion models.

The sequence processing is performed as follows: each frame is transformed into its corresponding Max-tree representation and each node $\mathcal{N}_k$ is analyzed. To check whether or not the pixels contained in a given node $\mathcal{N}_k$ are moving in accordance to the motion field $\Delta[n]$, a simple solution consists in computing the Mean Displaced Frame Difference (DFD) of this region with the previous frame:

$$\mathrm{DFD}_{f_t}^{f_{t-1}}(\mathcal{N}_k) = \sum_{n \in \mathcal{N}_k} |f_t[n] - f_{t-1}[n - \Delta[n]]| / \sum_{n \in \mathcal{N}_k} 1 \tag{20}$$

In practice, however, it is not very reliable to assess the motion on the basis of only two frames. The criterion should include a reasonable memory of the past decisions. This idea can be easily introduced in the criterion by adding a recursive term. Two mean DFD are measured: one between the current frame f_t and the previous frame f_{t-1} and a second one between the current frame and the previous filtered frame $\psi(f_{t-1})$ (ψ denotes the connected operator). The motion criterion is finally defined as:

$$Motion(\mathcal{N}_k) = \alpha \mathrm{DFD}_{f_t}^{f_{t-1}}(\mathcal{N}_k) + (1 - \alpha)\mathrm{DFD}_{f_t}^{\psi(f_{t-1})}(\mathcal{N}_k) \tag{21}$$

with $0 \le \alpha \le 1$. If α is equal to 1, the criterion is memoryless. Low values of α allow the introduction of an important recursive component in the decision process. In a way similar to recursive filtering schemes, the selection of a proper value for α depends on the application: if one wants to detect very rapidly any changes in motion, the criterion should be mainly memoryless ($\alpha \approx 1$), whereas if a more reliable decision involving the observation of a larger number of frames is necessary, then the system should rely heavily on the recursive part ($0 \le \alpha \ll 1$).

The motion criterion described by Eqs. 20 and 21 deals with one set of motion parameters. Objects that do not follow the given motion produce a high DFD and should be removed. The criterion is not increasing and the Viterbi algorithm

(a)   (b)   (c)   (d)

Fig. 19. Example of motion connected operator preserving fixed objects: (a) Original frame, (b) Motion connected operator ψ, (c) Dual operator: $\psi^*\psi(f)$ and (d) residue: $f - \psi^*(\psi(f))$

has to be used. This motion-oriented pruning strategy can be used on Max-tree, Min-tree or BPT representations.

A motion filtering example relying on a Max-tree is shown in Fig. 19. The operator goal is to remove all moving objects. The motion model is defined by: $\Delta[n] = (0,0), \forall n$. In this sequence, all objects are still except the ballerina behind the two speakers and the speaker on the left side who is speaking. The connected operator $\psi(f)$ removes all bright moving objects (Fig. 19(b)). The dual operator: $\psi^*(f) = -\psi(-f)$ removes all dark moving objects (Fig. 19(c)). The residue (the difference with the original image) presented in Fig. 19(d) shows what has been removed by the operator. As can be seen, the operator has very precisely extracted the ballerina and the (moving) details of the speaker's face.

The motion connected operator can potentially be used for a large set of applications. It permits in particular to different ways of handling the motion information. Indeed, generally, motion information is measured without knowing anything about the image structure. Connected operators take a different viewpoint by making decisions on the basis of the analysis of flat zones. By using motion connected operators, we can "inverse" the classical approach to motion and, for example, analyze simplified sequences where objects are following a known motion. Various connected operators involving nonincreasing criteria such as entropy, simplicity, perimeter can be found in [14, 15].

5.3 Pruning Strategies Involving Global Optimization Under Constraint

In this section, we illustrate a more complex pruning strategy involving global optimization under constraint. To fix the notations, let us denote by $\mathcal{C}$ the criterion that has to be optimized (we assume, without loss of generality, that the criterion has to be minimized) and by $\mathcal{K}$ the constraint. The problem is to minimize the criterion $\mathcal{C}$ with the restriction that the constraint $\mathcal{K}$ is below a given threshold $\mathcal{T}_\mathcal{K}$. Moreover, we assume that both the criterion and the constraint are additive over the regions represented by the nodes $\mathcal{N}_k$: $\mathcal{C} = \sum_{\mathcal{N}_k} \mathcal{C}(\mathcal{N}_k)$ and $\mathcal{K} = \sum_{\mathcal{N}_k} \mathcal{K}(\mathcal{N}_k)$. The problem is therefore to define a pruning strategy such that the resulting partition is composed of nodes $\mathcal{N}_i$ such that:

$$\text{Min} \sum_{\mathcal{N}_i} \mathcal{C}(\mathcal{N}_i) \text{ , with } \sum_{\mathcal{N}_i} \mathcal{K}(\mathcal{N}_i) \leq \mathcal{T}_\mathcal{K} \tag{22}$$

It has been shown [32] that this problem can be reformulated as the minimization of the Lagrangian: $\mathcal{L} = \mathcal{C} + \lambda\mathcal{K}$ where λ is the so-called Lagrange parameter. Both problems have the same solution if we find λ^* such that $\mathcal{K}$ is equal (or very close) to the constraint threshold $\mathcal{T}_\mathcal{K}$. Therefore, the problem consists in using the tree to find by pruning a set of nodes creating a partition such that:

$$\text{Min} \left(\sum_{\mathcal{N}_i} \mathcal{C}(\mathcal{N}_i) + \lambda^* \sum_{\mathcal{N}_i} \mathcal{K}(\mathcal{N}_i) \right) \tag{23}$$

Assume, in a first step, that the optimum λ^* is known. In this case, the pruning is done by a bottom-up analysis of the tree. If the Lagrangian value corresponding to a given node $\mathcal{N}_0$ is smaller than the sum of the Lagrangians of the children nodes $\mathcal{N}_i$, then the children are pruned:

$$\text{If } \mathcal{C}(\mathcal{N}_0) + \lambda^* \mathcal{K}(\mathcal{N}_0) < \sum_{\mathcal{N}_i} \mathcal{C}(\mathcal{N}_i)$$
$$+ \lambda^* \sum_{\mathcal{N}_i} \mathcal{K}(\mathcal{N}_i), \text{ prune the children nodes } \mathcal{N}_i. \tag{24}$$

This procedure is iterated up to the root node. In practice, the optimum λ^* is not known and the previous bottom-up analysis is embedded in a loop that searches for the best λ parameter. The computation of the optimum λ parameter can be done with a gradient search algorithm. The bottom-up analysis itself is not expensive in terms of computation since the algorithm has simply to perform a comparison of Lagrangians for all nodes of the tree. The part of the algorithm that might be expensive is the computation of the criterion and the constraint values associated to the regions. Note, however, that this computation has to be done once. Finally, the theoretical properties depend mainly on the criterion and on the constraint. In any case, the operator is connected and self-dual.

This type of pruning strategy is illustrated by two examples relying on a BPT. In the first example, the goal is to simplify the input image by minimizing the number of flat zones of the output image: $\mathcal{C}_1 = \sum_{\mathcal{N}_k} 1$. In the

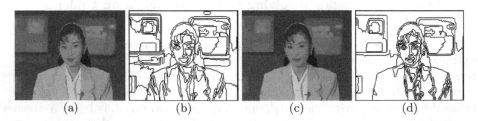

(a)   (b)   (c)   (d)

Fig. 20. Example of optimization strategies under a squared error constraint of 31 dB. (a) Minimization of the number of the flat zones, (b) contours of the flat zones of Fig. 20(a) (number of flat zones: 87, perimeter length: 4491), (c) Minimization of the total perimeter length, (d) contours of the flat zones of Fig. 20(b) (number of flat zones: 219, perimeter length: 3684).

second example, the criterion is to minimize the total length of the flat zones contours: $C_2 = \sum_{\mathcal{N}_k} Perimeter(\mathcal{N}_k)$. In both cases, the criterion has no meaning if there is no constraint because the algorithm would prune all nodes. The constraint we use is to force the output image to be a faithful approximation of the input image: the squared error between the input and output images $\mathcal{K} = \sum_{\mathcal{N}_k} \sum_{n \in \mathcal{N}_k} (\psi(f)(n) - f(n))^2$ is constrained to be below a given threshold. In the examples shown in Fig. 20, the squared error is constrained to be of at least 31 dB. Fig. 20(a) shows the output image when the criterion is the number of flat zones. The image is visually a good approximation of the original image but it involves a much lower number of flat zones: the original image is composed of 14335 flat zones whereas only 87 flat zones are present in the filtered image. The second criterion is illustrated in Fig. 20(c). The approximation provided by this image is of the same quality as the previous one (squared error of 31 dB). However, the characteristics of its flat zones are quite different. The total length of the perimeter of its flat zones is equal to 3684 pixels whereas the example of Fig. 20(a) involves a total perimeter length of 4491 pixels. The reduction of perimeter length is obtained at the expense of a drastic increase of the number of flat zones: 219 instead of 87. Figs 20(b) and 20(d) show the flat zone contours. As can be seen, the flat zone contours are more complex in the first example but the number of flat zones is higher in the second one.

This kind of strategy can be applied for a large number of criteria and constraints. Note that without defining a tree structure such as a Max-tree or a BPT, it would be extremely difficult to implement this kind of connected operators.

6   Conclusions

This paper has presented and discussed a region-based processing technique involving connected operators. There is currently an interest in defining processing tools that do not act on the pixel level but on a region level. Connected operators are examples of such tools that come from mathematical morphology.

Connected operators are operators that process the image by merging flat zones. As a result, they cannot introduce any contours or move existing ones. The two most popular approaches to create connected operators have been reviewed. The first one work on a pixel-based representation of the image and involves a reconstruction process. The operator involves first a simplification step based on a "classical" operator (such as morphological open, close, low-pass filter, median filter, etc) and then a reconstruction process. Three kind of reconstruction processes have been analyzed: anti-extensive, extensive and self-dual. The goal of the reconstruction process is to restore the contour information after the simplification. In fact, the reconstruction can be seen as a way to create a connected version of an arbitrary operator. Note that the simplification effect is defined and limited by the first step. The examples we have shown include simplification in terms of size or contrast.

The second second strategy to create connected operators involves three steps: in the first step, a region-based representation of the input image is constructed.

Three examples have been discussed: Max-tree, Min-tree and Binary Partition Tree. In the second step, the simplification is obtained by pruning the tree and, in the third step, the output image is constructed from the pruned tree. The tree creation defines the set of regions that the pruning strategy can use to create the final partition. It represents a compromise between flexibility and efficiency: on the one hand side, not all possible merging of flat zones are represented in the tree, but on the other hand side, once the tree has been defined complex pruning strategies can be defined. In particular, it is possible to deal in a robust way with nonincreasing criteria. Criteria involving the notions of area, motion and optimization under a quality constraint have been demonstrated.

References

1. S. L. Horowitz and T. Pavlidis. Picture segmentation by a directed split-and-merge procedure. In *Second Int. joint Conference on Pattern Recognition*, pages 424–433, 1974.
2. J. Serra and P. Salembier. Connected operators and pyramids. In SPIE, editor, *Image Algebra and Mathematical Morphology*, volume 2030, pages 65–76, San Diego (CA), USA, July 1993.
3. J. Crespo, J. Serra, and R.W. Schafer. Theoretical aspects of morphological filters by reconstruction. *Signal Processing*, 47(2):201–225, 1995.
4. H. Heijmans. Connected morphological operators and filters for binary images. In *IEEE Int. Conference on Image Processing, ICIP'97*, volume 2, pages 211–214, Santa Barbara (CA), USA, October 1997.
5. G. Matheron. Les nivellements. Technical report, Paris school of Mines, Center for Mathematical Morphology, 1997.
6. F. Meyer. From connected operators to levelings. In *Fourth Int. Symposium on Mathematical Morphology, ISMM'98*, pages 191–198, Amsterdam, The Netherlands, June 1998. Kluwer.
7. F. Meyer. The levelings. In *Fourth Int. Symposium on Mathematical Morphology, ISMM'98*, pages 199–206, Amsterdam, The Netherlands, June 1998. Kluwer.
8. J. Serra. Connectivity on complete lattices. *Journal of Mathematical Imaging and Vision*, 9:231–251, 1998.
9. C. Ronse. Set-theoretical algebraic approaches to connectivity in continuous or digital spaces. *Journal of Mathematical Imaging and Vision*, 8:41–58, 1998.
10. P. Maragos. Differential morphology. In S. Mitra and G. Sicuranza, editors, *Nonlinear Image Processing*, chapter 13. Academic Press, 2000.
11. L. Vincent. Morphological gray scale reconstruction in image analysis: Applications and efficients algorithms. *IEEE, Transactions on Image Processing*, 2(2):176–201, April 1993.
12. L. Vincent. Grayscale area openings and closings, their efficient implementation and applications. In J. Serra and P. Salembier, editors, *First Workshop on Mathematical Morphology and its Applications to Signal Processing*, pages 22–27, Barcelona, Spain, May 1993. UPC.
13. E. Breen and R. Jones. An attribute-based approach to mathematical morphology. In P. Maragos, R.W. Schafer, and M.A. Butt, editors, *International Symposium on Mathematical Morphology*, pages 41–48, Atlanta (GA), USA, May 1996. Kluwer Academic Publishers.

14. P. Salembier, A. Oliveras, and L. Garrido. Anti-extensive connected operators for image and sequence processing. *IEEE Transactions on Image Processing*, 7(4):555–570, April 1998.
15. P. Salembier and L. Garrido. Binary partition tree as an efficient representation for image processing, segmentation and information retrieval. *IEEE Transactions on Image Processing*, 9(4):561–576, April, 2000.
16. C. Gomila and F. Meyer. Levelings in vector space. In *IEEE International Conference on Image Processing, ICIP'99*, Kobe, Japan, October 1999.
17. P. Salembier and M. Kunt. Size-sensitive multiresolution decomposition of images with rank order based filters. *EURASIP Signal Processing*, 27(2):205–241, May 1992.
18. P. Salembier and J. Serra. Flat zones filtering, connected operators and filters by reconstruction. *IEEE Transactions on Image Processing*, 3(8):1153–1160, August 1995.
19. J. Crespo, R.W. Shafer, J. Serra, C. Gratin, and F. Meyer. A flat zone approach: a general low-level region merging segmentation method. *Signal Processing*, 62(1):37–60, October 1997.
20. V. Vilaplana and F. Marqués. Face segmentation using connected operators. In *Fourth Int. Symposium on Mathematical Morphology, ISMM'98*, pages 207–214, Amsterdam, The Netherlands, June 1998. Kluwer.
21. J. Serra. *Image Analysis and Mathematical Morphology*. Academic Press, 1982.
22. J. Serra. *Image Analysis and Mathematical Morphology, Vol II: Theoretical advances*. Academic Press, 1988.
23. J. Crespo. Space connectivity and translation-invariance. In P. Maragos, R.W. Schafer, and M.A. Butt, editors, *International Symposium on Mathematical Morphology*, pages 118–126, Atlanta (GA), USA, May 1996. Kluwer Academic Publishers.
24. H. Heijmans. Connected morphological operators for binary images. Technical Report PNA-R9708, CWI, Amsterdam, The Netherlands, 1997.
25. J.C. Klein. *Conception et réalisation d'une unité logique pour l'analyse quantitative d'images*. PhD thesis, Nancy University, France, 1976.
26. L. Vincent. Graphs and mathematical morphology. *EURASIP, Signal Processing*, 16(4):365–388, April 1989.
27. M. Grimaud. A new measure of contrast: the dynamics. In Serra Gader, Dougherty, editor, *SPIE Visual Communications and Image Processing'92*, volume SPIE 1769, pages 292–305, San Diego (CA), USA, July 1992.
28. O. Morris, M. Lee, and A. Constantinidies. Graph theory for image analysis: an approach based on the shortest spanning tree. *IEE Proceedings, F*, 133(2):146–152, April 1986.
29. L. Garrido, P. Salembier, and D. Garcia. Extensive operators in partition lattices for image sequence analysis. *EURASIP Signal Processing*, 66(2):157–180, April 1998.
30. A.J. Viterbi and J.K. Omura. *Principles of Digital Communications and Coding*. Mc Graw-Hill, New York, 1979.
31. F. Meyer, A. Oliveras, P. Salembier, and C. Vachier. Morphological tools for segmentation: connected filters and watersheds. *Annales des Télécommunications*, 52(7-8):366–379, July-Aug. 1997.
32. Y. Shoham and A. Gersho. Efficient bit allocation for an arbitrary set of quantizers. *IEEE Transactions on Acoustics, Speech, and Signal Processing*, 36:1445–1453, September 1988.

Exploiting High-Level Information Provided by ALISP in Speaker Recognition

Asmaa El Hannani[1,*] and Dijana Petrovska-Delacrétaz[1,2]

[1] DIVA Group, Informatics Dept., University of Fribourg, Switzerland
[2] Institut National des Télécommunications, 91011 Evry, France
asmaa.elhannani@unifr.ch, dijana.petrovska@int-evry.fr

Abstract. The best performing systems in the area of automatic speaker recognition have focused on using short-term, low-level acoustic information, such as cepstral features. Recently, various works have demonstrated that high-level features convey more speaker information and can be added to the low-level features in order to increase the robustness of the system. This paper describes a text-independent speaker recognition system exploiting high-level information provided by ALISP (Automatic Language Independent Speech Processing), a data-driven segmentation. This system, denoted here as ALISP n-gram system, captures the speaker specific information only by analyzing sequences of ALISP units. The ALISP n-gram system was fused with an acoustic ALISP-based Gaussian Mixture Models (GMM) system exploiting the speaker discriminating properties of individual speech classes. The resulting fused system reduced the error rate over the individual systems on the NIST 2004 Speaker Recognition Evaluation data.

1 Introduction

In recent years, research has expanded from only using the acoustic content of speech to trying to utilise high-level information, such as linguistic content, pronunciation and idiolectal word usage. Works examining the exploitation of high-level information sources have provided strong evidence that gains in speaker recognition accuracy are possible [1]. [2] explored the possibility of using word n-gram statistics for speaker verification. This technique although simple, gave encouraging results. Motivated by the work of [2], [3] applied similar techniques to phone n-gram statistics. This approach gave good results and was found to be a useful complementary features when used with short-term acoustic features. The research of [2] and [3] showed word and phone n-gram based model to be promising for speaker verification, however these techniques still based on human transcription of the speech data. The system we are proposing in this paper is inspired from the system described in [3], except that we used the automatic segmentation based on Automatic Language Independent Speech Processing (ALISP) tools [4] instead of the phonetic one. The ALISP-sequences, are automatically acquired from the output of the ALISP recognizer with no need

---

* Supported by the Swiss National Fund for Scientific Research, No. 2100-067043.01/1.

M. Faundez-Zanuy et al. (Eds.): NOLISP 2005, LNAI 3817, pp. 66–71, 2005.

of transcribed databases. In [5] we have built an ALISP-based GMM system exploiting the speaker discriminating properties of individual speech classes and we have shown that the ALISP segments could capture speaker information. In the ALISP n-gram system we are presenting here, speaker specific information is captured only by analyzing sequences of ALISP units. The ALISP-based GMM system and the ALISP n-gram system are combined to complement each other. The resulting fused system reduces the error rate over the individual systems.

The outline of this paper is the following: In Section 2 more details about the proposed method are given. Section 3 describes the database used and the experimental protocol. The evaluation results are reported in Section4. The conclusions and perspectives are given in Section 5.

2   System Description

2.1   ALISP Segmentation

The systems described bellow use in the first stage a data-driven segmentation Automatic Language Independent Speech Processing (ALISP) tools [4]. This technique is based on units acquired during a data-driven segmentation, where no phonetic transcription of the corpus is needed. In this work we use 64 classes. The modelling of the set of data-driven speech units, denoted as ALISP units, is achieved through the following stages. After the pre-processing step for the speech data, first Temporal Decomposition is used, followed by Vector Quantization providing a symbolic transcription of the data in an unsupervised manner. Hidden Markov Modeling is further applied for a better coherence of the initial ALISP units.

2.2   ALISP N-Gram System

The focus here is to capture high-level information about the speaking style of each speaker. Speaker specific information is captured by analyzing sequences of ALISP units produced by the data-driven ALISP recognizer. In this approach, only ALISP sequences are used to model speakers.

For the scoring phase each ALISP-sequence is tested against a speaker specific model and a background model using a traditional likelihood ratio. The speaker model, L_i, and the background model, L_{Bm}, are generated using a simple n-gram frequency count as follows:

$$L_i(k) = \frac{C_i(k)}{\sum_{n=1}^{N_i} C_i(n)} \tag{1}$$

$$L_{Bm}(k) = \frac{C_{Bm}(k)}{\sum_{n=1}^{N_{Bm}} C_{Bm}(n)} \tag{2}$$

where $C_i(k)$ and $C_{Bm}(k)$ represent the frequency count of the ALISP n-gram type, k, in the speaker data and world data, respectively. N_i and N_{Bm} are the number of all n-gram types in the speaker data and world data, respectively.

Then for each ALISP n-gram found in the test utterance a score is calculated using the log-likelihood ratio of the speaker likelihood to the background likelihood

$$S_{ti} = \frac{\sum_{k=1}^{K} (C_t(k).log\,[L_i(k) - L_{Bm}(k)])}{\sum_{k=1}^{M} C_t(k)} \qquad (3)$$

where $C_t(k)$ represents the number of occurrences of the ALISP n-gram type, k, in the test utterance t. The sums are over all of the ALISP n-gram types in the test segment.

Finally, the ALISP n-gram scores are fused together to generate an overall score for the test segment.

In this work three n-gram (1-gram, 2-gram and 3-gram) systems are built. The evaluation of their individual performances and their fusion is presented in section 4.

2.3 ALISP-Based GMM System

This system uses GMMs on a segmental level in order to exploit the different amount of discrimination provided by the ALISP classes [5]. In this segmental approach we represent each speaker by 64 GMMs each of them models an ALISP class. The speaker specific 64 models were adapted from the 64 gender and ALISP class dependent background models.

During the test phase, each test speech data is first segmented with the 64 ALISP HMM models. Then, each ALISP segment found in the test utterance is compared to the hypothesized speaker models and to the background model of the specific ALISP class.

Finally, and after the computation of a score for each ALISP segment, the segmental scores are combined together to form a single recognition score for the test utterance. A Multi-Layer Perceptrons (MLP) [6] is used to combine the individual scores for the ALISP segments.

2.4 Fusion

There are several scenarios for combining the decisions of multiple systems [7]. In [8] we have compared three fusion methods of speaker verification systems: the linear summation, the Logistic Regression (LR) and the Multi-Layer Perceptron (MLP). In this work we choose a Multi-Layer Perceptron to fuse the scores from the various systems. This perceptron has a layer consisting of inputs for each system, a hidden layer with 5 neurons, and an output layer using sigmoid as activation function.

3   Experimental Setup

All experiments are done on the NIST'2004 data which is split into two different subsets: the *Development-set* and the *Evaluation-set*, used to test the performance of the proposed system.

The speech parametrization is done with Mel Frequency Cepstral Coefficients (MFCC), calculated on 20 ms windows, with a 10 ms shift. For each frame a 15-element cepstral vector is computed and appended with first order deltas. Cepstral mean substraction is applied to the 15 static coefficients and only bands in the 300-3400 Hz frequency range are used. The energy and delta-energy are used in addition during the ALISP units recognition.

During the preprocessing step, after the speech parametrization, we separated the speech from the non-speech data. The speech activity detector is based on a bi-Gaussian modeling of the energy of the speech data [9]. Only frames higher than a certain threshold are chosen for further processing. Using this method, 56% of the original NIST 2004 data are removed.

In the ALISP-based GMM system[1], 64 ALISP-specific gender-dependent background models (with 32 Gaussians) are built and for each target speaker, 64 specific GMM with diagonal covariance matrices is trained via maximum a posteriori (MAP) adaptation of the Gaussian means of the matching gender background models. If an ALISP class does not occur in the training data for a target, the background model of this class becomes that target's model.

The gender dependent background models for the GMMs and the gender dependent ALISP recognizers, are trained on a total of about 6 hours of data from (1999 and 2001) NIST data sets. The MLP is trained on the development set.

4   Experimental Results

We present in this section results for "8sides-1side" NIST 2004 task on the evaluation data set, as defined in section 3. For this task we dispose of 40 minutes to

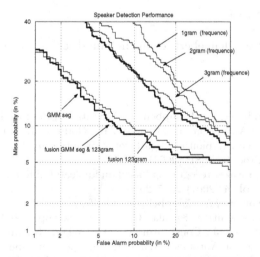

Fig. 1. Speaker verification results for the ALISP-based GMM system, the n-gram systems and their fusion on the evaluation data set (subset of NIST'04)

---

[1] Based on the BECARS package [10].

build the speaker model and 5 minutes for the test data (including silences). Performance is reported in term of the Detection Error Tradeoff (DET) curve [11]. Results are compared via Equal Error Rates (EER): the error at the threshold which gives equal miss and false alarm probabilities.

The Figure 1 shows DET curves of the fusion results. For reference, the four individual systems are also shown. Fusing the three ALISP n-gram (1-gram, 2-gram, 3-gram) systems lead to an improvement over the individual n-gram systems. These systems although worse compared to the ALISP-based GMM system, gave encouraging results.

In the next set of experiments, we fused the ALISP n-gram systems with the ALISP-based GMM system. Results are clearly showing that the new systems (ALISP n-gram) are supplying complementary information to the acoustic system (ALISP-based GMM).

5   Conclusions

In this paper we have presented a speaker verification system based on data-driven speech segmentation and exploiting high-level information. We have shown that the fusion of the acoustic ALISP-based GMM system with the n-gram systems treating high-level information (provided by the ALISP sequence), improve the speaker recognition accuracy. The great advantage of the proposed method is that it is not grounded on the usage of transcribed speech data. In other hand the ALISP data-driven segmentation can be used in different levels in speaker verification systems in order to extract complementary types of information.

Acknowledgement

Our thanks go to Jean Hennebert for his helpful discussions and MLP software.

References

1. Reynolds, D., Andrews, W., Campbell, J., Navratil, J., Peskin, B., Adami, A., Jin, Q., Klusacek, D., Abramson, J., Mihaescu, R., Godfrey, J., Jones, J., Xiang, B.: The supersid project: Exploiting high-level information for high-accuracy speaker recognition. In Proc. ICASSP (2003)
2. Doddington, G.: Speaker recognition based on idiolectal differences between speakers. Eurospeech vol. 4 (2001) 2517–2520
3. Andrews, W., Kohler, M., Campbell, J., Godfrey, J.: Phonetic, idiolectal, and acoustic speaker recognition. Speaker Odyssey Workshop (2001)
4. Chollet, G., Černocký, J., Constantinescu, A., Deligne, S., Bimbot, F.: Towards ALISP: a proposal for Automatic Language Independent Speech Processing. In Keith Ponting, editor, NATO ASI: Computational models of speech pattern processing Springer Verlag (1999)
5. El-Hannani, A., Petrovska-Delacrétaz, D.: Improving speaker verification system using alisp-based specific GMMs. submitted to AVBPA (2005)

6. Haykin, S.: Neural Networks: A Comprehensive Foundation. IEEE Computer society Press (1994)
7. Kittler, J., Hatef, M., Duin, R., Matas, J.: On combining classifiers. IEEE Transactions on Pattern Analysis and Machine Intelligence **vol. 20** (1998) 226–239
8. El-Hannani, A., Petrovska-Delacrétaz, D., Chollet, G.: Linear and non-linear fusion of alisp-based and GMM systems for text-independent speaker verification. In proc. of ODYSSEY04, The Speaker and Language Recognition Workshop (2004)
9. Magrin-Chagnolleau, I., Gravier, G., Blouet, R.: Overview of the 2000-2001 elisa consortium research activities. Speaker Odyssey Workshop (2001)
10. Blouet, R., Mokbel, C., Mokbel, H., Sanchez, E., Chollet, G., Greige, H.: Becars: A free software for speaker verification. Proc. Odyssey (2004)
11. Martin, A., Doddington, G., Kamm, T., Ordowski, M., Przybocki, M.: The det curve in assessment of detection task performance. Proc. Eurospeech'97 **vol. 4** (1997) 1895–1898

MLP Internal Representation as Discriminative Features for Improved Speaker Recognition

Dalei Wu, Andrew Morris, and Jacques Koreman

Institute of Phonetics, Saarland University, P.O. Box 15 11 50,
D-66041 Saarbrücken, Germany
{daleiwu, amorris, jkoreman}@coli.uni-saarland.de
http://www.coli.uni-saarland.de/~{daleiwu,amorris,koreman}

Abstract. Feature projection by non-linear discriminant analysis (NLDA) can substantially increase classification performance. In automatic speech recognition (ASR) the projection provided by the pre-squashed outputs from a one hidden layer multi-layer perceptron (MLP) trained to recognise speech sub-units (phonemes) has previously been shown to significantly increase ASR performance. An analogous approach cannot be applied directly to speaker recognition because there is no recognised set of "speaker sub-units" to provide a finite set of MLP target classes, and for many applications it is not practical to train an MLP with one output for each target speaker. In this paper we show that the output from the second hidden layer (compression layer) of an MLP with three hidden layers trained to identify a subset of 100 speakers selected at random from a set of 300 training speakers in Timit, can provide a 77% relative error reduction for common Gaussian mixture model (GMM) based speaker identification.

1 Introduction

Non-linear discriminant analysis (NLDA) based data enhancement by a multi-layer perceptron (MLP) has proved to be very effective for improving performance in automatic *speech* recognition (ASR) [6, 15]. This has been achieved by training an MLP with one output per phone to estimate phone posterior probabilities, and then using this MLP to project each data frame onto an internal representation of the data which the MLP has learnt (see Figure 1). This representation may be the net-input values to, or output values from, one of its hidden layers or the input to its output layer, i.e. the "pre-squashed MLP outputs" (see Figure 2).

The success of this simple data-driven approach to data enhancement in ASR has led to analogous procedures being attempted for *speaker* recognition. Since clustering in the acoustic space is mainly determined by the linguistic content (phones) of the speech, and speakers cause variation within these main clusters, the task of enhancing discrimination between the target classes (speakers) is more difficult in speaker recognition. Nevertheless, in [9] positive results with LDA based feature enhancement were obtained for speaker identification using Gaussian mixture models (GMMs), while in [8, 10] application of the more powerful NLDA based enhancement

M. Faundez-Zanuy et al. (Eds.): NOLISP 2005, LNAI 3817, pp. 72–80, 2005.

technique led to improved speaker verification rates, especially when used in combination with the original MFCC features.

In [8,10] the main aim of the application of NLDA preprocessing was to reduce training/test mismatch by training with data from different handset types. MLP feature enhancement is already known to be useful for dealing with training/test noise mismatch in ASR [15], but it is not yet established whether it will also do well if its only aim is to enhance discrimination between speakers. The present article therefore concentrates on speaker discrimination in clean speech only, where there is no strong mismatch of any kind in signal conditions.

In [8] (but not in [10]) there was a positive effect of MLP feature enhancement, except for short test utterances: MLP-enhanced features led to higher equal error rates than the MFCCs from which they were derived for 3 second test utterances, except when the MLP was trained on a large amount of training data (2-Session-full, cf. Table 2 in [8]). The present work was motivated by the development of speaker verification to run on a PDA, for which the enrolment and test data are very limited. We therefore concentrate on feature enhancement for limited training and test data, for which speaker recognition has not previously been improved by discriminatory preprocessing.

Further, this paper compares the MLP which was successfully applied in [8, 10] with several other, simpler architectures, to evaluate the gain in speaker identification accuracy obtained by adding extra layers.

In the present experiment the data enhancement is applied to a speaker identification task. The speakers with which the MLP is trained (which we refer to as the *speaker basis set*) are selected at random from the population while maintaining a representative balance of gender and dialect region, since this information is often available in a real system. As it was not possible to predict the number of speakers required to train the MLP, the size of the speaker basis set is varied..Results for several such random, non-overlapping selections are presented for the MLP which gives the best results. The assumption behind the speaker basis selection is that the feature projection learnt for a small number of speakers by the MLP is also effective when applied to any other speakers. In [11] it is shown that for any given number of basis speakers an automatic speaker selection can further enhance speaker identification.

Before training the speaker model for each new speaker to be enrolled into the GMM based speaker recognition system, and also before processing the data for a speaker to be recognised, each frame of speech data is now projected by the MLP onto its discriminative internal representation (see Fig.1).

The proposed approach to harness the discriminative power of MLPs for speaker recognition is a conceptually simple direct application of MLPs for data enhancement, as opposed to the application of an MLP in [7], where the primary task of the MLP was phone recognition, or in [10], where the MLP was mainly used to provide complementary data features.

In Section 2 we present the baseline GMM based speaker identification model whose performance we are aiming to improve [12]. In Section 3 we give the procedure used for the design and training of the MLP which we use for data enhancement. Section 4 describes the data features and procedures used for system testing and in Section 5 we present experimental results. These results show that the data enhancement procedure described can give significantly improved speaker recognition performance. This is followed by a discussion and conclusion.

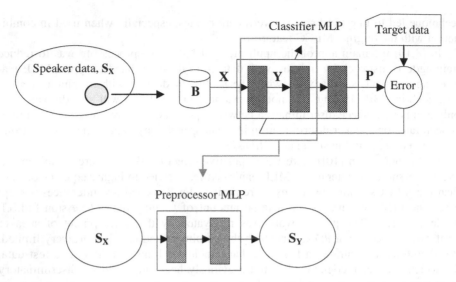

Fig. 1. Data enhancement procedure. A small random set of basis speakers, B, is selected from the training speakers. This is used to train an MLP with several hidden layers to estimate a-posteriori probabilities (P) only for speakers in B. All data S_X from speakers in the full closed set of speakers to be recognised is then passed through the first few layers of the trained MLP to produce new data features S_Y, with enhanced speaker discrimination.

2 Speaker Identification Baseline

In GMM based speaker identification a GMM data PDF $p(x|S)$ (1) is trained for each speaker for some fixed number M of Gaussians. This models the PDF for a single data frame, x, taking no account of the time order of the data frames in the full speech sample, X.

$$p(x|S) = \sum_{i=1}^{M} P(m_i) p(x|m_i, S) \tag{1}$$

When the speaker needs to be identified from speech data X, the speaker S_k is chosen as the most probable (2).

$$S_k = \arg\max_{S_k} P(S_k|X) \cong \arg\max_{S_k} P(X|S_k)$$
$$= \arg\max_{S_k} \log p(X|S_k) = \arg\max_{S_k} \sum_t \log p(x_t|S_k) \tag{2}$$

The GMM design, feature data and database used here (32 Gaussians, MFCC features, Timit) are taken from [14]. This simple model gives state-of-the-art speaker recognition performance. With Timit (though not with other databases, such as the CSLU speaker recognition database) no gain is found in training speaker models by adaptation from a global model.

As in [14], GMMs were trained by k-means clustering, followed by EM iteration. This was performed by the Torch machine learning API [3]. We used a variance threshold factor of 0.01 and minimum Gaussian weight of 0.05 (performance falling sharply if either was halved or doubled).

3   MLP Design and Training

The four MLP types tested are shown in Figure 2. Types *a, b, c* have previously been used successfully for data enhancement in ASR [6,15]. These are all feedforward MLPs in which each layer is fully connected to the next. The "neurons" in each layer comprise the usual linear net-input function followed by a non-linear squashing function, which is the sigmoid function for all layers except the output layer, which uses the softmax function to ensure that all outputs are positive and sum to 1 [1].

Also using Torch [3], each MLP is trained, by gradient descent, to maximise the cross entropy objective (i.e. the mutual information between the actual and target outputs). We trained in batch mode, with a fixed learning rate of 0.01. The data in each utterance was first normalised to have zero mean and unit variance. The estimated probabilities are often close to 0 or 1, and data with such a peaked distribution is not well suited as feature data. The enhanced features taken from the trained MLP of types *a* and *b* are therefore usually taken as the net input values in the output layer, prior to squashing. For type *c* they are normally taken as the squashed output from the last hidden layer (these values having less peaked distributions than the outputs from the output layer), but here we have used the net input to the second hidden layer as the enhanced features from MLPs *c* and *d*.

In ASR the MLP is trained to output a probability for each phoneme. In the model used here we select a random subset of the Timit speakers (but balanced for dialect region) available for training (the speaker basis set) and train the MLP to output a probability for each of these speakers. Although none of the MLPs *a-d* gave a high basis speaker classification score, the test results in Section 5 show that the speaker discriminative internal data representation which some of them learn can be very beneficial for GMM based speaker modelling.

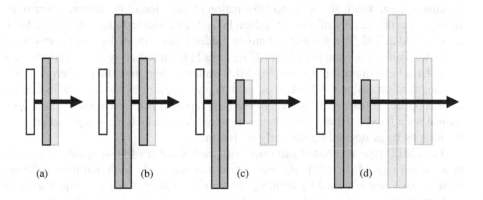

Fig. 2. Four MLP types *(a-d)* tested for data enhancement. Each active layer is shown as a (net-input function / non-linear activation function) sandwich. Only the dark sections of each MLP were used in data projection. The light parts were used only in training.

4 Test Procedure

Our baseline system is taken from the state of the art GMM based speaker identification system in [14], using the Timit speech database [5], GMMs with just 32 Gaussians, and 19 MFCC features.

4.1 Baseline Feature Processing

As in [14], all of the Timit signal data was first downsampled to 8 kHz, to simulate telephone line transmission (without downsampling, GMMs already achieve above 99.7% correct speaker identification using standard MFCC features). No further low- or high-pass filters were applied. Also as in [14], MFCC features, obtained using HTK [16], were used, with 20ms windows and 10ms shift, a pre-emphasis factor of 0.97, a Hamming window and 20 Mel scaled feature bands. All 20 MFCC coefficients were used except c0. On this database neither silence removal, cepstral mean subtraction, nor time difference features increased performance, so these were not used.

4.2 Test Protocol

Timit does not have a standard division into training, development and test sets which is suitable for work on speaker recognition. For this we first divided the 630 speakers in Timit into disjoint training, development and test speaker sets of 300, 162 and 168 speakers respectively. The speaker sets are all proportionally balanced for dialect region.

Data enhancement MLPs a-d (Figure 2) were trained using a speaker basis set of between 30 and 100 speakers, again proportionally balanced for dialect region. Within dialect region, the speakers are selected at random from the training set. Only one frame consisting of 19 MFCC features was used as input, in parallel to the GMM baseline system which also used no information of variation of the features over time. In each case the number of units in hidden layer 1, and also in hidden layer 3 in MLP d, was fixed at 100. The number of units in hidden layer 2 in MLPs c and d was fixed at 19 (the same as the number of MFCC features in the baseline system). Performance could have been improved by stopping MLP training when identification error on the development set (using GMMs trained on data preprocessed by the MLP in its current state) stopped increasing. However, in the tests reported here, each MLP was simply trained for a fixed number (35) of batch iterations, after which mean squared error on the training basis stopped significantly decreasing.

Each MLP type was tested just once with each number of basis speakers. For the best performing MLP (MLP d), test set tests were made with multiple different speaker basis sets obtained by dividing the training data into as many equal parts as each speaker basis size would permit. Because non-overlapping speaker basis sets were used, each speaker from the training data was used once for training the MLP.

Timit data is divided into 3 sentence types, SX_{1-5}, SI_{1-3} and SA_{1-2}. To make the speaker recognition system text-independent, a GMM for each speaker to be tested was trained on MLP projected sentences of type (SX_{1-2}, SA_{1-2}, SI_{1-2}) and tested on MLP projected sentences of type (SX_4, SX_5). Baseline GMMs were trained on MFCC features. The speaker identification procedure was as described in Section 2. Both training and testing used Torch [3].

5 Test Results

Test set speaker identification scores, for MLP types *a-d* against speaker basis size, are shown in Table 1 and Figure 3. The baseline identification error was 3.87%.

The best scoring MLP (MLP *d*) was then tested many times, for each number of basis speakers, also on the test set (Table 2). While results for different repetitions for each speaker basis size varied considerably, in 28 out of 30 tests the speaker identification error was lower than the baseline error. The optimal size of the speaker basis set used for training was 100, giving a relative error reduction of up to 77.0 %.

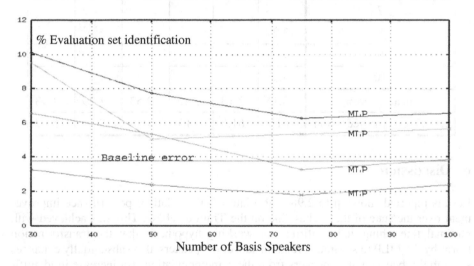

Fig. 3. Speaker identification error rate for the 168 speakers in the test set, for data enhancement using MLPs *a, b ,c, d*, with varying numbers of basis speakers

Table 1. Test set speaker identification error for MLPs *a-d* in Fig. 2 against speaker basis size

speaker basis size	30	50	75	100	best % relative error reduction
MLP *a*	10.10	7.74	6.25	6.55	-61.5
MLP *b*	9.52	5.06	5.36	5.65	-30.7
MLP *c*	6.55	5.36	3.27	3.87	15.5
MLP *d*	3.27	2.38	1.79	2.38	53.8

Table 2. MLP d speaker identification test set % error against speaker basis size. For each number of basis speakers, test-set tests were repeated, using disjoint speaker basis sets, as many times as were permitted by the number of available speakers (Baseline error 3.87%).

repetition \ spkr basis size	30	50	60	75	100	150
1	3.57	2.68	2.68	2.37	1.49	2.08
2	2.98	2.68	1.79	2.08	0.89	1.79
3	3.87	2.08	2.68	3.57	1.49	
4	2.08	2.08	1.79	2.08		
5	3.27	1.79	1.79			
6	4.76	1.49				
7	2.68					
8	3.27					
9	1.49					
10	3.57					
mean % error	3.15	2.13	2.15	2.53	1.29	1.93
best % rel. error reduction	61.5	61.5	53.7	46.3	77.0	53.7

6 Discussion

Results reported show up to 2.98% absolute (77.0% relative) performance improvement over the state of the art baseline on the Timit database. This was achieved with minimal fine-tuning and confirms our working hypothesis that the transformation learnt by the MLP to separate a random subset of speakers also substantially enhances separability between any speakers from the same population. An increase in identification accuracy has been found before with LDA when one output was trained for each speaker to be recognised [9]. By contrast, our MLP (a), which performs a linear separation equivalent to LDA [4], performs on average very badly. However, this could be because in our case none of the test speakers are used in training, so that the MLP is required to generalise to new speakers.

It appears that the ability of the features provided by the MLP to enhance speaker discrimination increases with the number of hidden layers. However, from the application viewpoint it would be advantageous to keep the MLP size and data transformation complexity to a minimum. It would be interesting to know whether the quality of data enhancement can be increased by dividing a given number of neurons into a greater number of layers, allowing for a more highly non-linear transformation.

Because of the large search space of possible MLP configurations, our search is still far from being optimised. Our decision to alternate large with small hidden layers is based on the intuition that the benefits of non-linear vector space expansion and data compression should possibly be balanced. Our choice of MLP types a-c for testing was also guided by what has been used successfully before in ASR [6,15], while MLP *d* was used in [8,10] for speaker recognition feature enhancement. The features it produced did not, however, consistently improve speaker verification for shorter test utterances, even though the quantity of training data used was at least 4 times larger than in our experiments. In future we could try varying layer sizes, and also test the discriminatory power of features from every compressive hidden layer, not just the second. So far we have seen performance always increasing with the number of hidden layers used in MLP training (while always using just three layers for data enhancement). We have yet to find the point where this benefit stops increasing.

To reduce the amount of experimentation required the number of MLP batch training iterations was fixed at 35, although it is well known that MLPs tend to overfit to training data after the learning curve begins to flatten out. In future we should use cross validation testing to permit training to stop when MLP preprocessing maximises speaker identification performance on the development set.

Results are only reported here for multiple *random* but balanced selections of each given number of basis speakers. While the number of speakers selected was always large enough to guarantee a fairly representative sample from the full speaker population, the somewhat erratic variation in identification performance resulting from different random speaker bases of the same size suggests that it would be instructive to see whether more principled methods could be used for basis speaker set selection. First results in this direction are reported in [11].

7 Conclusion

Test results reported show that the use of MLP based data enhancement for speaker identification using different handsets [8, 10] is also useful for speaker identification using very limited clean speech data. The number of target speakers which the MLP is trained to recognise must be small enough to avoid the classification problem becoming too difficult to train, but large enough to provide a feature basis sufficient to separate all speakers within a large population. The internal representation learnt by this MLP in separating the small set of basis speakers provides an enhanced feature vector which can improve GMM based speaker recognition performance. This form of data enhancement can be applied to speaker verification, as in [8, 10], as well as to speaker identification. It can be used with growing speaker sets, of unlimited size, with no need for further training as new speakers are added.

Acknowledgments

This work was supported by the EC SecurePhone project IST-2002-506883.

References

1. Bishop, C.M., Neural networks for pattern recognition, Oxford University Press, (1995).
2. Bengio, S., Bimbot, F., Mariethoz, j., Popovici, V., Poree, F., Bailly-Bailliere, E., Matas, G. & Ruiz, B., "Experimental protocol on the BANCA database", IDIAP-RR 02-05 (2002).

3. Collobert, R., Bengio, S. & Mariéthoz, J., "Torch: a modular machine learning software library", Technical Report IDIAP-RR 02-46 (2002).
4. Duda, O., Hart, P.E. & Stork, D.G., Pattern classification, Wiley (2001).
5. Fisher, W.M., Doddingtion, G.R. & Goudie-Marshall, K.M., "The DARPA speech recognition research database: Specifications and status", Proc. DARPA Workshop on Speech Recognition, pp. 93-99 (1986).
6. Fontaine, V., Ris, C. & Boite, J.-M., "Nonlinear Discriminant Analysis for improved speech recognition", Proc. Eurospeech'97, pp.2071-2074 (1997).
7. Genoud, D., Ellis, D. & Morgan, N., "Combined speech and speaker recognition with speaker-adapted connectionist models", Proc. ASRU (1999).
8. Heck, L., Konig, Y., Sönmez, K & Weintraub, M., "Robustness to telephone handset distortion in speaker recognition by discriminative feature design", Speech Communication 31, pp. 181-192 (2000).
9. Jin, Q. & Waibel, A., "Application of LDA to speaker recognition", Proc. ICSLP 2000 (2000).
10. Konig, Y., Heck, L., Weintraub, M. & Sönmez, K., "Nonlinear discriminant feature extraction for robust text-independent speaker recognition", Proc. RLA2C, ESCA workshop on Speaker Recognition and its Commercial and Forensic Applications, pp.72-75 (1998).
11. Morris, A.C, Wu, D. & Koreman, J., "MLP trained to separate problem speakers provides improved features for speaker identification", *IEEE Int. Carnahan Conf. on Security Technology (ICCST2005)*, Las Palmas (2005, accepted).
12. Reynolds, D.A., "Speaker identification and verification using Gaussian mixture speaker models", Speech Commun., 17 (1995), pp.91-108 (1995).
13. Reynolds, D.A., Doddington, D.R., Przybocki, M.A. & Martin, .F., "The NIST speaker recognition evaluation – overview, methodology, systems, results, perspective", Speech Communication, v.31, n.2-3, pp.225-254 (2000).
14. Reynolds, D.A., Zissman, M.A., Quatieri, T.F., O'Leary, G.C. & Carlson, B.A. "The effect of telephone transmission degradations on speaker recognition performance", Proc. ICASSP'95, pp.329-332 (1995).
15. Sharma, S., Ellis, D., Kajarekar, S., Jain, P. & Hermansky, H., "Feature extraction using non-linear transformation for robust speech recognition on the Aurora database", ICASSP 2000 (2000).
16. Young, S. et al. HTKbook (V3.2), Cambridge University Engineering Dept. (2002).

Weighting Scores to Improve Speaker-Dependent Threshold Estimation in Text-Dependent Speaker Verification

Javier R. Saeta[1], and Javier Hernando[2]

[1] Biometric Technologies, S.L.,
08007 Barcelona, Spain
j.rodriguez@biometco.com
[2] TALP Research Center, Universitat Politècnica de Catalunya (UPC),
08034 Barcelona, Spain
javier@talp.upc.es

Abstract. The difficulty of obtaining data from impostors and the scarcity of data are two factors that have a large influence in the estimation of speaker-dependent thresholds in text-dependent speaker verification. Furthermore, the inclusion of low quality utterances (background noises, distortion...) makes the process even harder. In such cases, the comparison of these utterances against the model can generate non-representative scores that deteriorate the correct estimations of statistical data from client scores. To mitigate the problem, some methods propose the suppresion of those scores which are far from the estimated scores mean. The tecnique results in a 'hard decision' that can produce errors especially when the number of scores is low. We propose here to take a 'softer decision' and weight scores according to their distance to the estimated scores mean. The Polycost and the BioTech databases have been used to show the effectiveness of the proposed method.

1 Introduction

The speaker verification is the process of deciding whether a speaker corresponds to a known voice. In speaker verification, the individual identifies her/himself by means of a code, login, card... Then, the system verifies her/his identity. It is a 1:1 process and it can be done in real-time. The result of the whole process is a binary decision. An utterance is compared to the speaker model and it is considered as belonging to the speaker if the Log-Likelihood Ratio (LLR) –the score obtained from the comparison- surpasses a predefined threshold and rejected if not.

In order to compare two systems, it is common to use the Equal Error Rate (EER), obtained when the False Acceptance Rate (FAR) and the False Rejection Rate (FRR) are equal. However, in real applications, a specific FAR or FRR is usually required. In this case, it is necessary to tune the speaker thresholds to achieve the desired rates.

In a typical Speaker Verification (SV) application, the user enrolls the system by pronouncing some utterances in order to estimate a speaker model. The enrollment procedure is one of the most critical stages of a SV process. At the same time, it

M. Faundez-Zanuy et al. (Eds.): NOLISP 2005, LNAI 3817, pp. 81–91, 2005.
© Springer-Verlag Berlin Heidelberg 2005

becomes essential to carry out a successful training process to obtain a good performance. The importance and sensitiveness of the process force us to pay special attention
on it. Consequently, it is necessary to protect the enrollment procedure by giving the
user some security mechanisms, like extra passwords or by providing a limited physical access. A general block diagram of an SV process can be found in Figure 1:

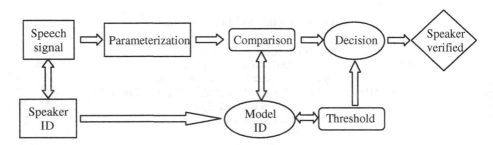

Fig. 1. Block diagram of a SV process

In real speaker verification applications, the speaker dependent thresholds should
be estimated a priori, using the speech collected during the speaker models training.
Besides, the client utterances must be used to train the model and also to estimate the
threshold because data is scarce. It is not possible to use different utterances for both
stages.

In development tasks, the threshold is usually set a posteriori. However, in real applications, the threshold must be set a priori. Furthermore, a speaker-dependent
threshold can sometimes be used because it better reflects speaker peculiarities and
intra-speaker variability than a speaker-independent threshold. The speaker dependent
threshold estimation method uses to be a linear combination of mean, variance or
standard deviation from clients and/or impostors.

Human-machine interaction can elicit some unexpected errors during training due
to background noises, distortions or strange articulatory effects. Furthermore, the
more training data available, the more robust model can be estimated. However, in
real applications, one can normally afford very few enrolment sessions. In this context, the impact of those utterances affected by adverse conditions becomes more
important in such cases where a great amount of data is not available. Score pruning
(SP) [1,2,3] techniques suppress the effect of non-representative scores, removing
them and contributing to a better estimation of means and variances in order to set the
speaker dependent threshold. The main problem is that in a few cases the elimination
of certain scores can produce unexpected errors in mean or variance estimation. In
these cases, threshold estimation methods based on weighting the scores reduce the
influence of the non-representative ones. The methods use a sigmoid function to
weight the scores according to the distance from the scores to the estimated scores
mean.

A theoretical approach of the state-of-the-art is reported on the next section. The
weighting threshold estimation method is developed in section 3. The experimental
setup, the description of the databases and the evaluation with empirical results are
shown in section 4, followed by conclusions in section 5.

2 Theoretical Approach

Several approaches have been proposed to automatically estimate a priori speaker dependent thresholds. Conventional methods have faced the scarcity of data and the problem of an a priori decision, using client scores, impostor data, a speaker independent threshold or some combination of them. In [4], one can find an estimation of the threshold as a linear combination of impostor scores mean (μ_I) and standard deviation from impostors σ_I as follows:

$$\Theta = \alpha \ (\mu_I - \sigma_I) + \beta \tag{1}$$

where α and β should be empirically obtained.

Three more speaker dependent threshold estimation methods similar to (1) are introduced in (2), (3) and (4) [5, 6]:

$$\Theta = \mu_I + \alpha \ \sigma_I^2 \tag{2}$$

where $\hat{\sigma}_{\bar{x}}^2$ is the variance estimation of the impostor scores, and:

$$\Theta = \alpha \ \mu_I + (1 - \alpha) \ \mu_c \tag{3}$$

$$\Theta = \Theta_{SI} + \alpha \ (\mu_c - \mu_I) \tag{4}$$

where μ_c is the client scores mean, Θ_{SI} is the speaker independent threshold and α is a constant, different for every equation and empirically determined. Equation (4) is considered as a fine adjustment of a speaker independent threshold.

Another expression introduced in [1] encompasses some of these approaches:

$$\Theta = \alpha \ (\mu_I + \beta \ \sigma_I) + (1 - \alpha)\mu_c \tag{5}$$

where α and β are constants which have to be optimized from a pool of speakers.

An approach introduced by the authors in [2] uses only data from clients:

$$\Theta = \mu_c - \alpha \ \sigma_c \tag{6}$$

where μ_c is the client scores mean, σ_c is the standard deviation from clients and α is a constant empirically determined. Equation (6) is very similar to (2), but uses standard deviation instead of variance and the client mean instead of impostors mean.

Some other methods are based on FAR and FRR curves [7]. Speaker utterances used to train the model are also employed to obtain the FRR curve. On the other hand, a set of impostor utterances is used to obtain the FAR curve. The threshold is adjusted to equalize both curves.

There are also other approaches [8] based on the difficulty of obtaining impostor utterances which fit the client model, especially in phrase-prompted cases. In these cases, it is difficult to secure the whole phrase from impostors. The solution is to use the distribution of the 'units' of the phrase or utterance rather than the whole phrase. The units are obtained from other speakers or different databases.

On the other hand, it is worth noting that there are other methods which use different estimators for mean and variance. With the selection of a high percentage of frames and not all of them, those frames which are out of range of typical frame likelihood values are removed. In [9], two of these methods can be observed, classified according to the percentage of used frames. Instead of employing all frames, one of the estimators uses 95% most typical frames discarding 2.5% maximum and minimum frame likelihood values. An alternative is to use 95% best frames, removing 5% minimum values.

Normalization techniques [10] can also be used for threshold setting purposes. Some normalization techniques follow the Bayesian approach while other techniques standardise the impostor score distribution. Furthermore, some of them are speaker-centric and some others are impostor-centric.

Zero normalization (Znorm) [11, 12, 13] estimates mean and variance from a set of impostors as follows:

$$S_{M,norm} = \frac{S_M - \mu_I}{\sigma_I} \tag{7}$$

where S_M are the client scores, μ_I is the estimated mean from impostors and σ_I the estimated variance from impostors [14].

We should also mention another threshold normalization technique such as Test normalization (Tnorm) [13, 15], which uses impostor models instead of impostor speech utterances to estimate impostor score distribution. The incoming speech utterance is compared to the speaker model and to the impostor models. That is the difference with regard to Znorm. Tnorm also follows the equation (7).

Tnorm has to be performed on-line, during testing. It can be considered as a test-dependent normalization technique while Znorm is considered as a speaker-dependent one. In both cases, the use of variance provides a good approximation for the impostor distribution.

Furthermore, Tnorm has the advantage of matching between test and normalization because the same utterances are used for both purposes. That is not the case for Znorm.

Finally, we can also consider Handset normalization (Hnorm) [16, 17, 18]. It is a variant of Znorm that normalizes scores according to the handset. This normalization is very important especially in those cases where there is a mismatch between training and testing.

Since handset information is not provided for each speaker utterance, a maximum likelihood classifier is implemented with a GMM for each handset [17]. With this classifier, we decide which handset is related to the speaker utterance and we obtain mean and variance parameters from impostor utterances. The normalization can be applied as follows:

$$S_{M,norm} = \frac{S_M - \mu_I(handset)}{\sigma_I(handset)} \tag{8}$$

where μ_I and σ_I are respectively the mean and variance obtained from the speaker model against impostor utterances recorded with the same handset type, and S_M are the client scores.

3 Threshold Estimation Based on Weighting Scores

A threshold estimation method that weights the scores according to the distance d_n from the score to the mean is introduced [19] in this section. It is considered that a score which is far from the estimated mean comes from a non-representative utterance of the speaker. The weighting factor w_n is a parameter of a sigmoid function and it is used here because it distributes the scores in a nonlinear way according to their proximity to the estimated mean. The expression of w_n is:

$$w_n = \frac{1}{1 + e^{-Cd_n}} \tag{9}$$

where w_n is the weight for the utterance n, d_n is the distance from the score to the mean and C is a constant empirically determined in our case.

The distance d_n is defined as:

$$d_n = |s_n - \mu_s| \tag{10}$$

where s_n are the scores and μ_s is the estimated scores mean.

The constant C defines the shape of the sigmoid function and it is used to tune the weight for the sigmoid function defined in Equation (9). A positive C will provide increasing weights with the distance while a negative C will give decreasing values. A typical sigmoid function, with C=1 is shown in Figure 2:

The average score is obtained as follows:

$$s_T = \frac{\sum_{n=1}^{N} w_n s_n}{\sum_{n=1}^{N} w_n} \tag{11}$$

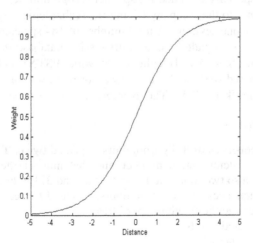

Fig. 2. Sigmoid function

where w_n is the weight for the utterance n defined in (9), s_n are the scores and s_T is the final score.

The standard deviation is also weighted in the same way as the mean. This method is called Total Score Weighting (T-SW).

On the other hand, it is possible to assign weights different from 1.0 only to a certain percentage of scores –the least representative- and not to all of them. This method is called Partial Score Weighting (P-SW). Normally, the farthest scores have in this case a weight different from 1.0.

4 Experiments

4.1 The Polycost Database

The Polycost database has been used for the experiments in this work. It was recorded by the participants of the COST250 Project. It is a telephone speech database with 134 speakers, 74 male and 60 female. The 85% of the speakers are between 20 and 35 years old. Almost each speaker has between 6 and 15 sessions of one minute of speech. Most speakers were recorded during 2-3 months, in English and in their mother tongue. Calls were made from the Public Switched Telephone Network (PSTN).

Each session contains 14 items: 4 repetitions of a 7-digit client code, five 10-digit sequences, 2 fixed sentences, 1 international phone number and 2 more items of spontaneous speech in speaker's mother tongue. For our experiments, we will use only digit utterances in English.

4.2 The BioTech Database

One of the databases used in this work has been recorded –among others- by the author and has been especially designed for speaker recognition. It is called the BioTech database and it belongs to the company Biometric Technologies, S.L. It includes landline and mobile telephone sessions. A total number of 184 speakers were recorded by phone, 106 male and 78 female. It is a multi-session database in Spanish, with 520 calls from the Public Switched Telephone Network (PSTN) and 328 from mobile telephones. One hundred speakers have at least 5 or more sessions. The average number of sessions per speaker is 4.55. The average time between sessions per speaker is 11.48 days.

Each session includes:

- different sequences of 8-digit numbers, repeated twice. They were the Spanish personal identification number and that number the other way round. There were also two more digits: 45327086 and 37159268.
- different sequences of 4-digit numbers, repeated twice. They were one random number and the fixed number 9014.
- different isolated words.
- different sentences.
- 1 minute long read paragraph.

- 1 minute of spontaneous speech, suggesting to talk about something related to what the user could see around, what (s)he had done at the weekend, the latest book read or the latest film seen.

4.3 Setup

In our experiments, utterances are processed in 25 ms frames, Hamming windowed and pre-emphasized. The feature set is formed by 12th order Mel-Frequency Cepstral Coefficients (MFCC) and the normalized log energy. Delta and delta-delta parameters are computed to form a 39-dimensional vector for each frame. Cepstral Mean Subtraction (CMS) is also applied.

Left-to-right HMM models with 2 states per phoneme and 1 mixture component per state are obtained for each digit. Client and world models have the same topology. The silence model is a GMM with 128 Gaussians. Both world and silence models are estimated from a subset of their respective databases.

The speaker verification is performed in combination with a speech recognizer for connected digits recognition. During enrolment, those utterances catalogued as "no voice" are discarded. This ensures a minimum quality for the threshold setting.

In the experiments with the BioTech database, clients have a minimum of 5 sessions. It yields 100 clients. We used 4 sessions for enrolment –or three sessions in some cases- and the rest of sessions to perform client tests. Speakers with more than one session and less than 5 sessions are used as impostors. 4- and 8-digit utterances are employed for enrolment and 8-digit for testing. Verbal information verification [20] is applied as a filter to remove low quality utterances. The total number of training utterances per speaker goes from 8 to 48. The exact number depends on the number of utterances discarded by the speech recognizer. During test, the speech recognizer discards those digits with a low probability and selects utterances which have exactly 8 digits. A total number of 20633 tests have been performed for the BioTech database, 1719 client tests and 18914 impostor tests.

It is worth noting that land-line and mobile telephone sessions are used indistinctly to train or test. This factor increases the error rates.

On the other hand, only digit utterances are used to perform tests with the Polycost database. After using a digit speech recognizer, those speakers with at least 40 utterances where considered as clients. That yields 99 clients, 56 male and 43 female. Furthermore, the speakers with a number of recognized utterances between 25 and 40 are treated as impostors. If the number of utterances does not reach 25, those speakers are used to train the world model. We use 40 utterances to train every client model.

In the experiments with the Polycost database, 43417 tests were performed, 2926 client tests and 40491 impostor tests. All the utterances come from landline phones in contrast with the utterances that belong to the BioTech database.

4.4 Results

In this section, the experiments show the performance of the threshold estimation methods proposed here. The following table shows a comparison of the EER for threshold estimation methods with client data only, without impostors and for the baseline Speaker-Dependent Threshold (SDT) method defined in Equation (6).

As it can be seen in Table 1, the T-SW method performs better than the baseline and even than the SP method. The P-SW performs better than the baseline too, but not than the SP. The results shown here correspond to the weighting of the scores which

Table 1. Comparison of threshold estimation methods in terms of Equal Error Rate

SDT	Baseline	SP	T-SW	P-SW
EER (%)	5.89	3.21	3.03	3.73

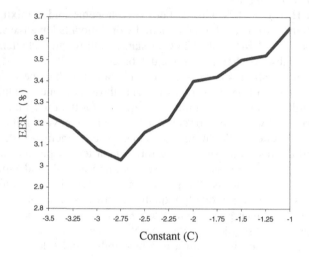

Fig. 2. Evolution of the EER with the variation of C

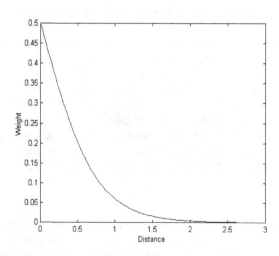

Fig. 3. Variation of the weight (w_n) with respecto to the distance (d_n) between the scores and the scores mean

distance to the mean is bigger than the 10% of the most distant score. It has been found that the minimum EER is secured when every one of the scores is weighted. It means that the optimal case for the P-SW method is the T-SW method.

In Figure 2, we can see the EER with respect to the constant C. It has been shown that the system performs better for a C = -2.75.

Figure 3 shows the function of the distance and the weight for the best C = -2.75. The weight exponentially decreases with the distance.

Table 2 shows the experiments with speaker-dependent thresholds using only data from clients following Equation (6).

The best EER is obtained for the Score Pruning (SP) method. The T-SW performs slightly worse and P-SW is the worst method. SP and SW methods improve the error rates with regard to the baseline. Results are given for a constant C = -3.0.

In Figure 4, the best EER is obtained for C = -3. This value is very similar to the one obtained for the BioTech database (C = -2.75).

Table 2. Comparison of threshold estimation methods for the Polycost database

SDT	Baseline	SP	T-SW	P-SW
EER (%)	1.70	0.91	0.93	1.08

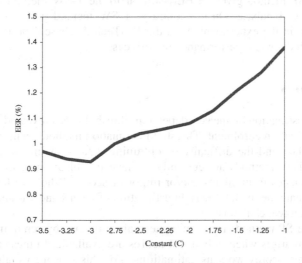

Fig. 4. Evolution of the EER with the variation of C

The comparison of the results obtained with both databases can be seen in Figure 5. First of all, EERs are lower for the Polycost database, mainly due to the fact that utterances are recorded from the PSTN while in the BioTech database calls come from the landline phones and the mobile phones. Furthermore, in the experiments with the Bio-Tech database, some clients are trained for example with utterances recorded from fixed-line phones and then tested with utterances from mobile phones and this random use of sessions decreases performance.

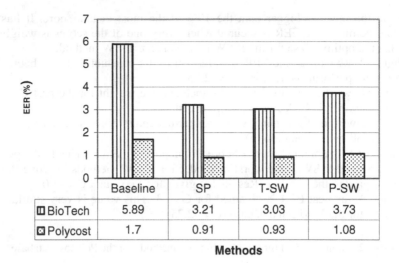

	Baseline	SP	T-SW	P-SW
▥ BioTech	5.89	3.21	3.03	3.73
▨ Polycost	1.7	0.91	0.93	1.08

Methods

Fig. 5. Comparison of EERs obtained for the BioTech and the Polycost databases

On the other hand, the improvement obtained with SP and SW methods is larger in experiments with the Polycost database where it almost reaches the 50%.

Otherwise, SP method gives an EER similar to the T-SW method in experiments with the Polycost database. On the contrary, T-SW method performs clearly better than SP method in the experiments with the BioTech database. The P-SW method is the method with the worst performance in both cases.

5 Conclusions

The automatic estimation of speaker dependent thresholds has revealed as a key factor in speaker verification enrolment. Threshold estimation methods mainly deal with the sparseness of data and the difficulty of obtaining data from impostors in real-time applications. These methods are currently a linear combination of the estimation of means and variances from clients and/or impostor scores. When we have only a few utterances to create the model, the right estimation of means and variances from client scores becomes a real challenge.

Although the SP methods try to mitigate main problems by removing the outliers, another problem arises when only a few scores are available. In these cases, the suppression of some scores worsens estimations. For this reason, weighting threshold methods proposed here use the whole set of scores but weighting them in a nonlinear way according to the distance to the estimated mean. Weighting threshold estimation methods based on a nonlinear function improve the baseline speaker dependent threshold estimation methods when using data from clients only. The T-SW method is even more effective than the SP ones in the experiments with the BioTech database, where there is often a mismatched handset between training and testing. On the contrary, with the Polycost database, where the same handset (landline network) is used, both of them perform very similar.

References

1. Chen, K.: Towards Better Making a Decision in Speaker Verification. Pattern Recognition, Vol. 36 (2003) 329-346
2. Saeta, J.R., Hernando, J.: Automatic Estimation of A Priori Speaker Dependent Thresholds in Speaker Verification. In: Proceedings 4th International Conference in Audio- and Video-based Biometric Person Authentication (AVBPA). Lecture Notes in Computer Science. Springer-Verlag (2003) 70-77
3. Saeta, J.R., Hernando, J.: On the Use of Score Pruning in Speaker Verification for Speaker Dependent Threshold Estimation. 2004: A Speaker Odyssey, The Speaker Recognition Workshop (2004) 215-218
4. Furui, S.: Cepstral Analysis for Automatic Speaker Verification. IEEE Trans. Speech and Audio Proc., vol. 29(2) (1981) 254-272
5. Lindberg, J., Koolwaaij, J., Hutter, H.P., Genoud, D., Pierrot, J.B., Blomberg, M., Bimbot, F.: Techniques for A Priori Decision Threshold Estimation in Speaker Verification. In: Proceedings RLA2C (1998) 89-92
6. Pierrot, J.B., Lindberg, J., Koolwaaij, J., Hutter, H.P., Genoud, D., Blomberg, M., Bimbot, F.: A Comparison of A Priori Threshold Setting Procedures for Speaker Verification in the CAVE Project. In: Proceedings ICASSP (1998) 125-128
7. Zhang, W.D., Yiu, K.K., Mak, M.W., Li, C.K., He, M.X.: A Priori Threshold Determination for Phrase-Prompted Speaker Verification. In: Proceedings Eurospeech (1999) 1203-1206
8. Surendran, A.C., Lee, C.H.: A Priori Threshold Selection for Fixed Vocabulary Speaker Verification Systems. In: Proceedings ICSLP vol. II (2000) 246-249
9. Bimbot, F., Genoud, D.: Likelihood Ratio Adjustment for the Compensation of Model Mismatch in Speaker Verification. In: Proceedings 2001: A Speaker Odyssey, The Speaker Recognition Workshop (2001) 73-76
10. Gravier, G. and Chollet, G.: Comparison of Normalization Techniques for Speaker Verification. In: Proceedings RLA2C (1998) 97-100
11. Auckentaler, R., Carey, M., Lloyd-Thomas, H.: Score Normalization for Text-Independent Speaker Verification Systems. Digital Signal Processing, Vol. 10 (2000) 42-54
12. Bimbot, F., Bonastre, F.J., Fredouille, C., Gravier, G., Magrin, I., Meignier, S., Merlin, T., Ortega-García, J., Petrovska, D., Reynolds, D.: A Tutorial on Text-Independent Speaker Verification. In: Proceedings Eusipco (2004) 430-451
13. Mirghafori, N., Heck, L.: An Adaptive Speaker Verification System with Speaker Dependent A Priori Decision Thresholds. In: Proceedings ICSLP (2002) 589-592
14. Navratil, J., Ramaswamy, G.N.: The Awe and Mystery of T-norm. In: Proceedings Eurospeech (2003) 2009-2012
15. Reynolds, D.: The Effect of Handset Variability on Speaker Recognition Performance: Experiments on the Switchboard Corpus. In: Proceedings ICASSP (1996) 113-116, 1996
16. Reynolds, D.A.: Comparison of Background Normalization Methods for Text-Independent Speaker Verification. In: Proceedings Eurospeech (1997) 963-966
17. Heck, L.P., Weintraub, M.: Handset Dependent Background Models for Robust Text-Independent Speaker Recognition. In: Proceedings ICASSP (1997) 1071-1074
18. Saeta, J.R., Hernando, J.: New Speaker-Dependent Threshold Estimation Method in Speaker Verification based on Weighting Scores. In Proceedings of the 3th Internacional Conference on Non-Linear Speech Processing (NoLisp) (2005) 34-41
19. Li, Q., Juang, B.H., Zhou, Q., Lee, C.H.: Verbal Information Verification. In: Proceedings Eurospeech (1997) 839-842

Parameter Optimization in a Text-Dependent Cryptographic-Speech-Key Generation Task

L. Paola García-Perera, Juan A. Nolazco-Flores, and Carlos Mex-Perera

Computer Science Department, ITESM, Campus Monterrey,
Av. Eugenio Garza Sada 2501 Sur, Col. Tecnológico,
Monterrey, N.L., México, C.P. 64849
{paola.garcia, carlosmex, jnolazco}@itesm.mx

Abstract. In this paper an improvement in the generation of the crypto-graphic-speech-key by optimising the number of parameters is presented. It involves the selection of the number of dimensions with the best performance for each of the phonemes. First, the Mel frequency cepstral coefficients, (first and second derivatives) of the speech signal are calculated. Then, an Automatic Speech Recogniser, which models are previously trained, is used to detect the phoneme limits in the speech utterance. Afterwards, the feature vectors are built using both the phoneme-speech models and the information obtained from the phoneme segmentation. Finally, the Support Vector Machines classifier, relying on an RBF kernel, computes the cryptographic key. By optimising the number of parameters our results show an improvement of 19.88%, 17.08%, 14.91% for 10, 20 and 30 speakers respectively, employing the YOHO database.

1   Introduction

The biometrics have been widely developed for access control purposes, but they are also becoming generators of cryptographic keys [14]. From all the biometrics voice was chosen for this research since it has the advantage of being flexible. For instance, if a user utters different phrases the produced key must be different. This means that by changing a spoken sentence or word the key automatically changes. Furthermore, the main benefit of using voice is that it can simultaneously act as a pass phrase for access control and as a key for encryption of data that will be stored or transmitted. Moreover, having a key generated by a biometric is highly desirable since the intrinsic characteristics that holds are unique for each individual, therefore, it will be difficult to guess.

Given the biometric information it is also possible to generate a private key and a public key. As an application we can propose the following scenario. A user utters a pass phrase that operates in two ways: as a generator of a private and public key and as a pass phrase for accessing his files. If an unauthorised user tries to access the files with a wrong pass phrase the access will be denied. But even if the pass phrase is correct the access will be denied since the phonetic features are not the ones that first generated the cryptographic key. With this

M. Faundez-Zanuy et al. (Eds.): NOLISP 2005, LNAI 3817, pp. 92–99, 2005.

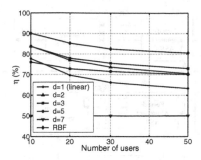

Fig. 1. η for different number of users and several types of kernels

Table 1. Value of η for different number of users, LPC and MFCC

Number of Users	Glob. Average LPC η	Glob. Average MFCC η
10	.8854	.927
20	.8577	.9088
30	.8424	.9004
50	.8189	.8871

example we can have a view of the potentiality of using the speech signal to generate such keys. Similar applications can be found in [10].

The results obtained in our previous work explained our proposed system architecture [4, 5, 7]. In those studies we tested different types of kernels, and as a result we obtained that the RBF kernel was superior than the linear and polynomial kernels [7], Figure 1.

We have also examined the parametrisation of the speech with different kinds of parameters; from here we concluded that the best feature vector for the SVM is based on the Mel frequency cepstral coefficients - as it is for speech recognition [4, 5], as shown in table 1.

Lastly, we also investigated the benefit of using the Gaussians' weights given by the model and the tuning of the user classifier per phoneme [6].

For the Automatic Speech Recognition (ASR) task is well known that the optimal number of parameters is around twelve. It is very common to use this number that SPHINX [15], one of the most prestigious softwares for automatic speech recognition employs it in its computations. Influenced by this trend, in our previous work, we had used this number. Consequently, the main purpose of this work is to improve the generation of the cryptographic-speech-key by optimising the number of parameters.

The system architecture is depicted in Figure 2 and will be discussed in the following sections. For a general view, the part under the dotted line shows the training phase that is performed offline. The upper part shows the online phase. In the training stage the *speech processing* and *recognition* techniques are used to obtain the model parameters and the segments of the phonemes

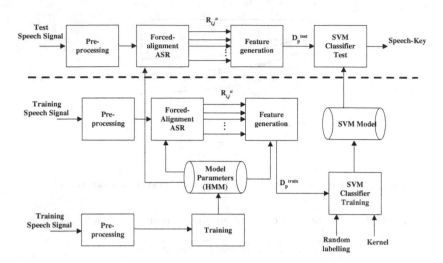

Fig. 2. System Architecture

in each user utterance. Afterwards, using the model parameters and the segments the feature generation is performed. Next, the *Support Vector Machine* (SVM) classifier produces its own models according to a specific kernel and bit specifications. From all those models, the ones that give the best results per phoneme are selected and compose the final SVM model. Finally, using the last SVM model the key is generated. The online stage is very much similar to the training and will repeatedly produce the same key if a user utters the same pass phrase.

2   Speech Processing and Phoneme Feature Generation

Firstly, the speech signal is divided into short windows and the *Mel frequency cepstral coefficients* (MFCC) are obtained. As a result an n-dimension vector, $(n - 1)$-dimension MFCCs followed by one energy coefficient is formed. To emphasize the dynamic features of the speech in time, the time-derivative (Δ) and the time-acceleration (Δ^2) of each parameter are calculated [13].

Afterwards, a forced alignment configuration of an ASR is used to obtain a model and the starts and ends of the phonemes per utterance. This ASR configuration is based on a three-state, left-right, Gaussian-based continuous Hidden Markov Model (HMM). For this research, the phonemes were selected instead of words since it is possible to generate larger keys with shorter length sentences.

Assuming the phonemes are modelled with a three-state left-to-right HMM, and assuming the middle state is the most stable part of the phoneme representation, let,

$$C_i = \frac{1}{K} \sum_{l=1}^{K} W_l G_l, \tag{1}$$

where G is the mean of a Gaussian, K is the total number of Gaussians available in that state, W_l is the weight of the Gaussian and i is the index associated to each phoneme.

Given the phonemes' starts and ends information, the MFCCs for each phoneme in the utterances can be arranged forming the sets $R_{i,j}^u$, where i is the index associated to each phoneme, j is the j-th user, and u is an index that starts in zero and increments every time the user utters the phoneme i.

Then, the feature vector is defined as

$$\psi_{i,j}^u = \mu(R_{i,j}^u) - C_i$$

where $\mu(R_{i,j}^u)$ is the mean vector of the data in the MFCC set $R_{i,j}^u$, and $C_i \in \mathcal{C}_P$ is known as the matching phoneme mean vector of the model. Let us denote the set of vectors,

$$D_p = \{\psi_{p,j}^u \mid \forall\, u, j\}$$

where p is a specific phoneme.

Afterwards, this set is divided in subsets: D_p^{tr} and D_p^{test}. 80% of the total D_p are elements of D_p^{tr} and the remaining 20% form D_p^{test}. Then, $D_p^{train} = \{[\psi_{p,j}^u, b_{p,j}] \mid \forall\, u, j\}$ where $b_{p,j} \in \{-1, 1\}$ is the key bit or class assigned to the phoneme p of the j-th user.

3   Support Vector Machine

The *Support Vector Machine* (SVM) *Classifier* is a method used for pattern recognition, and was first developed by Vapnik and Chervonenkis [1,3]. Although SVM has been used for several applications, it has also been employed in biometrics [12,11]. For this technique, given the observation inputs and a function-based model, the goal of the basic SVM is to classify these inputs into one of two classes. Firstly, the following set of pairs are defined $\{x_i, y_i\}$; where $x_i \in \mathbb{R}^n$ are the training vectors and $y_i = \{-1, 1\}$ are the labels. The SVM learning algorithm finds an hyperplane (w, b) such that,

$$\min_{x_i, b, \xi} \frac{1}{2} w^T w + C \sum_{i=1}^{l} \xi_i$$

$$\text{subject to } y_i(w^T \phi(x_i) + b) \geq 1 - \xi_i$$

$$\xi_i \geq 0$$

where ξ_i is a slack variable and C is a positive real constant known as a tradeoff parameter between error and margin.

To extend the linear method to a nonlinear technique, the input data is mapped into a higher dimensional space by function ϕ. However, exact specification of ϕ is not needed; instead, the expression known as kernel $K(x_i, x_j) \equiv \phi(x_i)^T \phi(x_j)$ is defined. There are different types of kernels as the linear, polynomial, radial basis function (RBF) and sigmoid. In this research, we study just

SVM techinque using radial basis function (RBF) kernel to transform a feature, based on a MFCC-vector, to a binary number (key bit) assigned randomly. The RBF kernel is denoted as $K(x_i, x_j) = e^{(-\gamma||x_i - x_j||^2)}$, where $\gamma > 0$.

The methodology used to implement the SVM training is as follows. Firstly, the training set for each phoneme (D_p^{train}) is formed by assigning a one-bit random label $(b_{p,j})$ to each user. Since a random generator of the values (-1 or 1) is used, the assignation is different for each user. The advantage of this random assignation is that the key entropy grows significantly. Afterwards, by employing a grid search the parameters C and γ are tuned.

Next, a testing stage is performed using D_p^{test}. This research considers just binary classes and the final key could be obtained by concatenating the bits produced by each phoneme. For instance, if a user utters two phonemes: /F/ and /AH/, the final key is $K = \{f(D_{/F/}), f(D_{/AH/})\}$, thus, the output is formed by two bits.

Finally, the SVM average classification accuracy is computed by the ratio

$$\eta = \frac{\alpha}{\beta}. \tag{2}$$

where α is the classification matches on test data and β is the total number of vectors in test data.

It is possible to choose the appropriate SVM model that corresponds to a specific phoneme by making the proper selection of the number of dimensions of the MFCCs. The SVM model should satisfy that the best average classification accuracy is obtained by all users in the SVM classifier outcome for that specific phoneme.

4   Experimental Methodology and Results

The YOHO database was used to perform the experiments [2, 8]. YOHO contains clean voice utterances of 138 speakers of different nationalities. It is a combination lock phrases (for instance, "Thirty-Two, Forty-One, Twenty-Five") with 4 enrollment sessions per subject and 24 phrases per enrollment session; 10 verification sessions per subject and 4 phrases per verification session. Given 18768 sentences, 13248 sentences were used for training and 5520 sentences for testing.

The utterances are processed using the Hidden Markov Models Toolkit (HTK) by Cambridge University Engineering Department [9] configured as a forced-alignment automatic speech recogniser. The important results of the speech processing stage are the twenty sets of mean vectors of the mixture of gaussians per phoneme given by the HMM and the phoneme starts and ends of the utterances. The phonemes used are: /AH/, /AX/, /AY/, /EH/, /ER/, /EY/, /F/, /IH/, /IY/,/K/, /N/, /R/, /S/, /T/, /TH/, /UW/, /V/, /W/. Following the method already described, the D_p sets are formed. It is important to note that the cardinality of each D_p set can be different since the number of equal phoneme utterances can vary from user to user. Next, subsets D_p^{train} and D_p^{test}

Table 2. η selection for different number of parameters, 10 users

Number of param.	10	12	14	16	18	20	22	selected coeffcient
/AH/	87.8096	92.4261	92.7356	93.6197	93.61	**93.7548**	93.6162	93.7548
/AO/	93.7447	94.8381	95.6992	95.8155	95.4674	95.8316	**96.0202**	96.0202
/AX/	93.749	94.1545	95.3309	95.2892	95.4443	95.2221	**95.7978**	95.7978
/AY/	97.281	97.7637	98.6195	**98.91**	98.7748	98.6099	98.6282	98.91
/EH/	94.2519	93.4383	95.2082	95.8032	96.3408	**96.5322**	96.1877	96.5322
/ER/	95.768	96.416	94.696	94.644	96.424	**97.348**	95.528	97.348
/EY/	87.778	88.9529	91.4003	**92.0903**	92.0439	91.8785	91.6377	92.0903
/F/	84.9242	85.6039	85.3135	86.0893	86.0493	86.5224	**87.1079**	87.1079
/IH/	93.1195	93.4377	94.9087	94.9783	94.3959	**95.009**	93.8153	95.009
/IY/	92.9352	93.1194	94.9087	95.2223	95.2971	**95.7253**	95.213	95.7253
/K/	86.692	86.046	87.136	86.458	**87.894**	85.216	86.614	87.894
/N/	96.4727	97.136	97.5795	97.9698	97.8956	97.9202	**98.1033**	98.1033
/R/	87.9715	86.6735	88.5042	89.7945	**90.0124**	89.9092	89.8415	90.0124
/S/	87.9832	88.6942	89.6029	**90.4101**	90.1797	90.065	90.0249	90.4101
/T/	90.1066	91.0542	**92.9159**	92.6551	92.629	92.3207	91.6112	92.9159
/TH/	84.1674	86.7468	87.7692	88.0125	90.7307	**91.8087**	89.2488	91.8087
/UW/	93.9652	95.1373	96.0497	96.0256	96.0958	95.5449	**96.2957**	96.2957
/V/	92.0974	94.8436	94.6873	95.4097	95.1399	95.3431	**95.5893**	95.5893
/W/	90.0719	91.1142	92.9558	92.4098	93.0521	**93.2115**	92.6727	93.2115
Average	91.0994	91.9787	92.9484	93.2424	93.5514	93.5670	93.3449	93.9247

are constructed. For the training stage, the number of vectors picked per user and per phoneme for generating the model is the same. Each user has the same probability to produce the correct bit per phoneme. However, the number of testing vectors that each user provided can be different.

Following the method a key bit assignation is required. For the purpose of this research, the assignation is arbitrary. Thus, the keys have liberty of assignation, therefore the keys entropy can be easily maximised if they are given in a random fashion with a uniform probability distribution.

The classification of D_p vectors was performed using SVMlight [16]. The behaviour of the SVM is given in terms of Equation 2.

The optimisation is accomplished as follows, see Table 2. The accuracy results η are computed for several number of dimensions. Afterwards, the models that develop the highest accuracy values are selected and compose a new set of results. Although Table 2 just shows the outcome for 10 users, the procedure was also executed for 20 and 30 users. The global results are depicted in Figure 3.

Table 3 shows the global values of η for different number of users, considering the selection. The statistics were computed as follows: 500 trials were performed for 10 and 20 users, and 1000 trails were performed for 30 and 50 users.

As shown in the Figure 3 the increment of the number of MFCC coefficients gives better results than just adjusting to speech known specifications.

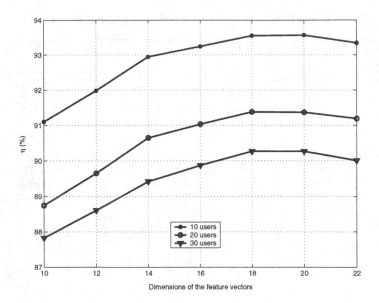

Fig. 3. η for different number of parameters

Table 3. Global average for η, after the best perfomance models

Number of users	η (%)
10	93.92474737
20	91.60074211
30	90.30173158

5   Conclusion

We have presented a method to improve the generation of a cryptographic key
from the speech signal based on the selection of the best performance for each
of the phonemes. With this method we obtained an improvement of 19.88%,
17.08%, 14.91% for 10, 20 and 30 speakers, from the YOHO database, respectively, compared with our previous results. In addition, it is important to note
that for this task the 18 and 20 dimension vector shows better performance
than 12 dimension vector which is the most common parameter number used in
speech recognition.

For future research, we plan to study the classification techniques, either improving the SVM kernel or by using artificial neural networks. Moreover, it is
important to study the robustness of our system under noisy conditions. Besides,
future studies on a M-ary key may be useful to increase the number of different keys available for each user given a fixed number of phonemes in the pass
phrase.

Acknowledgments

The authors would like to acknowledge the Cátedra de Seguridad, ITESM, Campus Monterrey and the CONACyT project CONACyT-2002-C01-41372 who partially supported this work.

References

1. Boser, B., Guyon I. and Vapnik V.: A training algorithm for optimal margin classifiers. In Proceedings of the Fifth Annual Workshop on Computational Learning Theory, (1992)
2. Campbell, J. P., Jr.: Features and Measures for Speaker Recognition. Ph.D. Dissertation, Oklahoma State University, (1992)
3. Cortes, C., Vapnik V.: Support-vector network. Machine Learning 20, (1995) 273-297
4. Garcia-Perera L. P., Mex-Perera C. and Nolazco-Flores J.A.: Multi-speaker voice cryptographic key generation. 3rd ACS/IEEE International Conference on Computer Systems and Applications (2005)
5. Garcia-Perera L. P., Mex-Perera C. and Nolazco-Flores J. A: Cryptographic-speech-key generation using the SVM technique over the lp-cepstra speech space. International School on Neural Nets, Lecture Notes on Computer Sciences (LNCS), Springer-Verlag, Vietri, Italy (2004).
6. Garcia-Perera L. P., Mex-Perera C., and Nolazco-Flores J.A. : Cryptographic-speech-key generation architecture improvements. J.S. Marques et al. (Eds.): IbPRIA 2005, LNCS 3523, Springer-Verlag Berlin Heidelberg (2005) 579585
7. Garcia-Perera L. P., Mex-Perera C. and Nolazco-Flores J. A.: SVM Applied to the Generation of Biometric Speech Key. A. Sanfeliu et al. (Eds.): CIARP 2004, LNCS 3287, Springer-Verlag Berlin Heidelberg (2004) 637-644
8. Higgins, A., J. Porter J. and Bahler L.: YOHO Speaker Authentication Final Report. ITT Defense Communications Division (1989)
9. Young,S., P. Woodland HTK Hidden Markov Model Toolkit home page. http://htk.eng.cam.ac.uk/
10. Monrose F., Reiter M. K., Li Q., Wetzel S.. Cryptographic Key Generation From Voice. Proceedings of the IEEE Conference on Security and Privacy, Oakland, CA. (2001)
11. E. Osuna, Freund R., and Girosi F.: Support vector machines: Training and applications. Technical Report AIM-1602, MIT A.I. Lab. (1996)
12. E. Osuna, Freund R., and Girosi F.: Training Support Vector Machines: An Application to Face Recognition, in IEEE Conference on Computer Vision and Pattern Recognition, (1997) 130-136
13. Rabiner L. R. and Juang B.-H.: Fundamentals of speech recognition. Prentice-Hall, New-Jersey (1993)
14. Uludag U., Pankanti S., Prabhakar S. and Jain A.K.: Biometric cryptosystems: issues and challenges, Proceedings of the IEEE , Volume: 92 , Issue: 6 (2004)
15. Lee K., Hon H., and Reddy R.: An overview of the SPHINX speech recognition system, IEEE Transactions on Acoustics, Speech and Signal Processing, Vol. 38, No. 1, (1990) 35 - 45
16. Joachims T., SVMLight: Support Vector Machine, SVM-Light Support Vector Machine http://svmlight.joachims.org/, University of Dortmund, (1999)

The COST-277 Speech Database

Marcos Faundez-Zanuy[1], Martin Hagmüller[2], Gernot Kubin[2],
and W. Bastiaan Kleijn[3]

[1] Escola Universitaria Politècnica de Mataró, Spain
[2] Graz University of Technology, Austria
[3] Department of Speech, Music and Hearing (KTH), Sweden
faundez@eupmt.es, hagmueller@tugraz.at, g.kubin@ieee.org,
bastiaan@speech.kth.se

Abstract. Databases are fundamental for research investigations. This paper presents the speech database generated in the framework of COST-277 "Nonlinear speech processing" European project, as a result of European collaboration. This database lets to address two main problems: the relevance of bandwidth extension, and the usefulness of a watermarking with perceptual shaping at different Watermark to Signal ratios. It will be public available after the end of the COST-277 action, in January 2006.

1 Introduction

Competitive algorithm testing on a database shared by dozens of research laboratories is a milestone for getting significant technological advances. Speaker recognition is one of these fields, where several evaluations have been conducted by NIST [1]. In this paper, we present the COST-277 database, generated by means of European collaboration between three European countries: Spain, Sweden and Austria. However, our purpose is not the collection of a new speech database. Rather than this, we have generated two new databases using a subset of an existing one [2], with the objective to study two new topics that can appear with recent technological advances:

1. The study of the relevance of bandwidth extension for speaker recognition systems.
2. The study of a watermark insertion for enhanced security on biometric systems.

A major advantage of database availability is also to set up the evaluation conditions that can avoid some common mistakes done in system designs [3]:

1. "Testing on the training set": the test scores are obtained using the training data, which is an optimal and unrealistic situation.
2. "Overtraining": The whole database is used too extensively in order to optimize the performance. This can be identified when a given algorithm gives exceptionally good performance on just one particular data set.

Thus, our database includes different material for training and testing in order to avoid the first problem. In addition, the availability of a new database helps to test the algorithms over new stuff and thus to check if the algorithms developed by a given

M. Faundez-Zanuy et al. (Eds.): NOLISP 2005, LNAI 3817, pp. 100–107, 2005.

laboratory can generalize their results, even in a new topic framework such as bandwidth extension and watermarked signals.

This paper is organized as follows: section 2 describes the database, and section three provides some experimental results as reference.

2 The COST-277 Database

We have generated two new databases using an existing one. Next section describes the original and new databases.

2.1 Original Database

Although the original database contains hundreds of speakers, several tasks (isolated digits, sentences, free text, etc.), recording sessions, microphones, etc., we have just picked up a small subset due to the procedure for database generation is time consuming and occupies a considerable amount of data (more than 2 DVD).

We have selected two subsets:

a) ISDN: 43 speakers acquired with a PC connected to an ISDN. Thus, the speech signal is A law encoded at a sampling rate fs=8kHz, 8 bit/sample and the bandwidth is 4kHz.

b) MIC: 49 speakers acquired with a simultaneous stereo recording with two different microphones (AKG C-420 and SONY ECM66B). The speech is in wav format at fs=16kHz, 16 bit/sample, and the bandwidth is 8kHz. We have just used the AKG microphone.

In both cases we have selected the following stuff for training and testing:

1. One minute of read text for training
2. Five different sentences for testing, lasting each sentence about 2-3 seconds.

All the speakers read the same text and sentences, so it is also possible to perform a text-dependent experiment.

2.2 Bandwidth Extended Database

A speech signal that has passed through the public switched telephony network (PSTN) generally has a limited frequency range between 0.3 and 3.4 kHz. This narrow-band speech signal is perceived as muffled compared to the original wide-band (0 – 8 kHz) speech signal. The bandwidth extension algorithms aim at recovering the lost low- (0 – 0.3 kHz) and/or high- (3.4 – 8 kHz) frequency band given the narrow-band speech signal. There are various techniques used for extending the bandwidth of the narrow-band. For instance, vector quantizers can be used for mapping features (e.g., parameters describing the spectral envelope) of the narrow-band to features describing the low- or high-band [4,5]. The method used in this database is based on statistical modelling between the narrow- and high-band [6].

The bandwidth extension algorithm has been directly applied to the ISDN original database, which is a real situation. However, it is interesting to have a reference of

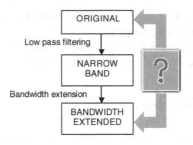

Fig. 1. General pattern recognition system

a "real" full band signal (see figure 1 for a conceptual diagram). For this purpose, we have generated a narrow band signal using the full band signal. We have used the *potsband* routine, which can be downloaded in [7]. This function meets the specifications of G.151 for any sampling frequency, and has a gain of −3dB at the passband edges.

The bandwidth extension algorithm has been tuned for speech signals with POTS (plain old telephone service) bandwidth, inside the range [300, 3400]. For this reason, we have created the following databases (see table 1):

Table 1. Speech databases, fs=sampling frequency (kHz), bps= bits per sample

Name	Bandwidth[kHz]	fs	bps	description
ISDN	[0, 4]	8	8	Original
ISDNb	[0.3, 3.4]	8	8	ISDN filtered with potsband
ISDNc	[0.1, 8]	8	8	ISDNb + BW extension
MIC	[0, 8]	16	16	Original
MICb	[0.3,3.4]	16	16	MIC filtered with potsband
MICc	[0.1, 8]	16	16	MICb + BW extension

Some experiments with these databases can be found in [8,9].

2.3   Watermarked Database

Watermarking is a possibility to include additional information in an audio signal channel without having to sacrifice bandwidth and without the knowledge of the listener. A widely know application of audio watermarking is digital rights management, where the watermark is used to protect copyrights.

Speech watermarking has been used to include additional information in the analog VHF communication channel between pilots and a air traffic controller [10]. Watermarking for biometric signal processing (e.g. speaker verification) can increase the security of the overall system.

Watermarking for speech signals is different than the usual audio watermarking due to the much narrower signal bandwidth. Compared to the 44.1 kHz sampling rate for CD-audio, telephony speech is usually sampled at 8 kHz. Therefore, compared CD-audio watermarking, less information can be embedded in the signal. For perceptual hiding usually the masking levels have to be calculated. The common algorithms used are optimized for CD-audio bandwidth and are computationally very expensive. Another difference is the expected channel noise. For CD-audio the channel noise is usually rather low. Speech on the other side is very often transmitted over noisy channels, in particular true for air traffic control voice communication. On the one hand, the channel noise is a disadvantage; on the other hand this allows much more power for the watermark signal since the channel noise will cover it anyway. The listener expects a certain amount of noise in the signal. A summary of the differences can be seen in table 2. Figure 2 shows an example of a speech frame spectrum with and without watermarking.

A more in depth explanation of the watermarking algorithm is beyond the scope of this paper and can be found in [10].

Our previous work [11] stated the convenience for a constant update in security systems in order to keep on being protected. A suitable system for the present time can become obsolete if it is not periodically improved. Usually, the combination of different systems and/ or security mechanisms is the key factor [12] to overcome some of these problems [13-14]. One application of speech watermarking is the combination of speaker recognition biometric system with a watermarking algorithm that will let to check the genuine origin of a given speech signal [15].

Table 2. Audio vs speech watermarking comparison

	CD-Audio Watermarking	Speech watermarking
Channel noise	Should be very low	Can be high
Bandwidth	Wideband (20 kHz)	Narrowband (4 kHz)
Allowed distortion	Should be not perceivable	Low
Processing delay	No issue	Very low (for real time communication)

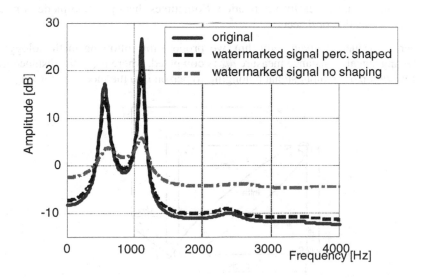

Fig. 2. Example of LPC spectrum envelope a speech fragment, with and without perceptual weighting compared with the original

Watermark floors higher than the SWR aren't included, since it is not useful.

We have watermarked the MICb database (see table 1) with the following signal to watermark ratios (SWR) and watermark floors (WM floor):

Table 3. Watermark levels (✔ :included, x: not included in the database)

SWR WM floor	0 dB	5 dB	10 dB	15 dB	20 dB
0 dB	✔	x	x	x	x
-5 dB	✔	✔	x	x	x
-10 dB	✔	✔	✔	x	x
-15 dB	✔	✔	✔	✔	x
-20 dB	✔	✔	✔	✔	✔
-25 dB	✔	✔	✔	✔	✔
-30 dB	✔	✔	✔	✔	✔

3 Algorithm Evaluation

Speaker recognition [16] can be operated in two ways:

a) Identification: In this approach no identity is claimed from the person. The automatic system must determine who is trying to access.

b) Verification: In this approach the goal of the system is to determine whether the person is who he/she claims to be. This implies that the user must provide an identity and the system just accepts or rejects the users according to a successful or unsuccessful verification. Sometimes this operation mode is named authentication or detection.

In order to evaluate a given algorithm, we propose the following methodology: for each testing signal, a distance measure d_{ijk} is computed, where d_{ijk} is the distance from the k realization of an input signal belonging to person i, to the model of person j.

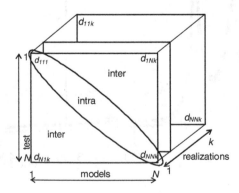

Fig. 3. Proposed data structure

The data can be structured inside a matrix. This matrix can be drawn as a three dimensional data structure (see figure 3). In our case, N=49 and k=5.

This proposal has the advantage of an easy comparison and integration of several algorithms by means of data fusion, with a simple matrix addition or more generally a combination. Once upon this matrix is filled up, the evaluation described in next sections should be performed.

3.1 Speaker Identification

The identification rate finds for each realization, in each raw, if the minimum distance is inside the principal diagonal (success) or not (error), and works out the identification rate as the ration between successes and number of trials (successes + errors):

```
for i=1:N,
    for k=1:#trials,
        if(d_iik<d_ijk)  ∀j?i, then success=success+1
        else error=error+1
        end
    end
end
```

3.2 Speaker Verification

Verification systems can be evaluated using the False Acceptance Rate (FAR, those situations where an impostor is accepted) and the False Rejection Rate (FRR, those situations where a speaker is incorrectly rejected), also known in detection theory as False Alarm and Miss, respectively. This framework gives us the possibility of distinguishing between the discriminability of the system and the decision bias. The discriminability is inherent to the classification system used and the discrimination bias is related to the preferences/necessities of the user in relation to the relative importance of each of the two possible mistakes (misses vs. false alarms) that can be done in speaker verification. This trade-off between both errors has to be usually established by adjusting a decision threshold. The performance can be plotted in a ROC (Receiver Operator Characteristic) or in a DET (Detection error trade-off) plot [17]. DET curve gives uniform treatment to both types of error, and uses a scale for both axes, which spreads out the plot and better distinguishes different well performing systems and usually produces plots that are close to linear. DET plot uses a logarithmic scale that expands the extreme parts of the curve, which are the parts that give the most information about the system performance. For this reason the speech community prefers DET instead of ROC plots. Figure 4 shows an example of DET plot, and figure 5 shows a ROC plot.

We can use the minimum value of the Detection Cost Function (DCF) for comparison purposes. This parameter is defined as [17]:

$$DCF = C_{miss} \times P_{miss} \times P_{true} + C_{fa} \times P_{fa} \times P_{false} \qquad (1)$$

Where C_{miss} is the cost of a miss (rejection), C_{fa} is the cost of a false alarm (acceptance), P_{true} is the a priori probability of the target, and $P_{false} = 1 - P_{true}$. $C_{miss} = C_{fa} = 1$.

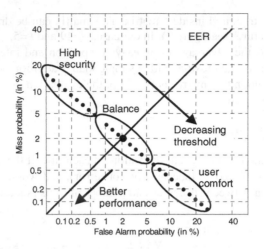

Fig. 4. Example of a DET plot for a speaker verification system (dotted line)

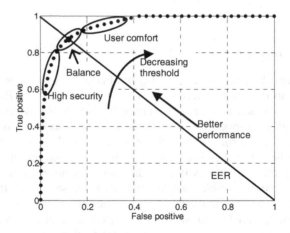

Fig. 5. Example of a ROC plot for a speaker verification system (dotted line)

Nevertheless, this parameter just summarizes the behaviour for a narrow range of operating points in the neighbourhood of the selected threshold. For this reason a whole DET or ROC plot is more interesting for system comparison purposes.

Using the data structure defined in figure 3, we can easily apply the DET curve analysis. We just need to split the distances into two sets: intra-distances (those inside the principal diagonal), and inter-distances (those outside the principal diagonal).

Acknowledgement

This work has been supported by FEDER and the Spanish grant MCYT TIC2003-08382-C05-02, and the European COST action 277 "Non-linear speech processing".

References

1. http://www.nist.gov
2. Ortega-García J., González-Rodríguez J., and Marrero-Aguiar V., "AHUMADA: A Large Speech Corpus in Spanish for Speaker Characterization and Identification". Speech communication Vol. 31 (2000), pp. 255-264, June 2000
3. Bolle R. M., Ratha N. K., Pankanti S., "Performance evaluation in 1:1 Biometric engines". Springer Verlag LNCS 3338, pp.27-46 S. Z. Li et al. (Eds.) Sinobiometrics 2004.
4. Enbom N., and Kleijn W. B., "Bandwidth expansion of speech based on vector quantization of the mel frequency cepstral coefficients," in IEEE Workshop on Speech Coding, Porvoo, Finland pp. 1953-1956, 1999
5. Epps J., and Holmes H.W., "A new technique for wideband enhancement of coded narrowband speech," in IEEE Workshop on Speech Coding, Porvoo, Finland pp. 174-176, 1999
6. Nilsson M., Kleijn W. B., "Avoiding over-estimation in bandwidth extension of telephony speech", IEEE ICASSP'2001, Salt Lake City, USA
7. http://www.ee.ic.ac.uk/hp/staff/dmb/voicebox/voicebox.html
8. Faundez-Zanuy M., Nilsson M., Kleijn W. B., "On the relevance of bandwidth extension for speaker identification". Vol. III pp.125-128, EUSIPCO'2002, Toulouse.
9. Faundez-Zanuy M., Nilsson M., Kleijn W. B., "On the relevance of bandwidth extension for speaker verification". Pp. 2317-2320. ICSLP'2002. Denver
10. Hagmüller M., Hering H., Kröpfl A., and Kubin G, "Speech watermarking for air traffic control", in Proc. of 12th European Signal Processing Conference, Vienna, Austria, Sept. 6-10, 2004, pp. 1653-1656.
11. Faundez-Zanuy M., "On the vulnerability of biometric security systems". IEEE Aerospace and Electronic Systems Magazine. Vol.19 n° 6, pp.3-8, June 2004.
12. Faundez-Zanuy M., "Data fusion in biometrics" IEEE Aerospace and Electronic Systems Magazine. IEEE Aerospace and Electronic Systems Magazine, Vol.20 n° 1, pp.34-38, January 2005.
13. Faundez-Zanuy, M., "Privacy issues on biometric systems". IEEE Aerospace and Electronic Systems Magazine, Vol.20 n° 2, pp.13-15, February 2005.
14. Faundez-Zanuy M., "Biometric recognition: why not massively adopted yet?". IEEE Aerospace and Electronic Systems Magazine. Vol.20 n° 8, pp.25-28, August 2005.
15. Faundez-Zanuy M., Hagmüller M., Kubin G. "Speaker identification security improvement by means of speech watermarking". Scientific Report to be published.
16. Faundez-Zanuy M., Monte-Moreno E., "state-of-the-art in speaker recognition". IEEE Aerospace and Electronic Systems Magazine. Vol.20 n° 5, pp 7-12, May 2005
17. Martin A., Doddington G., Kamm T., Ordowski M., and Przybocki M., "The DET curve in assessment of detection performance", V. 4, pp.1895-1898, Eurospeech 1997

Children's Organization of Discourse Structure Through Pausing Means

Anna Esposito[1,2]

[1] Seconda Università di Napoli, Dipartimento di Psicologia, Via Vivaldi 43, Caserta, Italy
anna.esposito@unina2.it, iiass.annaesp@tin.it
[2] IIASS, Via Pellegrino 19, 84019, Vietri sul Mare, INFM Salerno, Italy

Abstract. This study aims to investigate on how different kinds of pausing strategies, such as empty and filled pauses, and phoneme lengthening are used by children to shape the discourse structure in Italian, and to identify how many of the silent intervals can be attributed to the amount of given and added information the speaker is conveying in the speech flow. To this aim a cross-modal analysis (video and audio) of spontaneous narratives produced by male and female children (9 plus-minus 3 months years old) was performed. Empty speech pauses were divided into three categories according to their duration: a) short - from 0.150 up to 0.500 s long; b) medium - from 0.501 up to 0.900 s long; c) long - more than 0.900 s long. The analysis showed that each of the above categories seems to play a different role in the children discourse organization, Children pause, like adults, to recover from their memory the new information they try to convey. Higher is the recovery effort, longer is the pausing time. Longer are the pauses, lower is the probability that they can be associated to a given information Most of the long pauses (96% for female and 94% for male) are associated to a change of scene suggesting that long pauses are favored by children in signaling discourse boundaries. The consistency, among subjects, in the distribution of speech pauses seems to suggest that, at least in Italian, there is an intrinsic model of timing, probably a very coarse model, that speakers use to regulate speech flow and discourse organization.

1 Introduction

Spontaneous speech, as well as other types of speech, is characterized by the presence of silent intervals (empty pauses) and vocalizations (filled pauses) that do not have a lexical meaning. These pausing means play several communicative functions. In fact, it has been shown that their occurrence is determined by several factors such as build up tension or generate the listener's expectations about the rest of the story, assist the listener in his task of understanding the speaker, signal anxiety, emphasis, syntactic complexity, degree of spontaneity, gender, and educational and socio-economical information [1, 2, 13, 15, 16, 18].

Pauses are not only generate by psychological motivations but also as a linguistic mean for discourse segmentation. Speakers systematically signal changes in scene, time, and event structures using speech pauses. Empty pauses are more likely to coincide with boundaries, realized as a silent interval of varying length, at clause and paragraph level

M. Faundez-Zanuy et al. (Eds.): NOLISP 2005, LNAI 3817, pp. 108–115, 2005.

[4, 14, 17, 22]. This is particularly true for narrative structures where it has been shown that pausing marks the boundaries of narrative units [7, 8, 19, 20, 21, 22].

Several cognitive psychologists have suggested that pausing strategies reflect the complexity of neural information processing. Pauses will surface in the speech stream as the end product of a "planning" process that cannot be carried out during speech articulation and the amount and length of pausing reflects the cognitive effort related to lexical choices and semantic difficulties for generating new information [5, 8, 16].

Moreover, there is a practical interest in studying pause distribution along speech for application on automatic speech synthesis and recognition. Most of the current automatic speech systems ignore the effects of pausing, temporal timing, utterance-final vowel lengthening. Thus, a better knowledge of how pausing strategies affect spontaneous speech should be of support in the design of more natural speech synthesis systems and improve the performance of automatic speech recognition systems.

Along the above guidelines, the aim of the reported experiments was to investigate how different kind of pausing strategies, such as empty and filled pauses, and phoneme lengthening are used by children to shape the discourse structure in Italian and furthermore, to identify how many of the silent intervals can be attributed to the amount of "*given*" and "*added*" information the speaker is conveying in the speech flow.

2 Definitions

The present work interprets the concepts of "*given*" and "*added*" according to the definition proposed by Chafe [6], which considered as "*added*" *any verbal material that produces a modification in the listener's conscious knowledge*, and therefore "*given*" verbal material was intended as not to produce such a modification. Moreover, the label *not classified* is attributed to speech material such as filled pauses, and/or short interruptions (such as "*ap**") that follow empty pauses. Together with the above definitions it is necessary to introduce the concept of "*changes*" that in the present work are labels attributed to empty pauses identifying *changes in scene, time and event structures*. In this context, *changes* could be attributed to any pause independently of the kind of speech material (*given, added,* and *not classified*) that precedes them.

In this context, an *empty pause* (EP) is a *silent interval of more than 0.150 s*. Normally *filled pauses* (FP) are used to "hold the floor" i.e. preventing interruption by the listener while the speaker searches for a specific word [12] and different fillers may serve different functions, such as marking a successful memory search ("ah") or signaling the selection of an example ("oh"). However, in this context, filled pauses generally appear as "*hum, ehh*" because the task and the lack of an interlocutor preclude their use for other functions. Even though *phoneme lengthening* can appropriately be considered as a filled pause, such potential filled pauses were measured and analyzed on a separate ground. Moreover a "*clause*" is assumed to be "*a sequence of words grouped together on semantic or functional basis*" and a "*paragraph*" was considered as "*a sequence of several clauses connected together by the same subject or scene*".

3 Materials and Methods

The video recordings on which our analysis is based are of narrations by 10 female and 4 male children (9 plus-minus 3 months' years old). The children told the story of a 7-minute animated color cartoon they had just seen. The cartoon is of a type familiar to Italians children, involving a cat and a bird. The listener in each case was the child's teacher and other children. This kept out stranger-experimenter inhibitions from the elicitation setting; i.e., factors that could result in stress and anxiety. Limiting these factors allows us to rule out the "socio-psychological"-type of pause [3]. The cartoon has an episodic structure, each episode characterized by a "cat tries to get bird/is foiled" narrative arc. Because of the cartoon's episodic structure, typically children will forget entire episodes. Therefore, only two episodes were analyzed, the ones that all the children remembered. The data were recorded at the International Institute for Advanced Scientific Study, Vietri, Italy. None of the participants was aware that speech pauses were of interest. The video was analyzed using commercial video analysis software (VirtualDub™). The program allows viewing of video-shots in 3-D, and movement forward and backward through the shots. The speech waves, extracted from the video, were sampled at 16 kHz and digitalized at 16 bits. The audio was analyzed using Speechstation2™ from Sensimetrics. For the audio measurements the waveform, energy, spectrogram, and spectrum were considered together, in order to identify the beginnings and endings of utterances, filled and empty speech pauses and phoneme lengthening. The details of the criteria applied to identify the boundaries in the speech waveform are accurately described in Esposito and Stevens [9]. Both the video and audio data were analyzed perceptually as well, the former frame-by-frame and the latter *clause-by-clause*.

4 Preliminary Results

Tables 1 and 2 report, for each male and female child respectively, the absolute number of occurrences of various pausing means and their percentage (between brackets) over the two episodes. Tables 1 and 2 also report the percentage of empty pauses and the percentage of filled pauses and phoneme lengthening for each child.

Among children, S6 and S8 are whose that uses a higher percentage of empty pauses, which is compensated by a reduced number of filled pauses. Moreover, among empty pauses, short pauses (33%) are largely more frequent than medium ones (10%), which in turn are more frequent than long pauses (6%) suggesting that the three duration ranges play a different role in structuring the discourse. Since this is generally true also for each subject (except S5), it also suggests that children use a similar pause duration strategy to highlight different discourse units.

Table 1. *Female children:* absolute number of occurrence of empty (short, medium, long) pauses, filled pauses, and phoneme lengthening (% between brackets) broken down for per child

Female children	Short EP	Me-dium EP	Long EP	Filled	Vowel Length.	To-tal	% of EP	% of FP and Vowel Length.
S1	21 (22)	15 (16)	8 (9)	29 (31)	21 (22)	94	47	53
S2	25 (24)	18 (17)	10 (10)	33 (31)	18 (17)	104	53	48
S3	18 (22)	4 (5)	7 (9)	31 (38)	21 (26)	81	36	64
S4	61 (52)	2 (1)	1 (1)	33 (28)	21 (18)	118	54	46
S5	4 (12)	5 (15)	10 (30)	11 (34)	3 (9)	33	57	43
S6	39 (60)	2 (3)	3 (5)	10 (16)	10 (16)	64	68	32
S7	20 (26)	7 (10)	1 (1)	21 (27)	28 (36)	77	37	63
S8	10 (44)	6 (26)	0	6 (26)	1 (4)	23	70	30
S9	38 (36)	9 (8)	4 (4)	18 (17)	37 (35)	106	48	52
S10	19 (29)	7 (11)	3 (5)	24 (37)	12 (18)	65	45	55
Tot.	**255 (33)**	**75 (10)**	**47 (6)**	**216 (28)**	**172 (22)**	**765**	**51**	**49**

Table 2. *Male children*: absolute number of occurrence of empty (short, medium, long) pauses, filled pauses, and phoneme lengthening (% between brackets) broken down for per child

Male children	Short EP	Medium EP	Long EP	Filled	Vowel Length.	To-tal	% of EP	% of FP and Vowel Length.
S1	13 (28)	10 (21)	4 (8)	13 (28)	7 (15)	47	57	43
S2	20 (39)	9 (17)	2 (4)	8 (15)	13 (25)	52	60	40
S3	25 (45)	10 (18)	3 (5)	7 (12)	11 (20)	56	68	32
S4	18 (24)	13 (19)	7 (9)	21 (29)	15 (20)	74	52	49
Tot.	**76 (33)**	**42 (18)**	**16 (7)**	**49 (22)**	**46 (20)**	**229**	**59**	**41**

On overall, filled pauses are frequent as much as phoneme lengthening, even though there is a large intra-speaker and inter-speaker variability. Moreover, Tables 1 and 2 show that pausing means are differently used by different children with some child that make use of more filled pauses and vowel lengthening than empty pauses (as S3 and S7 for female children), other that does the opposite (as S6 and S8 for female children, and S3 for male children), yet other, along their speech, equally distribute empty and filled pauses.

Empty pauses, taken separately, are considerably more frequent than filled pauses and phoneme lengthening both for male (59% against 21% and 20%) and female (51% against 28% and 22%). Moreover, short pauses (33% both for male and female) are more frequent than medium ones (18% for male and 10% for female), which in turn are more frequent than long pauses (7% for male and 6% female). This is generally true for each subject (except female S5) suggesting that children use a similar strategy in distributing short, medium and long pauses in their narrations.

Table 3. Absolute number of occurrences of short, medium, and long empty pauses associated with a given, added, and not classified information, and changes of scene in the discourse structure. The percentage (between brackets) is computed over the number of pauses in each duration range. (Results are reported both for male and female children)

10 Female children	Short EP		Medium EP		Long_EP	
Given	9	(4%)	2	(3%)	0	(0)
Added	225	(88%)	67	(89%)	36	(77%)
Not_classified	21	(8%)	6	(8%)	11	(23%)
Changes	51	(20%)	61	(81%)	45	(96%)
4 Male children						
Given	5	(6%)	0		0	
Added	65	(86%)	37	(88%)	13	(81%)
Not_classified	6	(8%)	5	(12%)	3	(19%)
Changes	6	(8%)	14	(33%)	15	(94%)

To investigate if different duration ranges play a different role in structuring discourse units and if the need of pausing is due to the cognitive effort to recall from memory and lexicalize concepts that are not yet known by the listener, we evaluated the amount of *given*, *added* and *not classified* information that precedes each empty pause. Table 3 reports the number of short, medium, and long pauses that follow *given*, *added*, and *not classified* speech material. Here the label *not classified* is attributed to speech material such as filled pauses, and/or short interruptions (such as "*ap**").

The number of short, medium and long empty pauses that are associated to a change (*changes*) of scene or paragraph structure is also reported. Note that *changes* could happen independently of the kind of speech material (*given, added,* and *not classified*) that precedes them and therefore, should not be counted in the total percentage.

Children pause, like adults, to recover from their memory the new information they try to convey to the listeners, showing that higher is the recovery effort, longer is the pausing time. As it could be seen in Table 3, most of the pauses follow new added information, except for a few short (4% for female and 6% for male) and medium (3% only for female) pauses. Most of the long pauses (96% for female and 94% for male) are associated to a change of scene suggesting that long pauses are favoured by children in signalling discourse boundaries.

The relationship with the cognitive effort can be easily seen examining the amount of long pauses associated with changes of scene, time and event structure. In fact, an high percentage of medium (81% for female, 33% for male children) and long (96% for female and 94% for male children) pauses are made to signal these changes, whereas only a low percentage of short (20% for female and 8% for male children) pauses serve this purpose.

The above data also suggest a predictive scheme for the alternating pattern of cognitive rhythm in the production of spontaneous narratives. In this alternating pattern, long pauses account for the highest percentage (96% and 94% for female and male children respectively) of paragraphs or changes followed by medium pauses (81% and 33% for female and male children respectively). Even though they are more frequent than medium and long pauses, short pauses (no longer 0.500 s) have a low probability to signal a change of scene (20%) in the flow of the narration.

Tables 4 and 5 gives, for each child, the number of words (Wds), clauses (Cls) and paragraphs (Phs) marked by a pause (#Ps). In this case, filled and empty pauses are grouped together. Note that pauses that mark a word boundary can also mark a clause, a filler conjunction and a paragraph boundary.

The results in Tables 4 and 5 show that, on the average, 19% of the word boundaries are marked by a pauses independently of the number of words used. The pattern is still more reliable at clause and paragraph level where children mark with a pause 70% of the clause boundaries and 100% of the paragraph boundaries. There is a variability in the above pattern, mostly observed among the female children, since the female children S5, S6, S8, and S10 pause less than the others at word boundaries, and S7 and S9 use also other means (such as vowel lengthening) to signal paragraphs boundaries.

Table 4. Absolute number and percentage (between brackets) of words (Wds), clauses (Cls), and paragraphs (Phs) marked by a pause (#Ps) for female children

10 Female children	Wds	#Ps	(%)	Cls	#Ps	(%)	Phs	#Ps	(%)
S1	217	54	(25)	43	31	(72)	10	10	(100)
S2	249	49	(20)	56	37	(66)	11	11	(100)
S3	224	43	(19)	39	28	(72)	13	13	(100)
S4	307	74	(24)	49	36	(73)	15	15	(100)
S5	118	18	(15)	23	13	(57)	8	8	(100)
S6	324	47	(15)	68	36	(53)	15	15	(100)
S7	206	40	(19)	34	26	(76)	9	8	(89)
S8	149	15	(10)	25	14	(56)	6	6	(100)
S9	217	45	(21)	41	31	(76)	13	10	(77)
S10	187	29	(16)	34	23	(68)	11	11	(100)
Averaged total	2198	414	(19)	412	275	(68)	111	107	(96)

Table 5. Absolute number and percentage (between brackets) of words (Wds), clauses (Cls), and paragraphs (Phs) marked by a pause (#Ps) for male children

4 Male Children	Wds	#Ps	(%)	Cls	#Ps	(%)	Phs	#Ps	(%)
S1	142	29	(20)	21	15	(71)	9	9	(100)
S2	182	33	(18)	28	20	(71)	6	6	(100)
S3	218	35	(16)	34	23	(68)	8	8	(100)
S4	208	49	(24)	37	27	(73)	12	12	(100)
Averaged total	750	146	(19)	120	85	(71)	35	35	(100)

5 Conclusions

This study was devoted to investigate on the system of rules that underlie children pausing strategy and their psychological bases. The reported data show that empty pauses of short, medium, and long duration are largely used by children to signal new information to the listeners' conscious knowledge and only a few among the short (4% for female and 6% for male), and medium (3% only for female) empty pauses mark given information. This suggests that children pause, like adults [16, 20, 21, 22], to recover from their memory the new information they try to convey. Higher is the recovery effort, longer is the pausing time. Moreover, longer are the pauses, lower is the probability that they can be associated to a given information, supporting the hypothesis that pausing plays the functional role of indicating the cognitive effort needed for planning speech.

Pauses are not only generate by psychological motivations but also as a linguistic mean for discourse segmentation. Pauses are used by children to mark words, clause, and paragraph boundaries. The results show that a similar percentage of word (19% both for male and female children), clause (68% for female and 71% of male children) and paragraph (96% of female and 100% of male children) boundaries is marked by a pause. Short pauses (67.6% for females and 56.7 for males) than medium (19.9% for females and 31.4% for males), that in turn are more frequent than long pauses (12.5% for females and 11.9% for males). However, children systematically signal changes in scene, time, and event structures using medium (81% for female and 33% for male) and long pauses (96% for female and 94% for male) suggesting that only pauses longer than 0.5 s are favored by children to mark paragraph boundaries whereas short pauses rarely served for this function.

This result favors the hypothesis of an universal model for discourse structure, otherwise we would expect children, being less skilled in the use of the language's lexicon to make more pauses at word level than at the clause and paragraph level. This hypothesis is further on supported by the fact in a previous work [10-11] it has been shown that in 56% of the cases children's pauses occur right after the first word in a clause, i.e. right after a filler conjunction that signals a major transition in the speech flow and serves to plan the message content for the continuation of the discourse. The consistency among the subjects in the use of pausing means seems to suggest a very coarse and general timing model, that speakers use to regulate speech flow and discourse organization. More data are needed to make sense of how this model works, since it would be of great utility in the field of human-machine interaction, favoring the implementation of more natural speech synthesis and interactive dialog systems.

Acknowledgements

This work has been partially funded by the American project *IT's Enabled Intelligent and Ubiquitos Access to Educational Opportunities for Blind Student*, in collaboration with professor Nicholas Bourbakis, Information Technology Research Institute (ITRI), Wright State University, Dayton, OHIO, USA. The author would like to thank professor Maria Marinaro for her useful comments and suggestions, and Luisa Del Prete and Giulia Palombo who collected the video recordings. Miss Tina Marcella Nappi is acknowledged for her editorial help.

References

1. Abrams, K., Bever, T. G.: Syntactic Structure Modifies Attention during Speech Perception and Recognition. Quarterly Journal of Experimental Psychology, Vol. 21 (1969) 280-290
2. Bernstein, A.: Linguistic Codes, Hesitation Phenomena, and Intelligence. Language and Speech, Vol. 5 (1962) 31-46
3. Beaugrande, R.: Text Production. Norwood NJ: Text Publishing Corporation (1984)
4. Brotherton, P.: Speaking and not Speaking; Process for Translating Ideas into Speech. In A. Siegman, A., and Feldestein, S. (eds): Of Time and Speech. Hillsdale, Lawrence Erlbaum, N.J. (1979) 79-209
5. Butterworth, B. L.: Evidence for Pauses in Speech. In Butterworth B. L. (ed): Language Production: Speech and Talk, London Academic Press, Vol. 1 (1980) 155-176
6. Chafe, W. L.: Language and Consciousness. Language, Vol. 50 (1974) 111-133
7. Chafe, W. L.: The Deployment of Consciousness in the Production of a Narrative. In Chafe W.L.(cd): The Pear Stories, Norwood N.J., Ablex (1980) 9-50
8. Chafe, W. L.: Cognitive constraint on information flow. In Tomlin, R. (cd): Coherence and Grounding in Discourse, John Benjamins (1987) 20-51
9. Esposito, A., Stevens, K. N.: Notes on Italian Vowels: An Acoustical Study (Part I), RLE Speech Communication Working Papers, Vol. 10 (1995) 1-42
10. Esposito, A.: Pausing Strategies in Children. In Proceedings of the International Conference in Nonlinear Speech Processing, Cargraphics, Barcelona, SPAIN, 19-22 April (2005) 42-48
11. Esposito, A., Marinaro, M., Palombo, G.: Children Speech Pauses as Markers of Different Discourse Structures and Utterance Information Content. Proceedings of the International Conference From Sound to Sense: +50 years of discoveries in Speech Communication, MIT, Cambridge, June 10-13 (2004)
12. Erbaugh, M. S. (1987) A Uniform Pause and Error Strategy for Native and Non-Native Speakers. In Tomlin, R. (ed): Coherence and Grounding in Discourse, John Benjamins (1987) 109-130
13. Kowal, S., O'Connell, D. C., Sabin, E. J.: Development of Temporal Patterning and Vocal Hesitations in Spontaneous Narratives, Journal of Psycholinguistic Research, Vol. 4 (1975) 195-207
14. Gee, J. P., Grosjean, F.: Empirical Evidence for Narrative Structure. Cognitive Science, Vol. 8 (1984) 59-85.
15. Green, D. W.: The Immediate Processing of Sentence. Quarterly Journal of Experimental Psychology, Vol. 29 (1977) 135-146
16. Goldman-Eisler, F. Psycholinguistic: Experiments in Spontaneous Speech. London: New York, Academic press (1968)
17. Grosz, B., Hirschberg, J.: Some Intentional Characteristics of Discourse Structure. Proceedings. of International Conference on Spoken Language Processing, Banff (1992) 429-432
18. O'Connell, D. C., Kowal, S.: Pausology. Computers in Language Research 2, Vol.19 (1883) 221-301
19. Oliveira, M.: Prosodic Features in Spontaneous Narratives, Ph.D. Thesis, Simon Fraser University (2000)
20. Oliveira, M.: Pausing Strategies as Means of Information Processing Narratives. Proceeding of the International Conference on Speech Prosody, Ain-en-Provence (2002) 539-542
21. O'Shaughnessy D.: Timing Patterns in Fluent and Disfluent Spontaneous Speech. Proceedings of ICASSP Conference, Detroit (1995) 600-603
22. Rosenfield, B.: Pauses in Oral and Written Narratives. Boston, Boston University (1987)

F0 and Intensity Distributions of Marsec Speakers: Types of Speaker Prosody

Brigitte Zellner Keller[1, 2]

[1] Dept. Clinical Psychology & Rehabilitative Psychiatry,
University of Bern, Switzerland
[2] IMM, Lettres, University of Lausanne, Switzerland
Brigitte.ZellnerKeller@unil.ch

Abstract. Most research on F0 has attempted to model the behaviour of an entire linguistic community (e.g of speakers of US or UK English, French, Japanese etc). In this research, an attempt is made in two analyses to characterize some prosodic aspects of individual differences within the speaker community. For this, the statistical distributions of F0 and intensity parameters were examined. It was found in the first analysis (34 male speakers, nine speech styles) that F0 distributions showed a number of characteristic patterns while intensity distributions did not pattern in any particular fashion. F0 distributions fell into four patterns, suggesting four styles of F0 whatever the speech style is. This classification was confirmed in our second analysis (11 male speakers, one speech task). These various patterns of F0 distributions are discussed with regard to the speech task and to the speaker's style.

1 Introduction

It is a common assumption that speakers activate speech components in a similar fashion when they perform the same speech task. Differences among speakers are then interpreted in terms of paralinguistic and/or extralinguistic factors.

A number of current prosodic studies investigate the complex aspects of individual variations, in particular in studies related to the "family of emotions", attitudes and sociolinguistic parameters (see e.g., the proceedings of Speech Prosody 2004). These prosodic variations are mainly characterised by one, or more often several, specific patterns which are superimposed on the "neutral" expected prosodic profile. For example, in the case of fear, the mean values for speech rate and F0 are increased [3].

Other prosodic variations are related to the individual psychological profile (for example [4]). Such variations are distinguished by specific interplays between speech rate, intensity, F0 and pauses.

In this paper, statistical distributions of F0 and intensity are investigated. Possible differences at this level do not seem to have been considered in previous studies [1].

2 Methodology

For this study, the Machine-Readable Spoken English Corpus (MARSEC) see www.rdg.ac.uk/AcaDepts/ll/speechlab/marsec.) was used in two analyses: on the one

M. Faundez-Zanuy et al. (Eds.): NOLISP 2005, LNAI 3817, pp. 116–124, 2005.

hand, the intonation and the intensity of 34 male speakers recorded in different speech tasks were investigated. This permitted to examine the effect of speech style. On the other hand, the intonation and the intensity of 11 male speakers giving a commentary on the BBC were analysed. This permitted to examine the effect of individual variation within a specific speech style.

For the first study, 23'566 values of F0 for 34 male speakers were extracted from the database obtained from MARSEC by Keller [3]. The pitch periods of voiced sounds were determined by the position of the maximum of the autocorrelation function of the sound. All other parameters were kept at default (time step 10 ms, minimum pitch 70 Hz). Intensity values of these voiced sounds were calculated at a time step of 1 ms using the default values set in the Praat software. Average f0 and intensity values were calculated for each sound segment on the basis of these measures.

Nine styles of speech were represented in this database: address, fiction, lecture, market, news, poetry, religion, report and sports.

F0 values were converted into semitones by the formula given in Fant et al. [2]: 12[Ln(Hz/100)/Ln(2). This formula centers the semitone scale at 100 Hz = 0 semitones, and 67 Hz at about -7 semitones and 318 Hz at about 20 semitones. Frequency distributions were computed for each speaker with a stepsize of 0.5 semitones. The obtained frequency per semitone step was then weighted by converting the values into percentages of total observations per speaker.

Intensity values were squared-root transformed to better approximate a normal distribution. Distributions were computed for each speaker with a stepsize of 0.5 -i.e., 0.25 dB. Then the obtained frequency per intensity level was weighted by converting the values into percentages of total observations per speaker.

For the second study, signals were prealigned with an automatic algorithm written by Prof. Eric Keller (IMM, University of Lausanne), and then manually adjusted. Acoustic analyses were performed with the public software Praat. The pitch period of a sound was determined by the position of the maximum response of the autocorrelation function. The voicing threshold was set to 0.05 and the silence threshold was set to 0.15. All other parameters were kept at default values (time step 10 ms, minimum pitch 75 Hz).

F0 values were automatically extracted thanks to a program written by Eric Keller. It runs the Praat's F0 extraction of the input sound. On the basis of the TextGrids, the ouput provides F0 values for each voiced sound.

In both studies, the statistical analyses were performed with DataDesk 6.1, SPSS 11.0 and XLSTAT 5.1.

3 Results

3.1 First Analysis: F0 Results

Histograms computed in DataDesk, with the same window length and the same number of intervals show a considerable variation of distributions among speakers.

This variation was unexpected since the literature does not mention any particular issue in this domain. However, noticeable differences are observed by just looking at the graphs: some distributions are multimodal and others are not. Some distributions

Table 1. Statistical central values of F0 values for 34 male speakers in nine speech tasks

Statistics

| | N | | | | |
	Valid	Missing	Mean	Median	Mode
AMD	397	4787	1.829037	1.488125	7.4845
BP	901	4283	1.818313	1.454095	.0078[a]
BR	310	4874	3.338160	2.979429	-.2718[a]
CF	110	5074	3.080628	2.807319	-2.5968[a]
CL	1104	4080	4.246489	4.562503	-.9145[a]
CP	146	5038	4.697262	4.573999	2.0074[a]
DH	5181	3	1.396632	.911155	-2.6268[a]
DS	479	4705	5.884802	6.006810	6.8069[a]
GB	443	4741	5.152599	4.959489	4.7805
GF	3318	1866	6.350111	6.270877	7.0162
GL	590	4594	6.861170	6.853301	4.3922[a]
JB1	451	4733	7.779444	7.268890	13.9594
JB2	462	4722	2.637703	2.402813	-.7237[a]
JC	659	4525	6.940815	6.355093	8.9241
JH	345	4839	2.884157	2.526611	-4.2560[a]
JM	450	4734	.903066	.498188	-2.0857[a]
JS	400	4784	5.832542	5.839330	5.6475[a]
KG1	345	4839	7.661769	7.144898	15.0907
KG2	503	4681	6.449524	6.289607	5.1909[a]
LM	587	4597	5.329337	5.291426	3.7066[a]
MC	169	5015	4.391394	4.838427	-3.7520[a]
MF	128	5056	3.203771	2.950964	2.4201[a]
MJ	599	4585	3.920015	4.153611	13.4643
ms1	139	5045	1.817037	1.750320	3.6033
ms6	304	4880	3.765106	3.362408	.6912[a]
ms7	268	4916	.442596	.244106	-2.8346[a]
MW	111	5073	5.622323	5.555810	-1.0841[a]
PD	213	4971	3.833997	3.488348	1.9960[a]
PF	82	5102	.476567	.227052	1.2881[a]
PR	450	4734	8.118200	8.303180	6.8878[a]
RF	3300	1884	3.220368	2.730823	-.0904[a]
RSO	423	4761	.228776	.065835	-1.3390[a]
ST	71	5113	7.660508	7.489997	1.9399
VD	128	5056	5.387478	5.843171	-4.1178[a]

a. Multiple modes exist. The smallest value is shown

are strongly left-skewed, some others are rather centered. Some distributions are peaked, and others are rather flat.

The three central values computed in SPSS, mean, median and mode, (see Table 1) show various patterns. For example, the three central values for speaker JS are very close. For many other speakers, the lowest modal value differs considerably from the two other central values.

The computation of weighted frequency distributions permits first the graphical superposition of F0 distributions in order to illustrate these variations among speakers (see Figure 1). Secondly, this table was used for the computation in XLSTAT of an agglomerative hierarchical classification (AHC).

The AHC algorithm gathers the most similar observation pairs, and then progressively aggregates the other observations or observation groups according to their similarity until all observations are in a single group. The settings were as follows:

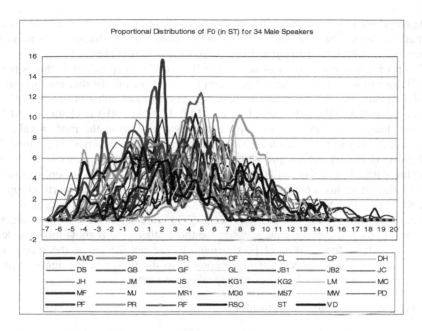

Fig. 1. Weighted frequency distributions for 34 male speakers' F0. The x axis represents semitones calulated by Fant's equation (2004). This equation sets 100 Hz = 0 st, 70 Hz = 6.17 st, 200 Hz= 12 st. The y axis represents the percentage of samples.

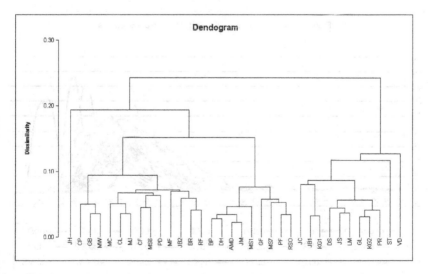

Fig. 2. Dendogram of the clustering of F0 distributions for 34 male speakers in nine speech tasks. The clustering suggests 4 groups.

spearman's dissimilarity and complete linkage. In that case, the dissimilarity between objects of A and B is the largest dissimilarity between an object of A and an object of

B. Aggregation using complete linkage tends to dilate the data space and to produce compact clusters. The AHC gives a clustering which classify the speakers according to their F0 distributions.It suggests four groups of speakers (Figure 2).

Group 1 is represented by a unique speaker JH in a unique speech style that is fiction. His F0 distribution has a strong left peak and then becomes fairly flat, multimodal and broad. Compared to the mean and median, his lowest modal value is very far-off. The second group (CP, GB, MW, MC, CL, PD, MF, JB2, BR, RF) is characterised by a "normal" height for male speakers with a mean value close to the median value, but both values being distant from the lowest modal value. The skewness is positive. The third group (BP, DH, AMD, JM, MS1, GF, MS7, PF, RSO) is similar to the second group but with a low register of F0. The skewness is positive. In this group, speaker GF is badly classified: his three central values are close and high. In the fourth group (JC, JB1, KG1, DS, JS,LM,GL, KG2, PR, ST, VD), at least two of the three central values are very high and the F0 range is fairly large. The kurtosis is close to 0 or negative. Apart from group 1 with a unique speaker, the three other groups show mixed styles of speech, i.e., these four classes of F0 distributions are not driven by the speech task. When hearing pairs of speech samples according to the leaves of the aggregation, an auditive impression of similarity between speakers emerges.

3.2 First Analysis: Intensity Results

The same procedure for the computation of histograms and weighted distributions were applied to intensity. The superposition of speakers' intensity distributions illustrate a similar pattern irrespective of speaker or speech style (see Figure 3).

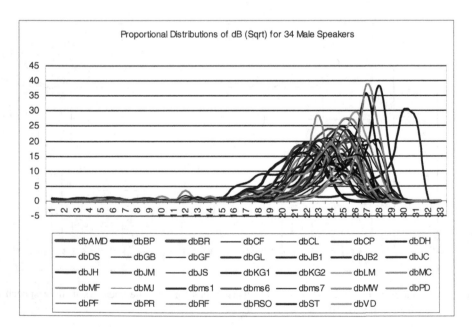

Fig. 3. Weighted frequency distributions for 34 male speakers' intensity. The x axis represents squared-root dB values. The y axis represents the percentage of samples.

3.3 Second Analysis: F0 Results

Although speakers perform the same speech task, the histograms computed in DataDesk, with the same window length and the same number of intervals again show a variation of distributions among speakers. Table 2 shows the central values of the 11 speakers. These variations are visible on the graphical superposition of F0

Table 2. Statistical central values of F0 values (in semitones) for 11 male speakers in one speech task

Statistics

	N		Mean	Median	Mode	Skewness	Std. Error of Skewness	Kurtosis	Std. Error of Kurtosis
	Valid	Missing							
ST2	340	436	1343696	9328647	3.42468a	.616	.132	.903	.264
ST3	440	336	9954147	9240932	5.20582a	-.018	.116	.578	.232
ST4	396	380	8925545	2177622	1.14639a	.292	.123	-.118	.245
ST5	349	427	0025711	2097615	-.73723	.854	.131	.359	.260
ST6	505	271	3760663	3272467	5.92541a	-.400	.109	-.735	.217
ST7	441	335	9693586	8377587	4.22985a	.340	.116	.881	.232
ST8	443	333	0122560	7621997	4.28062a	.304	.116	-.195	.231
ST9	394	382	0231211	9891787	2.04784a	.190	.123	-.212	.245
ST10	517	259	3608246	3251690	-1.83327a	1.013	.107	1.854	.214
ST11	345	431	8173770	9574606	2.33616a	-.352	.131	.390	.262
ST12	402	374	2137634	3857320	4.29279a	-.081	.122	.252	.243

a.Multiple modes exist. The smallest value is shown

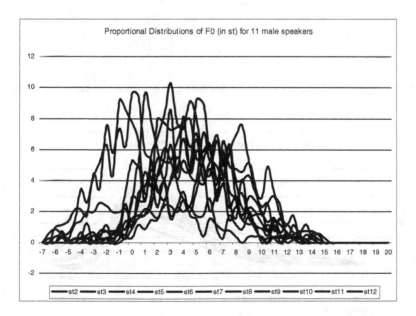

Fig. 4. Weighted frequency distributions for 11 male speakers' F0 in a commentary speech task. The x axis represent semi-tones calulated by Fant's equation (2004). The y axis represents the percentage of samples.

distributions (see Figure 4). The Kruskall Wallis test confirms that distributions of F0 (in semitones) differ significantly (Chi-square = 1055.890; df=10; p=.000).

The agglomerative hierarchical classification suggests three types of F0 distributions (Figure 5). The first group (ST5, ST10) has the largest positive skewness. Speakers in this group have a low F0 register. Group 2 (ST6, ST11) has the largest negative skewness. The median is higher than the mean and both are far-off the lowest modal value. Speakers in this group have a high register of F0. F0 distributions in the third group (ST4, ST3, ST2, ST8, ST9, ST12) has a skewness close to 0. All the distributions are multimodal. Again, when hearing pairs of speech samples according to the leaves of the aggregation, an auditive impression of similarity between speakers emerges.

3.4 Second Analysis: Intensity Results

Like in the first analysis, the superposition of speakers'intensity distributions illustrate a very similar pattern among the speakers in the commentary task (Figure 6).

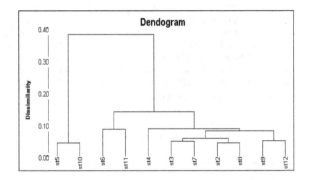

Fig. 5. Dendogram of the clustering of F0 distributions for 11 male speakers in a commentary speech task

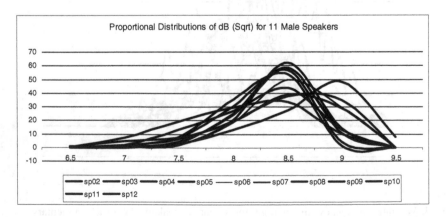

Fig. 6. Weighted frequency distributions for 11 male speakers' intensity. The x axis represents squared-root dB values. The y axis represents the percentage of samples.

4 Discussion

Variations in prosody within a linguistic community are triggered by a number of parameters, among others the speech task and the individual style. In this study, it is shown that raw data such as distributions of F0 and dB give interesting information in this area.

Our results show that F0 distributions among the speakers are not similar, whether the speech task is the same or not. The first difference is related to the heigth of the speaker's register and the way speakers use their register. Some speakers present left skewed distributions, meaning that their prefered F0 targets are in the lowest part of their register. Some other speakers have right-skewed distributions, meaning that they tend to favor F0 targets in the highest part of their register. Beyond the differences in terms of high and low register, we found that some speakers prefer activating their intonation in a multimodal way - with several preferred F0 targets - and some others activate only one preferred F0 target. The preferred F0 target(s) might be close to the two other central values (mean and median) or far-off. These differences sound differently and may characterise styles of intonation which are independant of the speech task.

Conversely, it was found in both analyses that intensity distributions are very similar among the speakers, whether they do perform the same speech task or not. Intensity distributions seem to be less sensitive to the individual characteristics of a speaker. Intensity curves are nearly perfectly superposed on each other and to a certain extent to the speech task. Intensity curves follow the same pattern despite the fact that they are not perfectly superposed on each other.

5 Conclusion

This paper is a contribution to the study of prosodic variations within the same linguistic community. The analyses of F0 and intensity distributions of 34 male speakers in nine speech tasks on the one hand, and the analysis of 11 male speakers in one speech task on the another hand show two interesting facts. Intensity distributions are speaker-independent and to a certain extent are also task-independent. Only one pattern of distribution emerges, whatever the speaker and the speech task are. F0 distributions are task-independent but speaker-dependent. At least fours types of F0 distribution were characterised, suggesting four types of speaker intonation.

Acknowledgments

My special thanks to Eric Keller (Lausanne) for his collaborative support and suggestions. This research is supported by a Swiss OFES grant in support of work performed under COST 277.

References

1. Baken and Orlikoff (2000). Clinical measurement of speech and voice. Singular Publishing Group, San Diego, California.
2. Fant, G., Kruckenberg, A., Gustafson K, & Liljencrants, J. (2002) A new approach to intonation analysis and synthesis of Swedish, Speech Prosody 2002, Aix en Provence.
3. Keller, E. (2003). Voice Characteristics of MARSEC Speakers. VOQUAL: Voice Quality: Functions, Analysis And Synthesis, Geneva - August 27-29, 2003.
4. Scherer, K. R. (2003). Vocal communication of emotion: A review of research paradigms. Speech Communication, 40, 227-256.
5. Zellner Keller, B. (2004). Prosodic Styles and Personality Styles: are the two interrelated? Proceedings of SP2004. (pp.383-386). Nara, Japan.

A Two-Level Drive – Response Model
of Non-stationary Speech Signals

Friedhelm R. Drepper

Zentralinstitut für Elektronik, Forschungszentrum Jülich GmbH,
Postfach 1913, D 52425 Jülich, Germany
f.drepper@fz-juelich.de

Abstract. The transmission protocol of voiced speech is hypothesized to be based on a fundamental drive process, which synchronizes the vocal tract excitation on the transmitter side and evokes the pitch perception on the receiver side. A band limited fundamental drive is extracted from a voice specific subband decomposition of the speech signal. When the near periodic drive is used as fundamental drive of a two-level drive-response model, a more or less aperiodic voiced excitation can be reconstructed as a more or less aperiodic trajectory on a low dimensional continuous synchronization manifold (surface) described by speaker and phoneme specific coupling functions. In the case of vowels and nasals the excitation can be described by a univariate coupling function, which depends on the momentary phase of the fundamental drive. In the case of other voiced consonants the coupling function may as well depend on a delayed fundamental phase with a phoneme specific time delay. The delay may exceed the length of the analysis window. The resulting long range correlation cannot be analysed or synthesized by models assuming stationary excitation.

1 Introduction

Speech signals are known to contain obviously non-stationary segments, which constitute a cue for stop consonants and which are characterized by isolated, non repetitive events with a duration of less than a couple of pitch periods. The present study is focussed on segments of speech, which cannot easily be classified as non-stationary, in particular on sustained voiced segments, which are characterized by repetitive time pattern. The vocal tract excitation of voiced speech is generated by a pulsatile airflow, which is strongly coupled to the oscillatory dynamics of the vocal fold. The excitation is created immediately in the vicinity of the vocal fold and/or delayed in the vicinity of a phoneme specific constriction of the vocal tract [1-3]. As has been pointed out by Titze [4], a mechanistic model of a dynamical system suitable to describe the self-sustained oscillations of the glottis cannot be restricted to state variables of the vocal fold alone, but has to be extended by state variables of the sub- and supraglottal aerodynamic subsystems.

Due to the strong nonlinearities of the coupled dynamics non-pathological, standard register phonation dynamics is characterized by a stable synchronization or

M. Faundez-Zanuy et al. (Eds.): NOLISP 2005, LNAI 3817, pp. 125–138, 2005.

mode locking of several oscillatory subsystems including the two vocal folds. The synchronization can furthermore be assumed to have the effect that some of these subsystems become topologically equivalent oscillators, whose states are one to one related by a continuous mapping with a continuous invers (conjugation) [5, 6]. Due to the pronounced mass density difference of about 1:1000 the coupling between the airflow and the glottal tissue is characterized by a dominant direction of interaction, such that the glottal oscillators can affirmatively be assumed to be a subset of those topologically equivalent oscillators. The (conjugation type) synchronization of the vocal folds has been described by kinematic and dynamic models [7, 8]. The glottal oscillators can be used to define a single glottal master oscillator, which enslaves (synchronizes) or drives the other oscillatory degrees of freedom including the higher frequency acoustic modes.

Time series of the electro-glottogram or of the sound pressure signal can more or less safely be used to reveal an oscillator, which is topologically equivalent to the glottal master oscillator. In the case of nonpathological voiced speech both types of observation reveal a unique frequency of voiced phonation, the so called fundamental frequency, which is also known to have a perceptional counterpart, the pitch. As has already been observed by Seebeck [9], human pitch perception does not rely on spectral components of the speech signal in the frequency range of the fundamental frequency. In spite of numerous attempts, the extraction of the momentary fundamental frequency out of the speech signal has not yet reached the generality, precision and robustness of auditive perception and of the analysis based on the electro-glottogram [2, 10 -12].

The time series of successive cycle lengths of oscillators, which are (implicitly assumed to be) equivalent to the glottal master oscillator show an aperiodicity with a wide range of relevant frequencies reaching from half of the pitch down to less than 0.1 Hz. Except at the high frequeny end the deviation of the glottal cycle lengths from the long term mean forms a non-stationary stochastic process. More or less distinct frequency bands or time scales have been described as: subharmonic bifurcation [8], jitter, microtremor and prosodic variation of the pitch [7, 12]. As a general feature, cycle length differences increase with the time scale, the relative differences ranging from less than 1 % up to more than 30%. In spite of the partially minor amplitudes of aperiodicity all or most of these frequency bands appear to be perceptionally relevant [13]. Some of them are known to play a major role for the non-symbolic information content of speech.

The relevant frequency range of the excitation of voiced speech extends at least one order of magnitude higher than the fundamental frequency. It is therefore common practice to introduce a time scale separation, which separates the high frequency acoustic phenomena of speech signals above the pitch from the subharmonic, subacoustic and prosodic ones below the pitch. A simple approach towards time scale separation starts with the assumption of a causal frequency gap, which separates the frequency range of the autonomous lower frequency degrees of freedom from the dependent degrees of freedom (modes) in the acoustic frequency range. In the main stream approach of speech analysis this has lead to the more or less explicit assumption that the voiced (and unvoiced) excitation is wide sense stationary in the analysis window, which is usually chosen as 20 ms [2, 3]. The assumption of wide sense stationarity is closely related to the assumption that the excitation process can be described as a sum of a periodic process and filtered white noise with a time invariant, finite impulse response filter. In the case of voiced excitation there exists multiple evidence that this assumption is not fulfilled [14, 15]. In a first step of improvement the voiced excitation has been described as stochastic process in the basin of attraction of a low dimensional nonlinear dynamical system [14, 15]. The assumption of a low dimensional dynamical system, however, is in contradiction to the observed non-stationary aperiodicity of the glottal cycle lengths.

The present study introduces an analysis of (sustained) speech signals, which does not assume a periodic fundamental drive nor an aperiodic drive, which obeys a low dimensional dynamics. The assumption of a causal frequency gap is avoided by treating the more or less aperiodic voiced broadband excitation as an approximately deterministic response of a near periodic, non-stationary fundamental drive, which is extracted continuously from voiced sections of speech with uninterrupted phonation [16, 17]. The extraction of the fundamental drive includes a confirmation that the drive can be interpreted as a topologically equivalent reconstruction of the glottal master oscillator which synchronizes the vocal tract excitation [16].

As an important property of non-pathological, standard register voiced speech the state of the fundamental drive is assumed to be described uniquely by a fundamental phase, which is related to pitch perception, and a fundamental amplitude which is related to loudness perception. Whereas the extraction of the fundamental phase is limited to voiced sections of speech, the fundamental amplitude can as well be used for the time scale separation of unvoiced sections. The (response related) state of the fundamental drive should not be confused with the state of the dynamical system, which describes the self-sustained oscillations of the glottis [4]. The phase of the glottal master oscillator should rather be compared with a phase, which is suited to describe a unique state on the limit cycle, which attracts the self sustained oscillations of the glottis.

As result of a detailed study of the production of vowels (with a sufficiently open vocal tract to permit the manipulation of airflow velocity sensors) Teager and Teager [18] pointed out that the conversion of the potential energy of the compressed air in the subglottal airduct to convective, acoustic and thermal energy happens in a highly organized cascade. They observed that the astonishingly complex convective airflow pattern within the vocal tract (flow separations, vortex rings, swirly vortices along the cavity walls, ...) show a degree of periodicity in time, which is comparable to the one of the corresponding far field acoustic response.

Also in the case of sustained voiced fricatives (and of vowel – voiceless fricative transitions) the far field acoustic response indicates a causal connection to the glottal dynamics [19]. It is therefore plausible to assume that at least a part of the frequency range of the convective flow pattern on the upstream side of the fricative specific constriction shows a vowel type periodicity. However, there is still a lot of speculation about the relevant delays of the cause and effect relationship between the primary response and the glottal dynamics. In the case of the fricative specific retarded excitation the delay may assume a large value, due to (comparatively slow) subsonic convective transport of the relevant action (trigger). The speculation refers in particular to the question, whether the subsonic transport is limited to the downstream side of the phoneme specific constriction [19] or applies to the whole distance starting from the glottis.

It cannot be excluded that the delay (or memory) of the subsonic excitation may reach the length of the conventional analysis window of 20 ms. In this case the resulting long range correlation cannot be analysed affirmatively by conventional methods assuming stationary excitation within the analysis window. The continuous reconstruction of the glottal master oscillator for segments of uninterrupted phonation opens the possibility to describe the excitation as superposition of a direct and a delayed phase locked response with correct long range correlation.

As has also been pointed out by Teager and Teager [18] there are many reasons to assume that the human auditory pathway uses analysis tools, which deviate from spectral analysis. Teager proposed a phenomenological approach, which is based on short term analysis of the distribution of energy in different frequency bands [21]. The present approach is focussed on a phenolmenological speech production model, which extends the validity range of the classical source and filter model, which is also grounded on evidence from speech physiology and psychoacoustics and which is suited to bring additional light to the complex airflow pattern of voiced consonants, which are extremely difficult to analyse in vivo [18], in vitro [19] and in silico [19].

2 Extraction of the Fundamental Drive

The amplitude and phase of the fundamental drive are extracted from subband decompositions of the speech signal. The decompositions use complex (4th order gammatone) bandpass filters with roughly approximate audiological bandwidths ΔF and with a subband independent analysis – synthesis delay as described in Hohmann [22].

The extraction of the fundamental phase ψ_t is based on an adaptation of the best (central) filter frequencies F_j of the subband decomposition to the momentary frequency of the glottal master oscillator (and its higher harmonics) [16, 17]. At the lower frequency end of the subband decomposition the best filter frequencies F_j are centred on the different harmonics of the analysis window specific estimate of the fundamental frequency. In the next higher frequency range the best filter frequencies are centred on pairs of neighbouring harmonics.

$$F_j = \left\{ \begin{array}{l} j F_1 \\ \sqrt{j(j-1)}\, F_1 \end{array} \right\} \qquad \text{for} \qquad \left\{ \begin{array}{l} 1 \le j \le 6 \\ 6 < j \le 12 \end{array} \right\} \tag{1a}$$

$$\Delta F_j = \left\{ \begin{array}{l} F_1 \\ 2\, F_1 \end{array} \right\} \qquad \text{for} \qquad \left\{ \begin{array}{l} 1 \le j \le 6 \\ 6 < j \le 12 \end{array} \right\}. \tag{1b}$$

As a second feature of human speech it is assumed that voiced segments of speech are produced with at least two subbands, which are not distorted by vocal tract resonances or additional constrictions of the airflow [17]. In the case of subbands with separated harmonics, $1 \le j \le 6$, the absence of a distortion is detected by nearly linear relations between the unwrapped phases of the respective subband states. For sufficiently adapted centre filter frequencies such subbands show an (n:m) phase locking. The corresponding phase relations can be interpreted to result from (n:1) and (m:1) phase relations to the fundamental drive. The latter ones are used to reconstruct the phase velocity of the fundamental drive. In the case of a subband with paired harmonics, $6 < j \le 12$, the phase relation to the fundamental drive is obtained by determining the Hilbert phase of the modulation amplitude of the respective subband.

The phase velocity of the fundamental drive is used to improve the centre filter frequencies. For voiced phones the iterative improvement leads to a fast converging fundamental phase velocity $\dot{\psi}_t$ with a high time and frequency resolution. Based on a, so far, arbitrary initial phase, successive estimates of $\dot{\psi}_t$ lead to a reconstruction of the fundamental phase ψ_t, which is uniquely defined for uninterrupted segments of confirmed topological equivalence [17]. The uninterrupted continuation of the fundamental phase can even be achieved in cases of a confirmation gap as long as there remains an overlap of confirmed analysis windows. The latter feature can e.g. be used for the analysis of vowel-nasal transitions (figures 5 and 6).

The extraction of the **fundamental amplitude** A_t is based on the assumption, that human auditive perception incorporates useful information on the dynamics of important sound sources of the human environment in particular on human speech. The relevant features of loudness perception concern the scaling of the loudness as function of the signal amplitude and the relative weights of the partial loudnesses of individual subbands [10]. The fundamental amplitude A_t is assumed to be related to loudness perception by a power law [17]. The exponent $1/\nu$ is chosen such that the fundamental amplitude represents a linear homogenous function of the time averaged amplitudes $\overline{A}_{i,t}$ of a synthesis suited set of subbands with approximately audiological bandwidths,

$$ A_t = \left(\sum_{j=1}^{N} (g_j \overline{A}_{j,t})^\nu \right)^{\frac{1}{\nu}} \qquad \text{with} \qquad \sum_{j=1}^{N} g_j^\nu = 1. \tag{2} $$

Zwicker, Feldtkeller [23] and Moore [10] give an exponent $\nu = 0.6$. Sottek [24] cites newer measurements, resulting in an exponent in the range of $\nu = 0.3$. The latter value has been adopted in the study. The weights g_j are proportional to inverse hearing thresholds. In the range up to 3 kHz they can be roughly approximated by the power law $g_j \approx h_j^\mu$, where h_j represents the (integer) centre harmonic number, which approximates the ratio F_j / F_1. The present study uses $\mu = 1$ [3, 23]. The synthesis suited set of subbands is generated by replacing the over-complete subband set $6 < j \le 12$ by a set $6 < j \le N$, which is spaced equidistantly on the logarithmic frequency scale with 4 filters per octave,

$$ F_j = 5 \cdot 2^{(j-5)/4} F_1, \qquad \Delta F_j = 2^{(j-5)/4} F_1. \tag{3} $$

The feasibility of the extraction of the fundamental drive as well as the validity of its interprettation as a reconstruction of a glottal master oscillator of voiced excitation is demonstrated with the help of simultaneous recordings of a speech signal and an electro-glottogram, which have been obtained from the pitch analysis database of

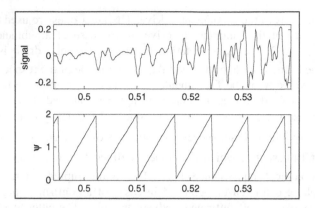

Fig. 1. Upper panel: 45 ms of a speech signal, which was taken from the /w/ in the word "wind" representing part of a publicly accessible pitch analysis data base [25]. The lower panel shows the reconstruction of the fundamental phase ψ in units of π. The time scale (in units of seconds) corresponds to the original one.

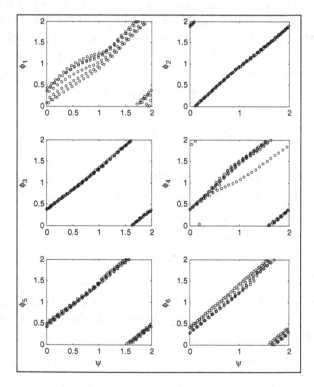

Fig. 2. Relation of the subband phases Φ_j, $(j = 1,2,...,6)$, obtained from the speech signal of figure 1, to the fundamental phase ψ. The subbands 2, 3 and 5 are characterized by near perfectly linear phase relations, whereas the other subbands are found to be unsuited for the reconstruction of the fundamental phase.

Keele University [25]. The upper panel of figure 1 shows the analysis window for a segment of the speech signal, which was taken from the /w/ in the first occurrence of the word "wind" spoken by the first male speaker. The lower panel shows the reconstruction of the fundamental phase (given in wrapped up form), based on the set of separable subbands with the harmonic numbers 2, 3 and 5.

The near perfectly linear phase locking of these subbands, which is used for the reconstruction of the drive, is demonstrated in figure 2. The subband phases Φ_j are given in a partially unwrapped form, depending on the respective centre harmonic number h_j. The enlarged range of the subband phases is normalized by the same centre harmonic number. Alternatively the fundamental phase can also be obtained from a subband decomposition of the electro-glottogram. The exchangeability of the two phase velocities is demonstrated in figure 3, which shows the relation between the two fundamental phases for the speech segment, which covers the "win" part of the word "wind", uttered by the first female speaker. The phase shift between the two phases did not change significantly during the 160 ms being covered.

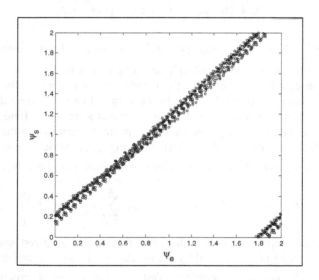

Fig. 3. Relation between the wrapped up fundamental phase ψ_s, obtained from the speech signal, and the fundamental phase ψ_e, obtained from the electro-glottogram. Both fundamental phases are extracted from 160 ms of uninterrupted voiced speech.

3 Entrainment of the Primary Response

In spite of the (temporary) arbitrariness of the initial fundamental phase, the reconstructed glottal master oscillator can be used as fundamental drive of a two level drive – response model, which is suited to describe voiced speech as secondary response [16, 26, 27]. The additional subsystem describes the excitation of the vocal tract as primary response of the fundamental drive and the classical secondary response subsystem describes the more or less resonant "signal forming" on the way through the vocal tract as action of a linear autoregressive filter, which (in a first approximation) is assumed as independent of the fundamental phase.

As a particular advantage of the two-level drive- response model the fundamental phase cannot only be interpreted as state variable of the fundamental drive. The unwrapped fundamental phase can also be assumed to be approximately proportional to time. As a characteristic simplifying assumption of the two-level drive-response model the near periodic time profile of the excitation is replaced by a precisely periodic fundamental phase profile [16, 26, 27].

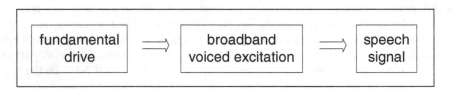

Fig. 4. The two-level drive – response model

In the context of the drive- response model this means that the excitation E_t at time t is assumed to be restricted (phase locked or entrained) to a generalized synchronization manifold (surface) in the combined state space of the fundamental drive and the primary response [28-30]. In the simplest case, the time dependence is replaced by a unique dependence on the simultaneous state of the fundamental drive [28,29]. More generally, the dependence of the primary response on the fundamental drive takes the form of a multi-valued mapping [30], which, however, can be expressed as a unique function of the unwrapped fundamental phase ψ_t,

$$E_t \;=\; A_t\, G_p(\psi_t) \;=\; A_t \sum_{k=0}^{K} c_k \exp(i\,k\,\frac{\psi_t}{p}). \qquad (4)$$

As part of the improved time scale separation the generalized synchronization manifold is assumed to be described by the product of the slowly variable fundamental amplitude A_t and the potentially fast varying complex coupling function $G_p(\psi_t)$, the real part of which describes the broadband excitation. To generate the mentioned multi-valued mapping (with unique branches) the coupling function $G_p(\psi_t)$ has to be assumed as a $2\pi\,p$ periodic function of the unwrapped fundamental phase ψ_t. The integer period number $p \geq 1$ defines the number of different branches of the multi-valued mapping. Coupling function $G_p(\psi_t)$ can thus be well approximated by the finite Fourier series of equation (4), which, however, has to be estimated for non-equidistant phases!

Voiced excitations can be represented by coupling functions with values of p, which are distinctly smaller than the number of fundamental cycles within the analysis window. The case $p = 1$ corresponds to the normal voice type characterized by a unique mapping and the case $p = 2$ corresponds to the period doubling voice type. The latter voice type can e.g. be identified by observing alternating period lengths, a feature, which is described as diplophonia in vocology. When p exceeds the number of fundamental cycles within the analysis window, equation (4) is able to describe a fully general excitation, including the unvoiced case. The real part of the coupling function of the normal voice type, $G_1(\psi_t)$, can be expressed as a polynomial in the harmonic functions $\cos(\psi_t)$ and $\sin(\psi_t)$. Similar polynomial coupling (or waveshaper) functions have been introduced by Schoentgen [31] to synthesize vowels with realistic vocal aperiodicities.

The excitation parameters c_k cannot be determined independently from the parameters which characterize the vocal tract resonances. In the standard approach the parameter estimation is performed hierarchically, by making the higher level assumption that the excitation has a nearly white (or tilted) spectrum. When dropping this assumption, special care has to be taken to avoid numerical instabilities in the case of a near periodic fundamental drive. To achieve a comparable numerical robustness, it is useful to perform separate parameter estimations for different frequency bands and to use optimally chosen (subband specific) time step lengths Δ for the autoregressive models. The bandwidths of the subbands should be chosen substantially broader than the bandwidths of the vocal tract resonances, which are most relevant for the respective subband. It is therefore advantageous to use a subband decomposition with larger bandwidths, than the ones used for the extraction of the fundamental drive. The band limitation can be used to reduce the number of resonances (poles of the autoregressive filter), which are relevant for the respective subband. Useful choices are two poles and one subband per octave or one pole and two subbands. For simplicity one pole and maximally two subbands of decomposition (1) and (3) are chosen. Interpreting excitation E_t of equation (4) as the aggregate of the set of subband specific excitations $E_{j,t}$ with subband specific coupling functions $G_{j,p}(\psi_t)$ and index sets $S_{j,p}$ of the Fourier type decomposition, we arrive at the following subband specific conditional stochastic process with a two-level drive – response model as deterministic part (skeleton) [16, 26, 27],

$$X_{j,t+\Delta} = -b_j X_{j,t} + A_t G_{j,p}(\psi_t) + A_t \sigma_j \xi_{j,t}, \qquad (5)$$

where $X_{j,t}$ denotes the complex state of the subband with index j, b_j the complex, subband specific resonator parameter, $\xi_{j,t}$ a $(0,1)$ Gaussian complex white noise process and $A_t \sigma_j$ the standard deviation, which for simplicity has been assumed to be not dependent on the fundamental phase. As an important computational advantage the estimation of the complex excitation and resonator parameters $c_{j,k}$ and b_j can be reduced to multiple linear regression. The subband specific summation index set

$S_{j,p}$ in equation (4) is chosen in accordance to the respective bandpass filter. To avoid a bad conditioning of the parameter estimation in the case of near periodic driving, the index set $S_{j,p}$ is pruned by the index, which equals the respective centre harmonic number h_j. Together with the option, to extend the analysis window due to the explicit reconstruction of the non-stationary part, these precautions lead to a precise and robust reconstruction of the voiced excitation.

When the speech signal of the respective analysis window can be described successfully by model (5) with a low periodicity $p \leq 2$, the speech signal has a high probability to belong to a vowel or a nasal. As is well known (and shown in figures 5 and 6) vowels and nasals are characterized by the fact that the time points of glottal closure can be detected as a unique pulse (or as a unique outstanding slope). Since there is no syllable without a vowel kernel, such kernels can be used to resolve the arbitrariness of the initial fundamental phase and to calibrate the wrapped up fundamental phase in terms of the time interval since the last glottal closure.

When the respective speech signal cannot be described successfully by a single low period coupling function, the unique reconstruction of the fundamental phase for uninterrupted segments of voiced phonation opens the possibility to extend model (5) by a retarded (subsonic) excitation which is suited to describe the delayed characteristic response of fricatives. According to the more detailed (aeroacoustic) view of speech production [18-20] the excitation of voiced fricatives (and of vowel – voiceless fricative transitions) should be extended by an additional or alternative coupling function, which depends on a delayed fundamental phase with a phoneme (and potentially speaker) specific delay τ ,

$$ X_{j,t+\Delta} = -b_j X_{j,t} + A_t G_{j,I}(\psi_t) + A_{t-\tau} G_{j,II}(\psi_{t-\tau}) + A_t \sigma_j \xi_{j,t} . \quad (6) $$

The average delay τ between the sonic and the subsonic excitation accounts for the additional time, which is needed for the (comparatively slow and quiet) subsonic transport of kinetic energy by convective airflow to the phoneme specific site of the vocal tract, where the enhanced transformation to acoustic (and thermal) energy takes place (typically at the teeth). Assuming a near optimal evolutionary adaptation of human speech production leading to a near optimal support of the distinction between the sonic coupling function $G_{j,I}(\psi_t)$ and the subsonic one $G_{j,II}(\psi_{t-\tau})$, a typical physiological tremor frequency of 7 Hz would correspond to a typical delay time of about 35 ms. For delay times in excess of 20 ms, the respective autocorrelation cannot be analysed by conventional methods, which assume uncorrelated excitation in non-overlapping analysis windows (of typically 20 ms length).

4 Long Range Correlation in a Vowel – Nasal Diphone

Contrary to the mainstream view, properties of the excitation can be used advantageously as additional cues for phoneme recognition. As a first example, where the long range correlation of a voiced speech signal represents a potential cue, the vowel- nasal diphone of figure 3 is selected, which represents the transition from the vowel to the nasal in the word "wind". Due to the difference in length and shape of the nasal tract compared to the vocal tract, a transition between a nasal and a vowel can be discerned by a sudden change of the phase position of the glottal pulse [32] relative to the normal position for vowel kernels. Fortunately the shift of the glottal pulse happens so fast, that the gap of the confirmation of the topological equivalence of the fundamental drive to the glottal master oscillator is short enough to permit an uninterrupted continuation of the fundamental phase. Figures 3, 5 and 6 are obtained with a 30 ms time window of analysis and an advancement step size of 5ms. That means that the gap of the confirmation of the topological equivalence was shorter than 5 advancement steps. Figure 3 confirms the successful continuation of the fundamental phase.

Figures 5 and 6 reveal a phase shift of about 1/12 of the fundamental cycle. Knowing the fundamental frequency of 230 Hz the phase shift can be translated to a time shift of about 0.36 ms and a distance shift of about 12 cm. As has been pointed out by Kawahara and Zolfaghari [32] the time or distance shift has to be interpreted as an effective shift, which includes a group delay difference, which results from differing vocal tract and nasal tract resonances. The latter ones are known to be increased by various sinuses, which are coupled to the nasal tract.

The reconstruction of a continuous fundamental phase for speech segments with uninterrupted phonation opens the possibility to complement the analysis of the

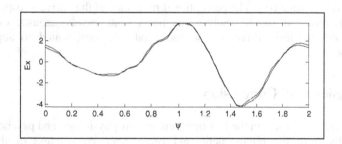

Fig. 5. Fundamental phase dependent coupling function $G_{j,2}(\psi)$ reconstructed with periodicity $p = 2$ for the vowel of the first occurrence of the word "wind" used in figure 3. The two curves correspond to the odd and even periods. The good agreement can be interpreted as a hint to the high robustness of the reconstruction of the excitation.

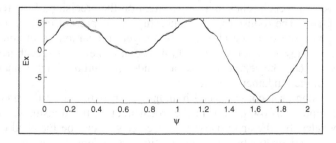

Fig. 6. Fundamental phase dependent coupling function $G_{j,2}(\psi)$ for the nasal of the word "wind" of figure 5

spectral properties of the speech signal by a run time analysis. The run time differences may refer either to different paths of the response to the early (sonic) acoustic excitation, which is created in the vicinity of the glottis, or to different speeds of the action of the fundamental drive on the retarded (subsonic) acoustic excitation, which is created in the vicinity of a phoneme specific constriction of the vocal tract. The delay of the retarded action results from subsonic transport of convective energy to the site of the enhanced production of acoustic energy. The delay depends in particular on the relative share of subsonic transport on the way from the glottis to the secondary site of acoustic excitation. In the case of the fricatives the high precision determination of the delay cannot only be achieved by parameter estimation of the delay τ in equation (6) but also by inspection of the fundamental phase profile of the primary (or secondary) response. As has been demonstrated by Jackson and Shadle [19] fricatives show characteristic delays of the amplitude (envelope) maximum of the subsonic (unharmonic) excitation. Both types of run time differences are potentially suited as additional cues for phoneme recognition.

In the case of the higher frequency subbands of voiced fricatives the interference between the responses of the sonic and the subsonic excitation may lead to a sensitive dependence on the recent history of the fundamental phase. This mechanism of deterministic amplification of the aperiodicity of the fundamental drive process is suited to explain (a part of) the typical aperiodicity of the higher frequency subbands of sustainable voiced fricatives. The perceptional relevance of the "drive – response chaos" of voiced fricatives should be analysed by appropriate psycho-acoustic experiments. The described two-level drive – response model is well suited to support such experiments.

5 Discussion and Conclusion

The present study is based on the link between speech-physiology and psycho-acoustics, which results from the phylogenetic and ontogenetic coevolution of the auditive pathway and the sound production system in an acoustic environment, which is strongly influenced by sound utterances of contemporary members of the own species. The assumption that the far field acoustic response of the pulsed turbulent airflow in the vocal tract can be described by a low dimensional synchronisation manifold, is a remarkable hypothesis, which should be interpreted as result of the ontogenetic adaptation of human speech production. The success of the proposed description of non-pathological voiced speech relies to a large extent on the precision of the reconstructed fundamental phase. The robustness and generality of the present method to extract the fundamental phase out of a speech signal is not yet comparable to the one of human pitch perception. However, the newly established link between speech-acoustics and psycho-acoustics can be exploited as a guide to future improvements of the reconstruction of the fundamental drive.

The transmission protocol of voiced human speech is based on the production and analysis of complex airflow pattern in the vocal tract of the transmitter. The present study demonstrates that the analysis on the receiver side can be focussed on the mode locking of the pulsed airflow by replacing the time dependent excitation of the classical source - filter model by a fundamental phase dependence which can be described by a low dimensional generalized synchronization manifold. In the simpler cases of vowels and nasals the manifold (surface) can be described by a single coupling function, which depends on a single fundamental phase. In the case of voiced consonants with a phoneme specific constriction of the vocal tract the excitation may have to be extended or replaced by a coupling function, which depends on a delayed fundamental phase. The evolution of speech has lead to many voiced phonemes, which can be distinguished by properties of these coupling functions and the closely related two-level drive - response models. To make the coupling functions visible (or audible) with increased precision, a voice specific subband decomposition of the speech signal has been proposed, which is suited to extract the phase of the fundamental drive with high precision. The extraction relies on the fact that non-pathological voiced speech leaves several subbands undistorted by vocal tract resonance or phoneme specific constriction of the airflow.

There have been numerous attempts to increase the precision of the spectral analysis of voiced speech by introducing a "dynamic time warping" preprocessing step [33] which enhances the proportionality between the (artificial) time and the fundamental phase. Such a preprocessing step, however, ignores the dynamic nature of the production of voiced speech which involves phoneme specific delays of the primary response and a fully dynamic secondary response. The dynamics of the secondary response may show a sensitive resonance behaviour with respect to changes in the time scale. A time warping of the speech signal, which enhances the visibility of the synchronization manifold of the primary excitation, can thus be expected to have a non-negligible corrupting effect on the spectrum of the secondary response.

In the case of vowel – nasal and vowel – fricative transitions, in particular, the response of the fundamental drive may show a long range correlation with a delay which exceeds the length of the conventional window of analysis. In these cases a phase vocoder, which is based on a continuously reconstructed fundamental drive process and a related two level drive – response model with appropriate time delays, is expected to solve some of the major coarticulation problems of present day phase vocoders, which, so far, have prevented them to replace concatenative synthesis in high quality speech reconstruction.

Acknowledgement. The author would like to thank V. Hohmann, B. Kollmeier and J. Nix, Oldenburg, M. Kob, C. Neuschaefer-Rube, Aachen, G. Langner, Darmstadt, N. Stollenwerk, Porto, J. Schoentgen, Brussels, J. Rouat, Montreal, P. Grassberger, M. Schiek and P. Tass, Jülich for helpful discussions.

References

[1] Fant G. *Acoustic theory of speech production*, Mouton, 'S-Gravenhage (1960)

[2] Vary P., U. Heute, W. Hess, *Digitale Sprachsignalverarbeitung*, B.G. Teubner Verlag, Stuttgart (1998)

[3] Schroeder M.R., *Computer Speech*, Springer (1999)

[4] Titze I.R., *Acta Acustica* **90**, 641-648 (2004)

[5] Kantz H., T. Schreiber, *Nonlinear time series analysis*, Cambridge Univ. Press (1997)

[6] Kocarev L., U. Parlitz, *Phys. Rev. Lett.* **76**, 1816 (1996)

[7] Schoentgen J., "Stochastic models of jitter", *J. Acoust. Soc. Am.* **109** (4): 1631-1650 (2001)

[8] Herzel H., D. Berry, I.R. Titze and I. Steinecke, „Nonlinear dynamics of the voice: Signal analysis and biomechanical modeling ", *Chaos* **5**, 30-34 (1995)

[9] Seebeck A., "Über die Sirene", *Annalen der Physik*, LX, 449 ff, ibid. LXIII, 353 ff and 368 ff (1843)

[10] Moore B.C.J., *An introduction to the psychology of hearing*, Academic Press (1989)

[11] De Cheveigné A. and H. Kawahara, "Comparative evaluation of F0 estimation algorithms", Eurospeech 2001, Alborg (2001)

[12] Winholtz W.S. and L.O. Ramig, "Vocal tremor analysis with the vocal demodulator", *J.Speech Hear. Res.* **35**, 562-573 (1992)

[13] Hanquinet J., F. Grenez and J. Schoentgen, "Synthesis of disordered voices", *this volume* (2005)

[14] Kubin G., "Nonlinear processing of speech," in *Speech Coding and Synthesis* (W. B. Kleijn and K. K. Paliwal, eds.), pp. 557–610, Amsterdam: Elsevier (1995)

[15] Moakes P. A. and S.W. Beet, "Analysis of non-linear speech generating dynamics," in *ICSLP 94*, Yokohama, pp. 1039–1042 (1994)

[16] Drepper F.R. in C. Manfredi (editor), *MAVEBA 2003*, Firenze University Press (2004)

[17] Drepper F.R., "Selfconsistent time scale separation of instationary speech signals", *Fortschritte der Akustik-DAGA'05* (2005)

[18] Teager H.M. and S.M. Teager, "Evidence for nonlinear sound production mechanisms in the vocal tract," in *Proc NATO ASI on Speech Production and Speech Modelling*, pp. 241–261 (1990)

[19] Jackson P.J.B. and C.H. Shadle, "Pitch scaled estimation of simultaneous voiced and turbulence-noise components in speech", *IEEE trans. speech audio process.*, vol. **9**, pp. 713-726 (2001)

[20] Maragos P., J.F. Kaiser and T.F. Quatieri, "Energy separation in signal modulations with application to speech analysis", IEEE Trans. Signal Processing, Vol. 41, pp. 3024-3051 (1993).

[21] Zhao, W., C. Zhang, S.H. Frankel and L. Mongeau, "Computational Aeroacoustics of Phonation, Part I: ... ", *J. Acoust. Soc. Am.*, Vol. **112**, No. 5, pp. 2134-2154 (2002)

[22] Hohmann V., *Acta Acustica* **10**, 433-442 (2002)

[23] Zwicker E. und Feldtkeller R., *Das Ohr als Nachrichtenempfänger*, Hirzel Verlag, (1967)

[24] Sottek R., *Modelle zur Signalverarbeitung im menschlichen Gehör*, Verlag M. Wehle, Witterschlick/Bonn (1993)

[25] ftp.cs.keele.ac.uk/pub/pitch

[26] Drepper F.R., "Rekonstruktion stationärer Mannigfaltigkeiten der Teilbanddynamik instationärer Sprachsignale" *Fortschritte der Akustik-DAGA'03* (2003)

[27] Drepper F.R., "Voiced excitation as entrained primary response of a reconstructed glottal master oscillator", *Fortschritte der Akustik-DAGA'05* (2005)

[28] Afraimovich V.S., N.N. Verichev, M.I. Rabinovich, *Radiophys. Quantum Electron.* **29**, 795 ff (1986)

[29] Rulkov N.F. , M.M. Sushchik, L.S. Tsimring, H.D.I. Abarbanel, *Phys. Rev. E* **51**, 980-994 (1995)

[30] Rulkov N.F. , V.S. Afraimovich, C.T. Lewis, J.R.Chazottes and A. Cordonet, *Phys. Rev. E* **64**, 016217 (2001)

[31] Schoentgen J., "Shaping function models of the phonatory excitation signal", J. Acoust. Soc. Am. **114** (5): 2906-2912 (2003)

[32] Kawahara H. and P. Zolfaghari, "Systematic F0 glitches around nasal-vowel transitions", *Eurospeech 2001* (2001)

[33] Graf J. T. and N. Hubing, "Dynamic time warping comb filter for the enhancement of speech degraded by white Gaussian noise," *Proc. ICASSP*, vol. **2**, pp. 339–342, (1993)

Advanced Methods for Glottal Wave Extraction

Jacqueline Walker and Peter Murphy

Department of Electronic and Computer Engineering,
University of Limerick,
Limerick, Ireland
{jacqueline.walker, peter.murphy}@ul.ie

Abstract. Glottal inverse filtering is a technique used to derive the glottal waveform during voiced speech. Closed phase inverse filtering (CPIF) is a common approach for achieving this goal. During the closed phase there is no input to the vocal tract and hence the impulse response of the vocal tract can be determined through linear prediction. However, a number of problems are known to exist with the CPIF approach. This review paper briefly details the CPIF technique and highlights certain associated theoretical and methodological problems. An overview is then given of advanced methods for inverse filtering: model based, adaptive iterative, higher order statistics and cepstral approaches are examined. The advantages and disadvantages of these methods are highlighted. Outstanding issues and suggestions for further work are outlined.

1 Introduction

Although convincing results for glottal waveform characteristics are reported in the literature from time to time, a fully automatic inverse filtering algorithm is not yet available. The benefits of an automatic inverse filtering technique are considerable. The separation of the speech signal into representative acoustic components that are feasible from a speech production point of view provides for a flexible representation of speech that can be exploited in a number of speech processing applications, including synthesis (e.g. the benefits of including glottal information in pitch modification schemes is highlighted in [25]), enhancement, coding [18] and speaker recognition [43]. Such an interactive source filter representation offers a compromise representation of speech lying somewhere between a detailed articulatory model on the one hand and a purely data driven approach on the other hand. Although a source filter representation is of potential benefit in a number of speech processing applications, one application of particular interest is the study of pathological voice where direct physical correlations to the acoustic waveform may be required. The paper is organized as follows: in Sect. 2 a review of the closed phase inverse filtering technique is given. In Sect. 3 a survey of advanced methods for glottal pulse extraction, highlighting advantages and disadvantages, is presented. Finally in Sect. 4, remaining problems and suggestions for further work are discussed.

M. Faundez-Zanuy et al. (Eds.): NOLISP 2005, LNAI 3817, pp. 139–149, 2005.

2   Closed Phase Glottal Inverse Filtering

Following the linear model for voice production, voiced speech can be represented as:

$$S(z) = AP(z)G(z)V(z)R(z) \ , \tag{1}$$

where A represents the overall amplitude, $P(z)$ is the Z transform of an impulse train, $p(n)$, $G(z)$ is the Z transform of the glottal pulse, $g(n)$, $V(z)$ is the Z transform of the vocal tract impulse response, $v(n)$ and $R(z)$ is the Z transform of the radiation load, $r(n)$. As shown in Fig. 1, glottal inverse filtering requires solving the equation:

$$G(z)P(z) = \frac{S(z)}{AV(z)R(z)} \ , \tag{2}$$

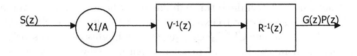

Fig. 1. Closed phase inverse filtering

that is, to determine the glottal waveform the influence of the vocal tract and the radiation load must be removed. The radiation load is due to the lip/open air interface: the unidirectional volume velocity at the lips is radiated in all directions and is recorded as sound pressure in the far field. Acoustically, the effect of radiation is a first-order differentiation of the volume velocity at the lips resulting in a zero at zero frequency. To invert this effect a first-order integrating filter is used with a pole placed just inside the unit circle to ensure stability. It is also possible to incorporate the differentiation into an effective driving pulse of the differentiated glottal flow:

$$G(z)P(z)R(z) = \frac{S(z)}{AV(z)} \tag{3}$$

Hence, the problem reduces to determining the inverse of the vocal tract transfer function as shown in Fig. 2.

To solve (3), it is assumed that $V(z)$ is purely minimum phase. Linear prediction is used to model the vocal tract impulse response as an $L-$order all-pole filter:

Fig. 2. Closed phase inverse filtering to obtain an effective driving function

$$V(z) = \frac{A}{1 - \sum_{i=1}^{L} b_i z^{-i}} . \tag{4}$$

Therefore the speech signal at time n can be written as:

$$s(n) = \sum_{i=1}^{L} b_i s_{n-i} + A(g(n) - g(n-1)) . \tag{5}$$

During the closed phase of the glottal cycle the input is assumed to be zero and the b_i's can be determined. The inverse of this filter is then used to de-convolve the speech signal resulting in a differentiated glottal flow signal. The filter coefficients are determined by minimizing the prediction error such that the filter provides an optimum match to the speech signal [23,53]. The model order must be chosen such that L is more than double the number of formants in the frequency range of interest. The covariance method of linear prediction is used to solve the linear system equation because it gives a better result with the reduced number of samples available from only considering the closed phase during a pitch period [36]. To guarantee that the system equation is well defined a frame length greater than 2ms is required ([19], [53] use 4.75ms intervals). A number of variations [11,43] exist for determining the closed phase region (or alternatively a region of formant stationarity which may not correspond exactly to the closed phase).

Although a number of studies (cited above) have demonstrated the feasibility of CPIF for use on male speakers in modal register the technique is as yet still not widely used in speech processing applications. A number of problems persist with the technique. For inverse filtering it is important that pole representations provide a match to actual formant data. However, the technique occasionally estimates poles where there are no formants and sometimes misses formants [19,29]. In addition, formants with very large bandwidths are sometimes falsely predicted. It has also been shown that the prediction error may be greater during the closed phase and hence the minimum of the prediction error does not reliably indicate the closed glottis interval [11]. Furthermore, the assumed closed phase interval may have non-zero excitation [22].

2.1 CPIF with a Second Channel

A primary challenge in CPIF is to identify precisely the instants of glottal clo-sure and opening. Some investigators have made use of the electroglottographic (EGG) signal to locate the instants of glottal closure and opening [28,29,33,50]. In particular, it is claimed that use of the EGG can better identify the closed phase in cases when the duration of the closed phase is very short as in higher fundamental frequency speech (females, children) or breathy speech [50]. Two-channel methods are not particularly useful for more portable applications of inverse filtering requiring minimal operator intervention. However, precisely be-cause they can identify the glottal closure more accurately, results obtained using the EGG can potentially serve as 'benchmarks' by which other approaches working with the acoustic pressure wave alone can be evaluated.

3   Advanced Approaches to Glottal Inverse Filtering

Given the difficulties outlined above regarding CPIF, alternative or supplemental methods for inverse filtering are required and a wide range of alternative methods has been developed. In the sections which follow, we will consider model-based approaches, heuristic adaptive approaches and approaches using more sophisticated statistical techniques such as the cepstrum or higher order statistics.

3.1   Model-Based Approaches

A more complete model for speech is as an ARMA (autoregressive moving average) process with both poles and zeros:

$$s\left(n\right) = \sum_{i=1}^{L} b_i s_{n-i} + \sum_{j=1}^{M} a_j g_{n-j} + g(n) \; . \tag{6}$$

Such an approach allows for more realistic modeling of speech sounds apart from vowels, particularly nasals, fricatives and stop consonants [37]. Many different algorithms for finding the parameters of a pole-zero model have been developed [9,15,30,31,37,45,46]. ARMA modeling approaches have been used to perform closed phase glottal pulse inverse filtering [49] giving advantages over frame-based techniques such as linear prediction by eliminating the influence of the pitch, leading to better accuracy of parameter estimation and better spectral matching [49].

If the input to the ARMA process described by (6) is modeled as a pulse train or white noise, the pole-zero model obtained will include the lip radiation, the vocal tract filter and the glottal waveform. The difficulty with this is that there is no definitive guide as to how to separate the poles and zeros which model these different features [35]. However, an extension of pole-zero modeling to include a model of the glottal source excitation can overcome the drawbacks of inverse filtering and produce a parametric model of the glottal waveform. In [28], the glottal source is modeled using the LF model [14] and the vocal tract is modeled as two distinct filters, one for the open phase, one for the closed phase [42]. Glottal closure is identified using the EGG. In [16,17], the LF model is also used in adaptively and jointly estimating the glottal source and vocal tract filter using Kalman filtering. To provide robust initial values for the joint estimation process, the problem is first solved in terms of the Rosenberg model [44]. One of the main drawbacks of model-based approaches is the number of parameters which need to be estimated for each period of the signal [28] especially when the amount of data is small e.g. for short pitch periods in higher pitched voices. To deal with this problem, inverse filtering may be used to remove higher formants and the estimates can be improved by using ensemble averaging of successive pitch periods.

Modeling techniques need not involve the use of standard glottal source models. Fitting polynomials to the glottal wave shape is a more flexible approach which can place fewer constraints on the result. In [33], the differentiated glottal waveform is modeled using polynomials (a linear model) where the timing of the glottis opening and closing is the parameter which varies. Initial values for the glottal source endpoints plus the pitch period endpoints are found using the EGG. The vocal tract filter coefficients and the glottal source endpoints are then jointly estimated across the whole pitch period. This approach is an alternative to closed phase inverse filtering in the sense that even closed phase inverse filtering contains an implied model of the glottal pulse [33], that is, the assumption of zero airflow through the glottis for the segment of speech from which the inverse filter coefficients are estimated. An alternative is to attempt to optimize the inverse filter with respect to a glottal waveform model for the whole pitch period [33]. Interestingly in this approach, the result is the appearance of ripple in the 'source-corrected' inverse filter during the closed phase of the glottal source, even for synthesized speech with zero excitation during the glottal phase, and which is clearly an analysis artefact due to the inability of the model to account for it [33]. (Note that the speech was synthesized using the Ishizaka-Flanagan model [24].) Improvements to the model are presented in [34,48], and the sixth-order Milenkovic model is used in GELP (Glottal Excited Linear Prediction) [10].

In terms of the potential application of glottal inverse filtering, the main difficulty with the use of glottal source models in glottal waveform estimation arises from the influence the models may have on the ultimate shape of the result. This is a particular problem with pathological voices. The glottal waveforms of these voices may diverge quite a lot from the idealized glottal models. As a result, trying to recover such a waveform using an idealized source model as a template may give less than ideal results. A model-based approach which partially avoids this problem is described in [43] where non-linear least squares estimation is used to fit the LF model to a glottal derivative waveform extracted by closed phase filtering (where the closed phase is identified by the absence of formant modulation). This model-fitted glottal derivative waveform is the 'coarse structure'. The fine structure of the waveform is then obtained by subtraction from the inverse filtered waveform.

3.2   Adaptive Inverse Filtering Approaches

The key to CPIF is to calculate the vocal tract filter impulse response free of the influence of the glottal waveform input. In the iterative adaptive inverse filtering method (IAIF-method) [3], a 2-pole model of the glottal waveform based on the characteristic 12dB/octave tilt in the spectral envelope [13] is used to remove the influence of the glottal waveform from the speech signal before estimating the vocal tract filter. The vocal tract filter estimate is used to inverse filter the original speech signal to obtain a glottal waveform estimate. The procedure is then repeated using a higher order parametric model of the glottal waveform obtained from the initial glottal waveform estimate. As the method removes the influence of the glottal waveform from the speech before estimating the vocal tract filter, it does not take a closed phase approach but utilises the whole pitch period. A flow diagram of the method is shown in Fig. 3. The method relies on linear prediction and due to the influence of the harmonic structure of the glottal source, incorrect formant estimation can occur [5]. In particular, the technique

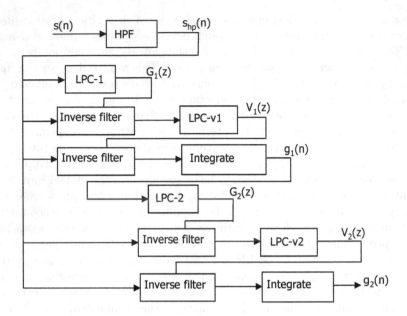

Fig. 3. The iterative adaptive inverse filtering method

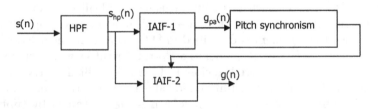

Fig. 4. The pitch synchronous iterative adaptive inverse filtering method

does not perform well for higher fundamental frequency voices [4]. Fig. 4 shows how IAIF was adapted to a pitch synchronous approach which was introduced in [5].

Comparing the results of the IAIF method with closed phase inverse filtering show that the IAIF approach seems to produce waveforms which have a shorter and rounder 'closed phase'. In [5] comparisons are made between original and estimated waveforms for synthetic speech sounds. It is interesting to note that pitch synchronous IAIF produces a closed phase ripple in these experiments (when there was none in the original synthetic source waveform). In [6] discrete all-pole modelling was used to avoid the bias given toward harmonic frequencies in the model representation. An alternative iterative approach is presented in [1]. The method de-emphasises the low frequency glottal information using high-pass filtering prior to analysis. In addition to minimising the influence of the glottal source, an expanded analysis region is provided in the form of a pseudo-closed phase. The technique then derives an optimum vocal tract filter function through applying the properties of minimum phase systems.

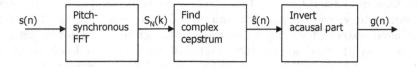

Fig. 5. A cepstral technique for inverse filtering

3.3 Higher Order Statistics and Cepstral Approaches

These approaches exploit the properties of newer statistical techniques such as higher order statistics which are theoretically immune to Gaussian noise [32,38]. The bispectrum (third-order spectrum) contains system phase information and many bispectrum-based blind deconvolution algorithms exist. The properties of the cepstrum have also been exploited in speech processing. Transformed into the cepstral domain, the convolution of input pulse train and vocal tract filter becomes an addition of disjoint elements, allowing the separation of the filter from the harmonic component [40]. The main drawback with bispectral and other higher order statistics approaches is that they require greater amounts of data to reduce the variance in the spectral estimates [21]. As a result, multiple pitch periods are required which would ordinarily be pitch asynchronous. This problem may be overcome by using the Fourier series and thus performing a pitch synchronous analysis [20] or possibly by performing ensemble averaging of successive pitch periods (as is done in [28]). Cepstral techniques also have some limitations including the requirement for phase unwrapping and the fact that the technique cannot be used when there are zeros on the unit circle [41].

It has been demonstrated that the higher order statistics approach can recover a system filter for speech, particularly for speech sounds such as nasals [20]. Such a filter may be non-minimum phase and when its inverse is used to filter the speech signal will return a residual which is much closer to a pure pseudo-periodic pulse train than inverse filters produced by other methods [8,20]. In [8], the speech input estimate generated by this approach is used in a second step of ARMA parameter estimation by an input-output system identification method. Similarly in [27], various ARMA parameter estimation approaches are applied to the vocal tract impulse response recovered from the cepstral analysis of the speech signal [39]. There are a few examples of direct glottal waveform recovery using higher order spectral or cepstral techniques. In [52], ARMA modelling of the linear bispectrum [12] was applied to speech for joint estimation of the vocal tract model and the glottal volume velocity waveform using higher-order spectral factorization [47]. Fig. 5 shows an approach to direct estimation from the complex cepstrum as suggested by [2] based on the assumption that the glottal volume velocity waveform may be modeled as a maximum phase system.

4 Discussion

One of the primary difficulties in glottal pulse identification is in the evaluation of the resulting glottal flow waveforms. There are several approaches which can be taken. One approach is to verify the algorithm which is being used for the glottal flow waveform recovery. Algorithms can be verified by applying the algorithm to a simulated system which may be synthesized speech but need not be [26,27]. In the case of synthesized speech, the system will be a known all-pole vocal tract model and the input will be a model for a glottal flow waveform. The success of the algorithm can be judged by quantifying the error between the known input waveform and the version recovered by the algorithm. This approach is most often used as a first step in evaluating an algorithm [4,5,49,52] and can only reveal the success of the algorithm in inverse filtering a purely linear time-invariant system.

It has been shown that the influence of the glottal source on the vocal tract filter during the open phase is to slightly shift the formant locations and widen the formant bandwidths [53], that is, the vocal tract filter is in fact time-varying. It follows then that inverse filtering with a vocal tract filter derived from the closed phase amounts to assuming the vocal tract filter is time-invariant. Using this solution, the variation in the formant frequency and bandwidth has to go somewhere and it ends up as a ripple on the open phase part of the glottal volume velocity (see for example Fig. 5c in [53]). Alternatively, one could use a time-varying vocal tract filter which will have different formants and bandwidths in closed and open phases and the result would be a glottal waveform independent of the vocal tract [7,29].

However, a common result in inverse filtering is a ripple in the closed phase of the glottal volume velocity waveform which is most often assumed to illustrate non-zero air flow in the closed phase: for example, in [50] where this occurs in hoarse or breathy speech. In [50], it is shown through experiments that this small amount of air flow does not significantly alter the inverse filter coefficients (filter pole positions change by < 4%) and that true non-zero air flow can be captured in this way. However, the non-zero air flow and resultant source-tract interaction may mean that the 'true' glottal volume velocity waveform is not exactly realized [50]. A similar effect is observed when attempting to recover source waveforms from nasal sounds. Here the strong vocal tract zeros mean that the inverse filter is inaccurate and so a strong formant ripple appears in the closed phase [50]. However, the phenomenon of closed phase ripple may also be an artefact as it often occurs where a time-invariant vocal tract filter has been derived over a whole pitch period and not from the closed phase only and may be due to formant localization error [4,28,52].

In addition to discovering an optimum glottal identification algorithm, which has been the primary focus of the present paper, a number of closely related issues remain to be addressed. Evaluating what is considered to be a good result remains largely unresolved - this can be determined precisely for synthesis (formant and bandwidth specification or least mean square of estimates compared to original glottal flow) but no method exists for testing the result of inverse filtering real speech. Some advance could come in the form of more detailed synthesis on the one hand and extracting more knowledge from real speech on the other hand e.g. investigating source-tract interaction, the time-varying open phase transfer characteristics and secondary excitation prior to attempting inverse filtering. Another consideration is what characteristics are perceptually relevant and what characteristics are physically relevant? In [51] some progress has been made on the former through examination of minimal perceivable differences in voice source parameters. For the latter, in correlations with physical entities such as glottal area, it may be preferable to derive the actual glottal flow as opposed to the effective glottal flow. Further work on parameterizing the glottal volume velocity and the voice source (derivative glottal volume velocity) is still required. An important advance in this direction is that the derived models must become physically constrained. Finally, on the practical side, general guidelines for appropriate recording conditions are required. These issues will be thoroughly reviewed in a follow-up study.

Acknowledgement

This work is supported by Enterprise Ireland Research Innovation Fund, RIF/2002/037.

References

1. Akande, O. and Murphy, P. J.: Estimation of the vocal tract transfer function for voiced speech with application to glottal wave analysis. Speech Communication, **46** (2005) 15–36
2. Alkhairy, A.: An algorithm for glottal volume velocity estimation. Proc. IEEE Int. Conf. Acoustics, Speech and Signal Processing. **1** (1999) 233–236
3. Alku, P., Vilkman, E., Laine, U. K.,: Analysis of glottal waveform in different phonation types using the new IAIF-method. Proc. 12th Int. Congress Phonetic Sciences, **4** (1991) 362–365
4. Alku, P.: An automatic method to estimate the time-based parameters of the glottal pulseform. Proc. IEEE Int. Conf. Acoustics, Speech and Signal Processing. **2** (1992) 29 32
5. Alku, P.: Glottal wave analysis with pitch synchronous iterative adaptive inverse filtering. Speech Communication. **11** (1992) 109–118
6. Alku, P., Vilkman, E.: Estimation of the glottal pulseform based on Discrete All-Pole modeling. Proc. Int. Conf. on Spoken Language Processing. (1994) 1619-1622
7. Ananthapadmanabha, T. V., Fant, G.: Calculation of true glottal flow and its components. Speech Communication. **1** (1982) 167–184
8. Chen, W.-T., Chi, C.-Y.: Deconvolution and vocal-tract parameter estimation of speech signals by higher-order statistics based inverse filters. Proc. IEEE Workshop on HOS. (1993) 51–55
9. Childers, D. G., Principe, J. C., Ting, Y. T. Adaptive WRLS-VFF for Speech Analysis. IEEE Trans. Speech and Audio Proc. **3** (1995) 209–213
10. Childers, D. G., Hu, H. T.: Speech synthesis by glottal excited linear prediction. J. Acoust. Soc. Amer. **96** (1994) 2026-2036
11. Deller, J. R.: Some notes on closed phase glottal inverse filtering. IEEE Trans. Acoust., Speech, Signal Proc. **29** (1981) 917-919

12. Erdem, A. T., Tekalp, A. M.: Linear Bispectrum of Signals and Identification of Nonminimum Phase FIR Systems Driven by Colored Input. IEEE Trans. Signal Processing. **40** (1992) 1469–1479
13. Fant, G. C. M.: Acoustic Theory of Speech Production. (1970) The Hague, The Netherlands: Mouton
14. Fant, G., Liljencrants, J., Lin, Q.: A four-parameter model of glottal flow. STL-QPR. (1985) 1–14
15. A recursive maximum likelihood algorithm for ARMA spectral estimation. IEEE Trans. Inform. Theory **28** (1982) 639–646
16. Fu, Q., Murphy, P. J.: Adapive Inverse filtering for High Accuracy Estimation of the Glottal Source. Proc. NoLisp'03. (2003)
17. Fu, Q., Murphy, P. J.: Robust glottal source estimation based on joint source-filter model optimization. Accepted for publication, IEEE Transactions on Speech and Audio Processing (2005)
18. Hedelin, P.: High quality glottal LPC-vocoding. Proc. IEEE Int. Conf. Acoustics, Speech and Signal Processing. (1986) 465-468
19. Hess, W.: Pitch Determination of Speech Signals: Algorithms and Devices. Springer, (1983).
20. Hinich, M. J., Shichor, E.: Bispectral Analysis of Speech. Proc. 17th Convention of Electrical and Electronic Engineers in Israel. (1991) 357–360
21. Hinich, M. J., Wolinsky, M. A.: A test for aliasing using bispectral components. J. Am. Stat. Assoc. **83** (1988) 499-502
22. Holmes, J. N.: Formant excitation before and after glottal closure Proc. IEEE Int. Conf. Acoustics, Speech and Signal Processing. **1** (1976) 39-42
23. Hunt, M. J., Bridle, J. S., Holmes, J. N.: Interactive digital inverse filtering and its relation to linear prediction methods. Proc. IEEE Int. Conf. Acoustics, Speech and Signal Processing. **1** (1978) 15-19
24. Ishizaka, K., Flanagan, J. L.: Synthesis of voiced sounds from a two mass model of the vocal cords. Bell Syst. Tech. J. **51** (1972) 1233–1268
25. Jiang, Y., Murphy,P. J.: Production based pitch modification of voiced speech Proc. Int. Conf. Spoken Language Processing. (2002) 2073-2076
26. Konvalinka, I. S., Mataušek, M. R.: Simultaneous estimation of poles and zeros in speech analysis and ITIT-iterative inverse filtering algorithm. IEEE Trans. Acoust., Speech, Signal Proc. **27** (1979) 485–492
27. Kopec, G. E., Oppenheim, A. V., Tribolet, J. M.: Speech Analysis by Homomorphic Prediction IEEE Trans.Acoust., Speech, Signal Proc. **25** (1977) 40–49
28. Krishnamurthy, A. K.: Glottal Source Estimation using a Sum-of-Exponentials Model. IEEE Trans. Signal Processing. **40** (1992) 682–686
29. Krishnamurthy, A. K., Childers, D. G.: Two-channel speech analysis. IEEE Trans. Acoust., Speech, Signal Proc. **34** (1986) 730–743
30. Lee, D. T. L., Morf, M., Friedlander, B.: Recursive least squares ladder estimation algorithms. IEEE Trans. Acoust., Speech, Signal Processing. **29** (1981) 627–641
31. Makhoul, J.: Linear Prediction: A Tutorial Review. Proc. IEEE. **63** (1975) 561–580
32. Mendel, J. M.: Tutorial on Higher-Order Statistics (Spectra) in Signal Processing and System Theory: Theoretical Results and Some Applications. Proc. IEEE. **79** (1991) 278–305
33. Milenkovic, P.: Glottal Inverse Filtering by Joint Estimation of an AR System with a Linear Input Model. IEEE Trans. Acoust., Speech, Signal Proc. **34** (1986) 28–42
34. Milenkovic, P. H.: Voice source model for continuous control of pitch period. J. Acoust. Soc. Amer. **93** (1993) 1087-1096

35. Miyanaga, Y., Miki, M., Nagai, N.: Adaptive Identification of a Time-Varying ARMA Speech Model. IEEE Trans. Acoust., Speech, Signal Proc. **34** (1986) 423–433
36. Moore, E., Clements, M.: Algorithm for automatic glottal waveform estimation without the reliance on precise glottal closure information. Proc. IEEE Int. Conf. Acoustics, Speech and Signal Processing. **1** (2004) 101–104
37. Morikawa, H., Fujisaki, H.: Adaptive Analysis of Speech based on a Pole-Zero Representation. IEEE Trans. Acoust., Speech, Signal Proc. **30** (1982) 77–87
38. Nikias, C. L., Raghuveer, M. R.: Bispectrum Estimation:A Digital Signal Processing Framework. Proc. IEEE. **75** (1987) 869–891
39. Oppenheim, A. V.: A speech analysis-synthesis system based on homomorphic filtering. J. Acoust., Soc. Amer. **45** (1969) 458–465
40. Oppenheim, A. V., Schafer, R. W.: Discrete-Time Signal Processing. Englewood Cliffs:London Prentice-Hall (1989)
41. Pan, R., Nikias, C. L.: The complex cepstrum of higher order cumulants and non-minimum phase system identification. IEEE Trans. Acoust., Speech, Signal Proc. **36** (1988) 186–205
42. Parthasarathy, S., Tufts, D. W.: Excitation-Synchronous Modeling of Voiced Speech. IEEE Trans. Acoust., Speech, Signal Proc. **35** (1987) 1241–1249
43. Plumpe, M. D., Quatieri, T. F., Reynolds, D. A.: Modeling of the Glottal Flow Derivative Waveform with Application to Speaker Identification. IEEE Trans. Speech and Audio Proc. **7** (1999) 569–586
44. Rosenberg, A.: Effect of the glottal pulse shape on the quality of natural vowels. J. Acoust. Soc. Amer. **49** (1971) 583–590
45. Steiglitz, K.: On the simultaneous estimation of poles and zeros in speech analysis. IEEE Trans. Acoust., Speech, Signal Proc. **25** (1977) 194–202
46. Steiglitz, K., McBride, L. E.: A technique for the identifcation of linear systems. IEEE Trans. Automat. Contr., **10** (1965) 461–464
47. Tekalp, A. M., Erdem, A. T.: Higher-Order Spectrum Factorization in One and Two Dimensions with Applications in Signal Modeling and Nonminimum Phase System Identification. IEEE Trans. Acoust., Speech, Signal Proc. **37** (1989) 1537–1549
48. Thomson, M. M.: A new method for determining the vocal tract transfer function and its excitation from voiced speech. Proc. IEEE Int. Conf. Acoustics, Speech and Signal Processing. **2** (1992) 23–26
49. Ting, Y., T., Childers, D. G.: Speech Analysis using the Weighted Recursive Least Squares Algorithm with a Variable Forgetting Factor. Proc. IEEE Int. Conf. Acoustics, Speech and Signal Processing. **1** (1990) 389–392
50. Veeneman, D. E., BeMent, S. L.: Automatic Glottal Inverse Filtering from Speech and Electroglottographic Signals. IEEE Trans. Acoust., Speech, Signal Proc. **33** (1985) 369–377
51. van Dinther, R., Kohlrausch, A. and Veldhuis,R.: A method for measuring the perceptual relevance of glottal pulse parameter variations Speech Communication **42** (2004) 175–189
52. Walker, J.: Application of the bispectrum to glottal pulse analysis. Proc. NoLisp'03. (2003)
53. Wong, D. Y., Markel, J. D., Gray, A. H.: Least squares glottal inverse filtering from the acoustic speech waveform. IEEE Trans. Acoust., Speech, Signal Proc. **27** (1979) 350–355

Cepstrum-Based Estimation of the Harmonics-to-Noise Ratio for Synthesized and Human Voice Signals

Peter J. Murphy and Olatunji O. Akande

Department of Electronic and Computer Engineering,
University of Limerick, Limerick, Ireland
{peter.murphy, olatunji.akande}@ul.ie

Abstract. Cepstral analysis is used to estimate the harmonics-to-noise ratio (HNR) in speech signals. The inverse Fourier transformed liftered cepstrum approximates a noise baseline from which the harmonics-to-noise ratio is estimated. The present study highlights the cepstrum-based noise baseline estimation process; it is shown to analogous to the action of a moving average filter applied to the power spectrum of voiced speech. The noise baseline, which is taken to approximate the noise excited vocal tract is influenced by the window length and the shape of the glottal source spectrum. Two existing estimation techniques are tested systematically using synthetically generated glottal flow and voiced speech signals with *a priori* knowledge of the HNR. The source influence is removed using a novel harmonic pre-emphasis technique. The results indicate accurate HNR estimation using the present approach. A preliminary investigation of the method with a set of normal/pathological data is investigated.

1 Introduction

The cepstrum is used to estimate the harmonics-to-noise ratio (HNR) in speech signals [1], [2]. The basic procedure presented in [1] is as follows; the cepstrum is produced for a windowed segment of voiced speech. The rahmonics are zeroed and the resulting liftered cepstrum is inverse Fourier transformed to provide a noise spectrum. After performing a baseline correction procedure on this spectrum (the original noise estimate is high), the logarithm of the summed energy of the modified noise spectrum is subtracted from the logarithm of the summed energy of the original harmonic spectrum in order to provide the harmonics-to-noise ratio estimate (Fig.1), (the need for baseline shifting with this approach is clearly explained in the Method section).

A modification to this technique, [2], illustrates problems with the baseline fitting procedure and hence does not adjust the noise baseline but calculates the energy and noise estimates at harmonic locations only (Fig.2). In addition, rather than zeroing the rahmonics, the cepstrum is low passed filtered to provide a smoother baseline (the reason the baseline shifting is not required is due to the window length used as detailed under Method). However the noise baseline estimate is shown to deviate from the actual noise level at low frequencies (Fig. 2). Each of these approaches, [1], [2],

M. Faundez-Zanuy et al. (Eds.): NOLISP 2005, LNAI 3817, pp. 150–160, 2005.

provide useful analysis techniques and data for studies of voice quality assessment, however, to date, neither method has been tested on synthesis data with *a priori* knowledge of the harmonics-to-noise ratio. The present study uses known amounts of random noise added to the glottal source to systematically test these techniques.

When such source signals are convolved with the vocal tract impulse response and radiation load the HNR is altered. However, an *a priori* HNR can still be estimated in the time domain by using a synthesized speech signal convolved with a noisy glottal source and one convolved with a noise-free source. The influence of the source spectrum on the noise baseline estimate is highlighted and is corrected for using a pre-emphasis technique.

An alternative cepstral-based approach for extracting a HNR from speech signals is estimated in [3]. However, this involves directly estimating the magnitude of the cepstral rahmonic peaks, leading to a geometric-mean harmonics-to-noise ratio

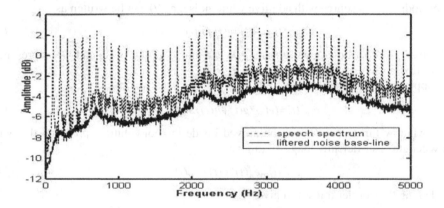

Fig. 1. HNR estimation using de Krom [1] cepstral baseline technique using a window length of 1024 points. The noise level is underestimated due to the baseline shifting process, which detects minima at between-harmonic locations.

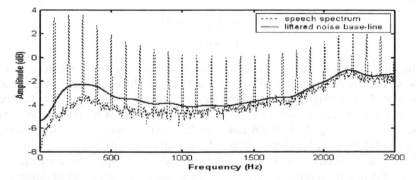

Fig. 2. HNR estimation using Qi and Hillman [2] cepstral baseline technique (window length 3200 points)

(i.e. an average of the dB harmonics-to-noise ratios at a specific frequency locations), which is quite distinct from traditional harmonics-to-noise ratio estimators which reflect the average signal energy divided by the average noise energy, expressed in dB. This is shown in eqtn. 1 for an N-point DFT, giving harmonic amplitudes, $|S|_i$, and noise estimates, $|N|_i$.

$$HNR = 10 \log 10 \left\{ \frac{\sum\limits_{i}^{N/2} |S_i|^2}{\sum\limits_{i=1}^{N/2} |N_i|^2} \right\} \qquad (1)$$

2 Method

A periodic glottal source with additive white noise, $g_{en}(t)$ can be written as

$$g_{en}(t) = e(t)*g(t)+n(t) \qquad (2)$$

where $e(t)$ is a periodic impulse train, $g(t)$ is a single glottal pulse and $n(t)$ represents aspiration noise.

Applying a Hanning window, (w)

$$g_{en}{}^w(t)=(e(t)*g(t)+n(t))\times w(t) \qquad (3)$$

The window function, $w(t)$, can be moved inside the convolution, provided the window length is sufficiently long [4], to give

$$g_{en}{}^w(t)=e_w(t)*g(t)+n_w(t) \qquad (4)$$

Taking the Fourier transform gives

$$G_{en}{}^w(f)=E_w(f)\times G(f)+N_w(f) \qquad (5)$$

Taking the logarithm of the magnitude squared values and approximating the signal energy at harmonic locations, $log|G_{en}{}^w|_h{}^2$ and at between-harmonic locations, $log|G_{en}{}^w|_{bh}{}^2$, gives

$$log|G_{en}{}^w|_h{}^2=log|E_w(f)\times G(f)|^2 \qquad (6)$$

$$log|G_{en}{}^w|_{bh}{}^2=log|N_w(f)|^2 \qquad (7)$$

Although the noise spectrum is broadband, its estimation in the presence of a harmonic signal can be concentrated at between-harmonic locations i.e. in the spectrum of the glottal source signal energy dominates at harmonic locations and noise energy dominates at between-harmonic locations. This approximation becomes more exact if the spectra are averaged in which case the harmonics approach the true harmonic values and the between-harmonics approach the true noise variance [5].

The cepstral technique is described with reference to Fig.3. An estimate of the HNR is obtained by summing the energy at harmonic locations and dividing by the sum of the noise energy. Extracting the noise energy baseline via the cepstral technique can be viewed as an attempt to estimate the noise level for all frequencies, including harmonic locations. It is noted that the cepstrum can be applied to periodic glottal source signals, separating the slowly varying glottal spectral tilt from the fast variation due to harmonic structure. The baseline is estimated via either of the methods [1], [2] outlined in the Introduction; in either approach the rahmonics are removed in the cepstrum and the resulting cepstrum is inverse Fourier transformed to provide the noise baseline i.e. in these analyses it is assumed that the cepstral liftering provides an estimate of the noise level. As shown in Fig.3, however, the estimated noise baseline diverges from the true noise floor at low frequencies. Considering the present theoretical derivation (eqtns. (6) and (7) and empirical data (Fig. 3) it can be seen that the baseline estimate is in fact influenced by both the source noise and source harmonic spectral tilt.

The above analysis applied to a windowed segment of voiced speech gives

$$s_{en}^{w}(t) = [(e_w(t)*g(t))+n_w(t)]*v(t)*r(t) \tag{8}$$

where $v(t)$ and $r(t)$ represent, respectively, the impulse response of the vocal tract and the radiation load, gives

$$log|S_{en}^{w}(f)|_h^2 = log|E_w(f)\times G(f)|^2 + log|V_R(f)|^2 \tag{9}$$

$$log|S_{en}^{w}(f)|_{bh}^2 = log|N_w(f)|^2 + log|V_R(f)|^2 \tag{10}$$

where $V_R(f)$ is the Fourier transform of $v(t)$ and $r(t)$ combined.

Again for the speech signal, HNR is estimated by summing the energy at harmonic locations and dividing by the summed noise energy estimated via the cepstral baseline technique. Now, the noise baseline (which is equivalent to a traditional vocal tract transfer function estimate via the cepstrum) is influenced by the glottal source excited vocal tract and by the noise excited vocal tract (Fig.2). It is the interpretation of the noise baseline as a MA filter that explains the need for baseline fitting in [1]; the

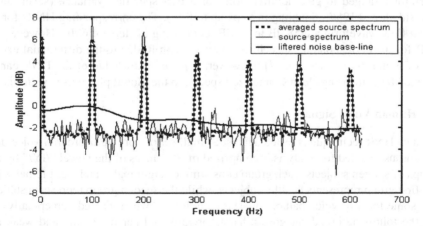

Fig. 3. Spectrum of glottal source with 1% additive noise. The solid line represents a single spectral estimate. The dashed line represents an average of n spectral estimates. The liftered noise baseline is also shown.

liftered spectral baseline does not rest on the actual noise level but interpolates the harmonic and between harmonic estimates and hence resides somewhere between the noise and harmonic levels. As the window length increases (as per [2], for example) the contribution of harmonic frequencies to the MA cepstral baseline estimate decreases. However, the glottal source still provides a bias in the estimate.

To remove the influence of the source, pre-emphasis is applied to the glottal source, $g_e^w(t)$ for the glottal signals and to $s_e^w(t)$ for the voiced speech signals (i.e. noiseless signals). $|G_{en}^w(f)|_h$ and $|S_{en}^w(f)|_h$ are estimated using periodogram averaging to provide estimates for $g_e^w(t)$ and $s_e^w(t)$ respectively (Fig.5).

A pre-emphasis filter,

$$h(z)=1-0.97z^{-1} \tag{11}$$

is applied to these estimates in the frequency domain by multiplying each harmonic value by the appropriate pre-emphasis factor.

3 Analysis

3.1 Synthesis Parameters

In order to evaluate the performance of the existing techniques along with the newly proposed method, synthesized glottal source and vowel /AH/ waveforms are generated at five fundamental frequencies (f0s) beginning at 80 Hz increasing in four steps of 60 Hz up to 320 Hz, covering modal register. The model described in [6] is adopted to synthesize the glottal flow waveform while the vocal tract impulse response is modeled with a set of poles. Lip radiation is modeled by a first order difference operator $R(z)=1-z^{-1}$. A sampling rate of 10 kHz is used for synthesis. Noise is introduced by adding pseudo-random noise to the glottal pulse via a random noise generator arranged to give additive noise of a user-specified variance (seven levels from std. dev. 0.125%, doubling in steps up to 8 %). The corresponding HNRs for the glottal flow waveform are 58 dB to 22 dB, decreasing in steps of 6 dB. However, the HNR for the corresponding speech signals vary with f0 due to the differential excitation of glottal harmonics, c.f. [7]. However, *a priori* knowledge of the HNR can be obtained by comparing clean synthesized speech to the glottal plus noise synthesis.

3.2 Human Voice Signals

The new HNR technique is tested against a small set of normal/disordered voice data. The corpus selected for analysis is comprised of utterances of the vowel "/IY/" by two groups of sixteen subjects, each group consisting of eight males and eight females [8]. The first group represents healthy subjects, while the second group represents subjects with some form of voice pathology. The pathologies, when labeled perceptually, fall into the following broad categories: harsh, breathy and harsh, breathy and weak and vocal fry. All recordings were performed in an anechoic chamber with a high quality microphone Bruel and Kjaer condenser microphone place 6 cm from subject's lips. In all case signals were sampled at 10 kHz sample rate.

3.3 Analysis Procedure

The procedures in [1] and [2] are implemented as outlined in the Introduction. In [1] window lengths of 1024, 2048 and 4096 are chosen, while in [2] a window length of 3200 is chosen, as per the original algorithm descriptions. In the proposed approach, the spectrum (2048-point FFT) of the test signal is computed using an analysis window (Hanning) of 2048 points overlapped by 1024 points. The analysis is applied to a 1.6 second segment of synthesized speech, providing 14 spectral estimates. The resulting power spectra are averaged (in order to reduce the noise variance at harmonic locations) to give a single 2048-point FFT. Harmonic peaks and bandwidths in the averaged spectrum are identified, and are modified using a pre-emphasis filter. The between-harmonics, which are not pre-emphasized, approach the noise variance in the averaged spectrum. The cepstrum is applied to the log spectrum of the pre-emphasized harmonics with the non pre-emphasized between-harmonics. The noise floor is extracted using a rectangular low-pass liftering window to select the first 40 cepstral coefficients. In order to calculate the noise energy, the extracted baseline is transformed back to a linear power spectrum and summed at the harmonic points. A sum, representing the signal energy, is taken of the harmonic peaks in the power spectrum of the signal (without pre-emphasis). The HNR is calculated as per eqtn.1.

4   Results

In order to illustrate the improvement offered by the new method over the existing cepstrum-based techniques, the liftered noise baseline is plotted together with the spectrum of the glottal source signal (with 1% additive noise) with (Fig.5) and without (Fig.4) pre-emphasis. It can be seen from Fig.4 that without pre-emphasis the estimated noise baseline deviates from the true noise floor as a result of the source influence on the liftered noise baseline. The result of removing the source influence before extracting the noise base line from the cepstrum is depicted in Fig.5 where the estimated noise baseline provides a much improved fit to the actual noise floor.

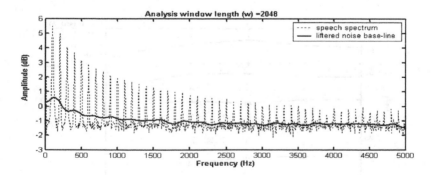

Fig. 4. Spectrum of the glottal source and liftered noise baseline, where pre-emphasis is not applied in the baseline estimation procedure (not shown). The spectrum is calculated with an analysis window length of 2048 points.

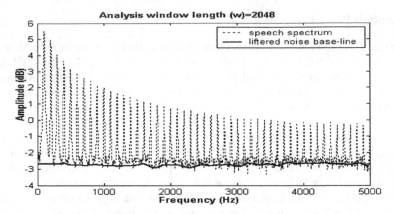

Fig. 5. Spectrum of the glottal source and liftered noise baseline, where pre-emphasis is applied in the baseline estimation procedure (not shown). The spectrum is calculated with an analysis window length of 2048 points.

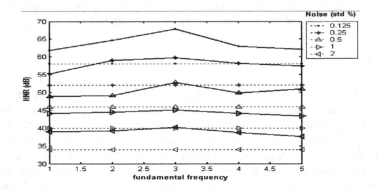

Fig. 6. Estimated HNR (solid line, de Krom [1]) versus f0 for synthesized glottal source waveforms (dotted line – actual HNR)

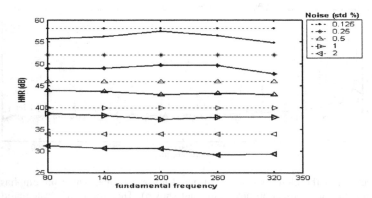

Fig. 7. Estimated HNR (solid line, Qi and Hillman [2]) versus f0 for synthesized glottal source waveforms (dotted line – actual HNR)

The HNR plotted against f0 is shown for (a) de Krom [1] (Fig.6) (b) Qi and Hillman [2] (Fig.7) and (c) the new approach (Fig.8). The results of the HNR measurement for the synthesized vowel /AH/ with the new method is shown in Fig.9. In order to evaluate the performance of a method, the estimated HNR is compared to the original HNR (dotted curve) in the figures.

5 Discussion

Increasing the window length moves the baseline closer to the actual noise level. The de Krom technique [1] tends to underestimate the baseline due to the fact that minima are estimated in the baseline fitting procedure. The Qi and Hillman approach [2] cannot match the noise level at low frequencies due to the influence of the source.

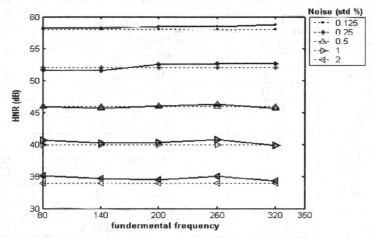

Fig. 8. Estimated HNR (solid line, with the new method) versus f0 for synthesized glottal source waveform (dotted line – actual HNR)

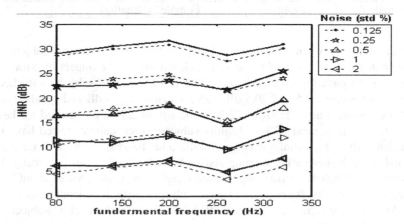

Fig. 9. Estimated HNR (with the new method) versus f0 for synthesized vowel /AH/ (dotted line – actual HNR)

Similar over- and under- estimates of the HNR for synthesized speech are also found (not shown). The effect of the source on the liftered noise floor is reduced by pre-emphasizing the harmonics of the test signal. HNR estimated with the new method tracks the corresponding input HNRs with marginal error as illustrated in Fig.8 and Fig.9.

Finally, the new harmonic pre-emphasis, cepstral-based HNR estimator, which has been validated with synthesis data, is evaluated against a preliminary set of normal and disordered voice data.

Table 1. Estimated HNR values for healthy and disordered voiced speech (vowel /IY/)

Normal Subjects	HNR (dB)	Subjects with voice pathology	HNR (dB)
Male 1	18.7	Male 1 (Breathy)	9.6
Male 2	17.92	Male 2 (Breathy and weak)	8.4
Male 3	17.59	Male 3 (Harsh)	8.90
Male 4	19.2	Male 4 (Harsh)	7.15
Male 5	20.51	Male 5 (Breathy)	3.76
Male 6	18.6	Male 6 (Breathy and weak)	7.39
Male 7	19.5	Male 7(Breathy)	6.02
Male 8	18.33	Male 8 (harsh)	6.68
Female 1	20.23	Female 1(Breathy and weak)	6.97
Female 3	26.2	Female2 (Vocal fry)	8.83
Female 3	26.47	Female 3 (weak)	7.97
Female 4	22.17	Female4 (Vocal fry)	9.36
Female 5	24.44	Female 5 (Breathy k)	5.11
Female 6	19.41	Female2 (Vocal fry)	3.27
Female 7	21.8	Female 1(harsh)	5.71
Female 8	19.46	Female2 (Breathy)	3.07

The difference between the estimated HNR for healthy subjects and subjects with voice pathology, shown in Table 1 and Fig., clearly shows the impact of noise in the spectrum of the pathologic voice corpus. The HNR for male subjects with healthy voice ranges between 17.5 and 20.5 dB with a mean of 18.79 dB and standard deviation of 0.93, while a mean HNR of 22.5dB and std. of 2.78 are observed for healthy female subjects. In general 95% of healthy subjects voice samples tested have HNR greater that 17dB. The pathologic corpus on the other hand is observed to have lower HNR and much higher variance compared with the healthy corpus. Estimated HNR ranges between 3-9.6dB for pathologic voice samples. A mean HNR of about 7.23dB and 6.2 dB are recorded for male and female subjects (with disordered voice) respectively. An average difference of about 10dB HNR separates the healthy subjects from the disordered subjects.

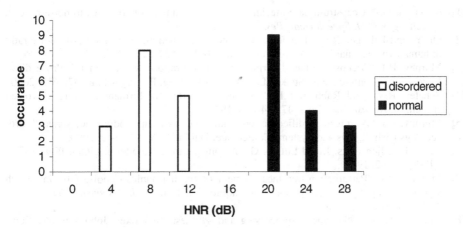

Fig. 10. Histogram of the estimated HNR values for normal and disordered corpus

6 Conclusion

Two existing cepstral–based HNR estimation techniques are evaluated using synthesized glottal waveforms and speech signals with *a priori* knowledge of the HNR for these signals. The methods provide reasonably consistent estimates of the HNR, however, HNR tends to be overestimated in [1] due to the baseline fitting procedure underestimating the noise levels and [2] tends to over-estimate the HNR due to the underestimate of noise levels due to the influence of the glottal source on the noise baseline. A combination of appropriate window length and harmonic pre-emphasis is shown to remove the bias due to the glottal source, providing an accurate noise baseline from which to estimate the HNR. A slight systematic error results due to the processing gain of the window. This can be simply adjusted by multiplication by the appropriated listed factor for that window. The normal and disordered data for real speech are completely separated using the new HNR estimator. However, as stated previously the HNR for voiced speech signals will in general have a fundamental frequency dependence that needs to be taken into consideration. Further work will apply the technique to a larger set of human voice signals and will investigate adapting the technique for use with shorter analysis windows with a view to analyzing continuous speech. In addition, a more detailed characterization of noise in voiced speech signals will be investigated.

Acknowledgements

This work is supported by Enterprise Ireland Research Innovation Fund, RIF/037.

References

[1] de Krom, G. "A cepstrum based technique for determining a harmonics-to-noise ratio in speech signals". *J. Speech Hear. Res.* 36(2):254-266, 1993.

[2] Qi, Y. and Hillman, R.E. "Temporal and spectral estimations of harmonics-to-noise ratio in human voice signals". *J. Acoust. Soc. Amer.* 102(1):537-543, 1997.

[3] Murphy, P.J. "A cepstrum-based harmonics-to-noise ratio in voice signals", Proceedings International Conference on Spoken Language Processing, Beijing, China, 672-675, 2000.

[4] Schafer, R.W. and Rabiner, L.R. "System for automatic formant analysis of voiced speech," *J. Acoust. Soc. Amer.* 47:634-648, 1970.

[5] Murphy, P.J. "Averaged modified periodogram analysis of aperiodic voice signals", Proceedings Irish Signals and Systems Conference, Dublin, 266-271, June, 2000.

[6] Fant, G., Liljencrants, J. and Lin, Q. G. "A four parameter model of glottal flow", STL-QPSR 4, 1-12, 1985.

[7] Murphy, P.J. "Perturbation-free measurement of the harmonics-to-noise ratio in speech signals using pitch-synchronous harmonic analysis", *J. Acoust. Soc. Amer.* 105(5):2866:2881, 1999.

[8] Childers, D.G., 1999. Speech processing and synthesis toolboxes, John Wiley & Sons, Inc., New York.

Pseudo Cepstral Analysis of Czech Vowels

Robert Vích

Institute of Radio Engineering and Electronics,
Academy of Sciences of the Czech Republic,
Chaberská 57 CZ-182 52 Prague 8, Czech Republic
vich@ure.cas.cz

Abstract. Real generalized cepstral analysis is introduced and applied to speech deconvolution. Real pseudo cepstrum of the vocal tract model impulse response is defined and applied to the analysis of Czech vowels. The energy concentration measure of the real pseudo cepstrum of the vocal tract model impulse response is introduced and evaluated for Czech vowels pronounced by male and female speakers. The goal of this investigation is to find a robust and more reliable method of vocal tract modeling also for voices with high fundamental frequency, i.e. for female and child voices. From the investigation follows that vowel and speaker dependent generalized cepstral analysis can be found which is more robust in speech modeling than cepstral and LPC analysis.

1 Introduction

In the papers [1-3] a parametric speech modeling approach based on homo-morphic signal processing [4] using spectral analysis was presented and applied to speech synthesis.

In 1979 Lim [5] suggested a new nonlinear signal transformation, which converts the convolution of two signals, an excitation in the form of a train of pulses and a model impulse response, into another convolution, in which the transformed impulse response is shorter than the original one and better separated from the excitation. Using this transformation it is possible to deconvolve a speech signal, i.e. to extract the transformed impulse response from the transformed speech signal by applying a suitable window and after inverse transformation to obtain the impulse response of the vocal tract model with greater accuracy than in classical homomorphic deconvolution. Let us call this nonlinear signal transformation generalized homomorphic approach.

In papers [6-8] the principle of generalized homomorphic signal analysis was applied to speech deconvolution and to vocal tract modeling. A comparison of the computational complexity and of the memory requirements of cepstral IIR and FIR vocal tract models may be found in [9].

In this paper the principle of generalized homomorphic signal analysis is briefly described. We call it real pseudo cepstral analysis and apply it to speech cepstral deconvolution and to vocal tract modeling. Further the generalized signal analysis is used for analysis of Czech vowels uttered by male and female speakers and is compared with results obtained by homomorphic signal analysis.

M. Faundez-Zanuy et al. (Eds.): NOLISP 2005, LNAI 3817, pp. 161–173, 2005.

2 Generalized Cepstral Analysis

At first the procedure proposed by Lim is briefly summarized and applied to speech analysis. The voiced speech signal $s(n)$ may be described by the convolution

$$s(n) = p(n) * h(n) .$$ (1)

$p(n)$ is the sequence of the vocal tract excitation impulses with the fundamental frequency period L and $h(n)$ is the impulse response of the vocal tract model. The parameter L is the fundamental frequency period expressed by the number of speech samples.

Fourier transform of the convolution (1) leads to

$$S(\omega) = P(\omega) \cdot H(\omega) .$$ (2)

$S(\omega)$ represents the speech signal spectrum, $P(\omega)$ is the spectrum of the excitation signal $p(n)$ and $H(\omega)$ is the frequency response of the vocal tract model. For spectrum calculation we use fast Fourier transform with the dimension N_F.

A new transformation is searched for, which converts the convolution (1) into another convolution

$$\tilde{s}(n) = \tilde{p}(n) * \tilde{h}(n)$$ (3)

with shorter "impulse response" $\tilde{h}(n)$, which is better recognizable in the transformed speech signal $\tilde{s}(n)$. The sequences $\tilde{s}(n)$, $\tilde{p}(n)$ and $\tilde{h}(n)$ are the transformed sequences of the corresponding signals $s(n)$, $p(n)$ and $h(n)$ in (1).

This generally nonlinear transformation is performed in the frequency domain followed by inverse Fourier transform. The aim is to find a suitable function $f(S(\omega))$ for speech spectrum transformation.

In homomorphic signal analysis the function $f(S(\omega)) = \ln S(\omega)$ is applied in the definition of the complex cepstrum. In real cepstrum computation $f(S(\omega)) = \ln|S(\omega)|$ is used and for estimation of the autocorrelation sequence in the time domain we use $f(S(\omega)) = |S(\omega)|^2$. In [10] a unifying view on cepstral and correlation analysis was presented. Lim proposed for the spectrum transformation the function

$$f(S(\omega)) = (S(\omega))^\gamma, \ -1 \le \gamma \le 1 .$$ (4)

In this contribution we shall not use the complex spectrum $S(\omega)$ as the argument of the function $f(\cdot)$, like in the definition of the complex speech cepstrum, but the magnitude speech spectrum $|S(\omega)|$ as in the computation of the real cepstrum. The transformation in the frequency domain is therefore defined as

$$f(S(\omega)) = |S(\omega)|^\gamma .$$ (5)

The application of this transformation to (2) results in

$$\tilde{S}_\gamma(\omega) = |S(\omega)|^\gamma = |P(\omega)|^\gamma \cdot |H(\omega)|^\gamma = \breve{P}_\gamma(\omega) \cdot \breve{H}_\gamma(\omega) . \tag{6}$$

The symbols $\tilde{S}_\gamma(\omega)$, $\breve{P}_\gamma(\omega)$ and $\breve{H}_\gamma(\omega)$ are introduced for the Fourier transforms of the magnitude spectra transformed with the parameter γ.

By inverse Fourier transform the convolution in the form of (3) is obtained, but with new sequences $\tilde{s}_\gamma(n)$, $\breve{p}_\gamma(n)$ and $\breve{h}_\gamma(n)$, i.e.

$$\tilde{s}_\gamma(n) = \breve{p}_\gamma(n) * \breve{h}_\gamma(n) . \tag{7}$$

We call the transformed sequences *real pseudo cepstra* corresponding to the signals $s(n)$, $p(n)$ and $h(n)$, respectively they could be called *pseudo correlation sequences*. Lim calls them *spectral root cepstra*.

The real pseudo cepstra are *two sided*; they have a *causal* and an *anticipative* part. Since the Fourier transforms in (6) of the real pseudo cepstra are real, for the sequences in (7) hold

$$\tilde{s}_\gamma(n) = \tilde{s}_\gamma(-n), \quad \breve{p}_\gamma(n) = \breve{p}_\gamma(-n), \quad \breve{h}_\gamma(n) = \breve{h}_\gamma(-n).$$

The pseudo cepstrum $\breve{p}_\gamma(n)$ of the periodic excitation contains a quasi periodical component with the fundamental period L of the voiced excitation. In the following $\breve{p}_\gamma(n)$ will not be examined.

As already mentioned in Chapter 1, the aim of the pseudo cepstral approach is the robust extraction of the pseudo cepstrum $\breve{h}_\gamma(n)$ by windowing the speech pseudo cepstrum $\tilde{s}_\gamma(n)$.

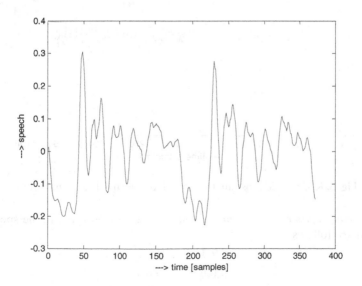

Fig. 1. Vowel "a" – male voice

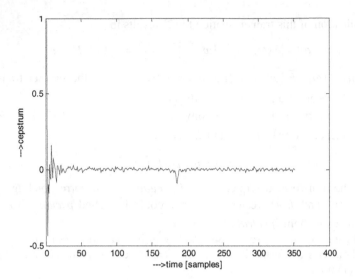

Fig. 2. Real cepstrum of the vowel "a" – male voice

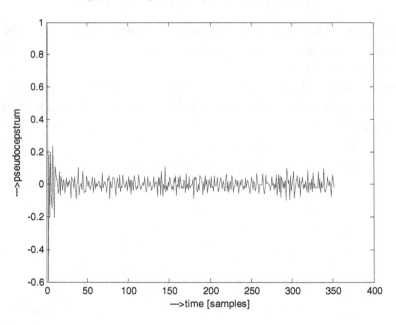

Fig. 3. Real pseudo cepstrum for $\gamma = -1$ of the vowel "a" – male voice

The *inverse pseudo cepstral transformation* is performed again in the spectral domain. From (6) follows

$$\begin{aligned}
|S(\omega)| &= \breve{S}_\gamma^{1/\gamma}(\omega), \\
|P(\omega)| &= \breve{P}_\gamma^{1/\gamma}(\omega), \\
|H(\omega)| &= \breve{H}_\gamma^{1/\gamma}(\omega).
\end{aligned} \qquad (8)$$

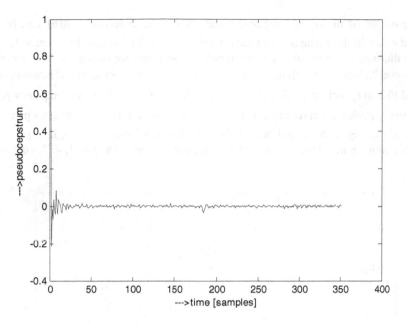

Fig. 4. Real pseudo cepstrum for $\gamma = -0.2$ of the vowel "a" – male voice

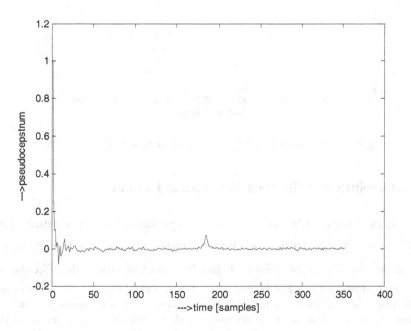

Fig. 5. Real pseudo cepstrum for $\gamma = +0.2$ of the vowel "a" – male voice

The design of the pseudo cepstral vocal tract impulse response $h(n)$ can be performed either in the frequency or time domain. This will be shown in Chapter 4.

We illustrate the generalized homomorphic signal transformation by an example of male vowel "a" sampled with the sampling frequency $F_s = 16$ kHz. The dimension of the applied FFT in speech spectral analysis is $N_F = 1024$. The fundamental frequency period of the male speaker is approximately $L = 186$ ($F_0 = 86$ Hz). In Fig.1 we see a part of the vowel "a", in Fig. 2 the causal part of the corresponding real cepstrum, in Figs. 3, 4, 5 and 6 the causal parts of the corresponding real pseudo cepstra for $\gamma = -1, -0.2, +0.2, +1$.

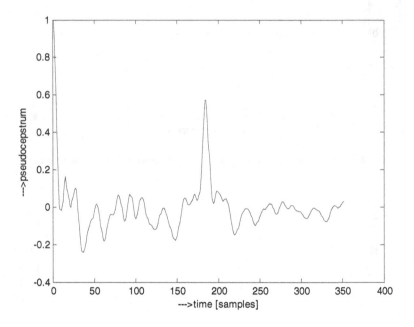

Fig. 6. Real pseudo cepstrum for $\gamma = +1$ of the vowel "a" – male voice

3   Deconvolution in the Pseudo Cepstral Domain

As shortly mentioned in Chapter 2, the pseudo cepstrum $\tilde{h}_\gamma(n)$ can be extracted from the transformed speech signal $\tilde{s}_\gamma(n)$ using a suitable window. The window type depends on the aim of the application of $\tilde{h}_\gamma(n)$. For the evaluation of the effective duration of the transformed impulse response, i.e. for the calculation of its energy concentration, which is described in chapter 5, we preferably use a rectangular window. If we want to design a finite impulse response model of the vocal tract, we use a finite window smoothly tapered to zero, like the Hamming or Hann window. In this case we can also use as a window the frequency response of a convenient magnitude spectrum smoothing filter. This type of windowing will be treated in the following.

To simplify the windowing, it is convenient to replace the symmetric two sided sequence $\breve{s}_\gamma(n)$ by a *one sided causal complex pseudo cepstral sequence* corresponding to a *minimum phase signal*. Then the windowing just described can be performed on the causal part of the complex pseudo cepstral sequence, but using the right part of the considered window only.

Let us designate the one sided causal complex pseudo cepstrum corresponding to $\breve{s}_\gamma(n)$ as $\breve{s}_{\gamma,m}(n)$. Then holds

$$\breve{s}_{\gamma,m}(n) = \breve{s}_\gamma(0) = \frac{1}{2\pi} \int_{-\pi}^{\pi} |S(\omega)|^\gamma \, d\omega \qquad \text{for } n = 0$$

$$\breve{s}_{\gamma,m}(n) = 2\breve{s}_\gamma(n) \qquad\qquad\qquad \text{for } n \geq 0 \qquad\qquad (9)$$

$$\breve{s}_{\gamma,m}(n) = 0 \qquad\qquad\qquad\qquad \text{for } n \leq 0.$$

The design of the vocal tract impulse response $h(n)$ starts now with windowing the causal sequence $\breve{s}_{\gamma,m}(n)$ by a window of the length $M \leq L$. In vocal tract modeling it is appropriate to apply a window given by the square of the magnitude response of a filter with rectangular finite impulse response, i.e. of a FIR filter of the length $K = \text{int}(N_F / L)$. This corresponds to the magnitude response of a triangular or Bartlett filter with the length $(2K - 1)$, where N_F is the dimension of the applied Fast Fourier Transform (FFT) and L is the fundamental period expressed in samples. The result of the pseudo cepstral windowing of $\breve{s}_{\gamma,m}(n)$ is the transformed impulse response, the pseudo cepstrum $\tilde{h}_{\gamma,m}(n)$.

4 FIR Vocal Tract Model Design

The aim of the generalized homomorphic deconvolution is the design of the impulse response of the vocal tract model. The cepstral vocal tract model defined in [1-3] was of the type *infinite impulse response*, shortly *IIR*. In [9] the *finite impulse response* – *FIR*- cepstral vocal tract model was proposed. In this paper we shall concentrate on the always stable *pseudo cepstral FIR* vocal tract model.

The determination of the vocal tract model impulse response is performed by the *inverse pseudo cepstral transform*. This inverse transform may be realized in two domains:

- In the *spectral domain* using the discrete Fourier transform. This procedure results in the *FIR vocal tract model with linear phase*.
- In the *time domain* using a recursive relation. In this case we obtain the *FIR vocal tract model with minimum phase*.

4.1 Spectral Domain

Suppose we have estimated the windowed pseudo cepstrum $\tilde{h}_{\gamma,m}(n)$. The corresponding magnitude spectrum $\tilde{H}_\gamma(\omega)$ is given by Fourier transform and further using (8) the magnitude frequency response of the vocal tract model follows

$$\left|H_\gamma(\omega)\right| = \breve{H}_\gamma^{1/\gamma}(\omega). \tag{10}$$

By inverse Fourier transform we obtain the two sided symmetric impulse response $h_\gamma(n)$ of the vocal tract model. The transfer function of the pseudo cepstral vocal tract model can then be obtained by application of the well known design of FIR filters by windowing [4]. The transfer function obtained in this way is of the type *FIR with linear phase*.

The design of the impulse response $h_\gamma(n)$ of the FIR vocal tract model may be performed directly in the frequency domain without the pseudo cepstrum $\breve{h}_{\gamma,m}(n)$. It starts with smoothing, i.e. with filtering of the transformed speech spectrum $\left|S(\omega)\right|^\gamma$ with a FIR filter. In our case we use double filtering with a rectangular filter of the length $K = \text{int}(N_F / L)$. This double filtering with a rectangular filter corresponds to smoothing with a triangular – Bartlett filter - of the length $(2K - 1)$. The result of this filtering operation is the approximation of transformed spectrum

$$\breve{H}_\gamma(\omega) = \left|H(\omega)\right|^\gamma, \tag{11}$$

from which the frequency response of the vocal tract model (10) results. This approach may be called *deconvolution in the frequency domain*. The frequency response is then the starting point for the FIR filter design by windowing as already mentioned above following Eq. (10).

4.2 Time Domain

The second approach of inverse transform of the causal pseudo cepstrum in the time domain is based on the application of a recursive formula. The *minimum-phase impulse response* $h_{\gamma,m}(n)$ of the pseudo cepstral speech model is related with the real pseudo cepstrum $\breve{h}_{\gamma,m}(n)$ for $n \geq 1$ by the recursive formula

$$h_{\gamma,m}(n) = \frac{1}{\gamma \breve{h}_{\gamma,m}(0)} \{h_{\gamma,m}(0)\breve{h}_{\gamma,m}(n) +$$
$$+ \sum_{k=1}^{n-1}\left(\frac{k}{n}\right)[\breve{h}_{\gamma,m}(n)h_{\gamma,m}(n-k) - \gamma\breve{h}_{\gamma,m}(n-k)h_{\gamma,m}(k)]\} \tag{12}$$

The first value of the impulse response is

$$h_{\gamma,m}(0) = \breve{h}_{\gamma,m}^{\frac{1}{\gamma}}(0),$$

with

$$\breve{h}_{\gamma,m}(0) = \frac{1}{2\pi}\int_{-\pi}^{\pi}\left|S(\omega)\right|^\gamma d\omega.$$

For $n < 0$ holds $h_{\gamma,m}(n) = 0$.

This impulse response is *one sided, causal* and *infinite* and must be truncated by a proper finite smoothly to zero tapered window. The corresponding pseudo cepstral transfer function of the vocal tract model is of the type *FIR* with *minimum phase*.

5   Concentration Measure of the Transformed Impulse Response

For evaluation of the effective duration of the transformed impulse response $\breve{h}_\gamma(n)$ of the vocal tract model we define, according to Lim, the *energy concentration measure* $d_M(\gamma)$ for $M = Lp$, where L is the fundamental frequency period of the voiced excitation and p is a chosen constant, $0 \le p \le 1$. In our experiments we use $p = 0.95$, i.e. we apply for windowing of the transformed impulse response $\breve{h}_\gamma(n)$ a rectangular window of length $M = 0.95L + 1$.

The concentration measure is given as

$$d_M(\gamma) = \frac{\sum_{n=1}^{M} \breve{h}_\gamma^2(n)}{\sum_{n=1}^{N_F/2} \breve{h}_\gamma^2(n)} . \tag{13}$$

The low summation index is set $n = 1$, since $\breve{h}_\gamma(0)$ corresponds to the mean value of the transformed spectral function $\breve{S}_\gamma(\omega) = |S(\omega)|^\gamma$ and it is not convenient to consider it in the concentration measure $d_M(\gamma)$. For weighting the sequence $\breve{s}_{\gamma,m}(n)$ we use a rectangular window $w(n) = 1$, $n = 1, ... M$. In the following example we evaluate $d_M(\gamma)$ for several values of γ in the interval $-1 \le \gamma \le 1$.

6   Concentration Measure for Male and Female Voices

In the experiment we evaluate $d_M(\gamma)$ for Czech vowels uttered by a male and a female for several values of γ in the interval $-1 \le \gamma \le 1$. We use the stationary parts of the sounds *a, e, i, o, u* sampled with the sampling frequency $F_s = 16\,\text{kHz}$. The dimension of the applied FFT in speech spectral analysis is $N_F = 1024$. The fundamental frequency period of the first male speaker is approximately $L = 186$ ($F_0 = 86\,\text{Hz}$), for the second male speaker $L = 145$ ($F_0 = 110\,\text{Hz}$), for the female speaker $L = 91$ ($F_0 = 176\,\text{Hz}$). For the second speaker also the voiced consonants *m* and *n* and the unvoiced fricative *s* are analyzed. The coefficient $p = 0.95$ and the frame length for male and female voices were set $N = N_F$. For spectrum analysis Hamming windowing was applied.

In Fig. 7 and 8 we see the energy concentration measures $d_M(\gamma)$ for the two male voices and in Fig. 9 for the female voice, as functions of the parameter γ. The concentration measure for the real logarithmic cepstrum $d_M(0)$ is shown by an asterisk on all curves. It corresponds to the value $\gamma = 0$.

It can be seen that the curves $d_M(\gamma)$ have a maximum in the neighborhood of $\gamma = 0$. The positions of the maxima depend on the ratio of the numbers of poles and zeros of the corresponding speech models, which was already stated in the paper by Lim using an experimental signal model. For a system with only zeros in its transfer function, i.e. for a finite impulse response system (FIR), the maximum of $d_M(\gamma)$ is located in the neighborhood of $\gamma = 1$. In the case of an all pole transfer function, the maximum lies at $\gamma = -1$. For an infinite impulse response system (IIR) with equal number of poles and zeros the maximum of $d_M(\gamma)$ is positioned at $\gamma = 0$. This statement is not peremptory, the maximizing value of γ depends also on the mutual position of the formants and on the fundamental frequency period L, i.e. it is speaker dependent.

When comparing Fig. 7 and Fig. 9 it seems that the speech models for the male and female voice have different number of poles and zeros for the same vowel. Further the maximum values $d_M(\gamma)$ for the female voice are mostly smaller than that for the male voice, which is given by the shorter fundamental frequency period L of the female voice in relation to the effective length of its transformed speech model impulse response $\bar{h}_\gamma(n)$.

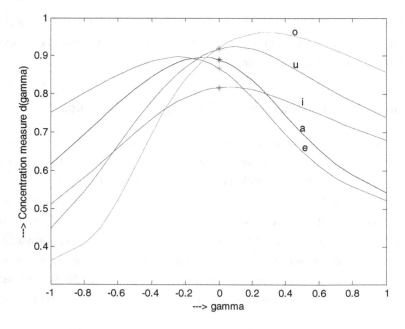

Fig. 7. Energy concentration measure $d_M(\gamma)$ for the first male voice

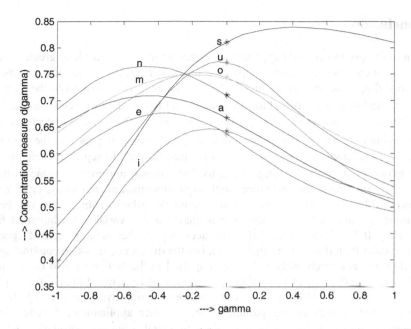

Fig. 8. Energy concentration measure $d_M(\gamma)$ for the second male voice. Beside the vowels also the consonants *m, n* and *s* are analyzed.

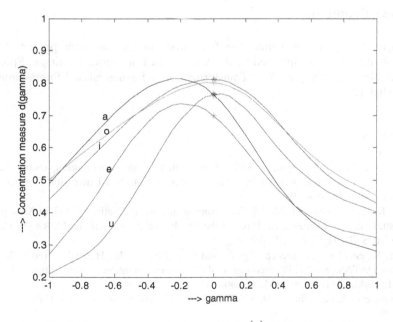

Fig. 9. Energy concentration measure $d_M(\gamma)$ for the female voice

7 Conclusion

The aim of the pseudo cepstral approach to speech analysis is to achieve greater accuracy of the vocal tract model in comparison to the accuracy obtained using cepstral speech modeling. The optimum value of the parameter γ is in relation to the number of formants and antiformants and to the fundamental frequency of voiced sounds. The pseudo cepstral speech model for $\gamma = 0$ corresponds to the cepstral speech model. The approximation error of the pseudo cepstral speech model depends on the parameter γ and on the length and type of the window used for pseudo cepstrum weighting.

The second goal of this investigation is to find a robust and more reliable method of vocal tract modeling also for voices with high fundamental frequency, i.e. for female and child voices. The signal transformation described in this paper has been tested using a synthetic signal with three formants and a variable fundamental frequency [11]. It has been verified that the accuracy of the pseudo cepstral speech model is greater than that of the cepstral one, but the difference is not so convincing.

Pseudo cepstral speech analysis has been applied in the last years also in recognition of noisy speech, e.g. in papers by Alexandre and Lockwood [12], Zühlke [13] and Chilton and Marvi [14]. In these contributions an improvement was registered, but for different values of the parameter γ. Another application of generalized homomorphic speech analysis is its use for fundamental frequency estimation, which is summarized in [15].

Acknowledgements

This paper was prepared within the framework of the research project AVOZ 20670512 and has been supported by the Ministry of Education, Youth and Sports of the Czech Republic OC 277.001 "Transformation of Segmental and Suprasegmental Speech Models".

References

1. Vích, R.: Cepstrales Sprachmodell, Kettenbrüche und Anregungsanpassung in der Sprachsynthese. Wissenschaftliche Zeitschrift der Technischen Universität Dresden, Vol. 49 No 4/5 (2000) 119-121
2. Vích R.: Cepstral Speech Model, Padé Approximation, Excitation and Gain Matching in Cepstral Speech Synthesis. In: Proc. of the 15th Biennial International EURASIP Conference BIOSIGNAL, Brno (2000) 77-82
3. Vích, R., Smékal, Z.: Speech Signals and their Models. In: Horová, I. (ed.): Summer School DATASTAT'01, Proceedings, Folia Facultatis Scientiarum Naturalium Universitatis Masarykianae Brunensis, Mathematica 11 (2002) 275-289
4. Oppenheim, A.V., Schafer, R.W.: Digital Signal Processing. Prentice-Hall, N. Jersey (1989)
5. Lim, J. S.: Spectral Root Homomorphic Deconvolution System. IEEE Transactions on Acoustics Speech, and Signal Processing, Vol. ASSP-27, No. 3 (1979) 223-233

6. Vích, R.: Experimente mit der Anwendung der Pseudokorrelation bei der Vokal-traktmodellierung. In: R. Hoffmann (Ed.): Tagungsband der 13. Konferenz Elektronische Sprachsignalverarbeitung, Dresden, Studientexte zur Sprachsignalverabeitung, Vol. 24 (2002) 253-260

7. Vích, R.: Pseudocepstrale Sprachanalyse und Konstruktion eines Vokaltraktmodells mit endlicher Impulsantwort. In: D. Wolf (Ed.) Signaltheorie und Signalverarbeitung, Akustik und Sprachakustik, Informationstechnik. Arild Lacroix zum 60. Geburtststag, Studientexte zur Sprachkommunikation, Vol. 29 (2003) 126-132

8. Vích, R.: Pseudo Cepstral Speech Analysis. In: Horová, I. (ed.): Summer School DATASTAT'03, Proceedings, Folia Facultatis Scientiarum Naturalium Universitatis Ma-sarykianae Brunensis, Mathematica 15, (2004) 277-291

9. Vondra, M., Vích, R.: Design of FIR Vocal Tract Models with Linear and Nonlinear Phase. In: R. Vích (Ed.): Proc. of the 12th Czech-German Workshop on Speech Processing Prague (2002) 28-32

10. 10 Vích, R., Horák, P., Schwarzenberg, M.: Korrelation von Sprachsignalen im Zeit-und Frequenzbereich. In: Hoffmann, R., Ose, R. (Eds.): Tagungsband der 6. Konferenz Elektro-nische Sprachsignalverarbeitung, Wolfenbüttel, (1995) 10-13

11. Vích, R., Plšek, M.: New Methods for Speech Spectrum Smoothing and Formant Estima-tion. In: Tagungsband des 48. Internationalen Wissenschaftlichen Kolloquiums der TU Ilmenau, Reihe 3.3 Sprachverarbeitung (2003) CD ROM

12. Alexandre, P., Lockwood, P.: Root Cepstral Analysis: A Unified View. Application to Speech Processing in Car Noise Environments. Speech Communication Vol. 12 (1993) 277-288

13. Zühlke, W.: Vergleich der Pseudokorrelationsbereiche mit dem Cepstralbereich. In: Kon-vens 2000, Sprachkommunikation, Ilmenau (2000) 141-144

14. Chilton, E., Marvi, H.: Two-Dimensional Root Cepstrum as Feature Extraction Method for Speech Recognition. Electronics Letters, Vol. 39 No.10 (2003) 815-816

15. Hess, W.: Pitch Determination of Speech Signals. Algorithms and Devices. Springer-Verlag, Berlin Heidelberg New York Tokyo (1983)

Bispectrum Estimators for Voice Activity Detection and Speech Recognition

J.M. Górriz, C.G. Puntonet, J. Ramírez, and J.C. Segura

E.T.S.I.I., Universidad de Granada,
C/ Periodista Daniel Saucedo, 18071 Granada, Spain
gorriz@ugr.es

Abstract. A new Bispectra Analysis application is presented is this paper. A set of bispectrum estimators for robust and effective voice activity detection (VAD) algorithm are proposed for improving speech recognition performance in noisy environments. The approach is based on filtering the input channel to avoid high energy noisy components and then the determination of the speech/non-speech bispectra by means of third order auto-cumulants. This algorithm differs from many others in the way the decision rule is formulated (detection tests) and the domain used in this approach. Clear improvements in speech/non-speech discrimination accuracy demonstrate the effectiveness of the proposed VAD. It is shown that application of statistical detection test leads to a better separation of the speech and noise distributions, thus allowing a more effective discrimination and a tradeoff between complexity and performance. The algorithm also incorporates a previous noise reduction block improving the accuracy in detecting speech and non-speech. The experimental analysis carried out on the AURORA databases and tasks provides an extensive performance evaluation together with an exhaustive comparison to the standard VADs such as ITU G.729, GSM AMR and ETSI AFE for distributed speech recognition (DSR), and other recently reported VADs.

1 Introduction

Speech/non-speech detection is an unsolved problem in speech processing and affects numerous applications including robust speech recognition [1], discontinuous transmission [2, 3], real-time speech transmission on the Internet [4] or combined noise reduction and echo cancellation schemes in the context of telephony [5]. The speech/non-speech classification task is not as trivial as it appears, and most of the VAD algorithms fail when the level of background noise increases. During the last decade, numerous researchers have developed different strategies for detecting speech on a noisy signal [6, 7] and have evaluated the influence of the VAD effectiveness on the performance of speech processing systems [8]. Most of them have focussed on the development of robust algorithms with special attention on the derivation and study of noise robust features and decision rules [9, 10, 11]. The different approaches include those based on energy

M. Faundez-Zanuy et al. (Eds.): NOLISP 2005, LNAI 3817, pp. 174–185, 2005.

thresholds [9], pitch detection [12], spectrum analysis [11], zero-crossing rate [3], periodicity measure [13], higher order statistics in the LPC residual domain [14] or combinations of different features [3, 2].

This paper explores a new alternative towards improving speech detection robustness in adverse environments and the performance of speech recognition systems. The proposed VAD proposes a noise reduction block that precedes the VAD, and uses Bispectra of third order cumulants to formulate a robust decision rule. The rest of the paper is organized as follows. Section 2 reviews the theoretical background on Bispectra analysis and shows the proposed signal model. Section 3 analyzes the statistical tests used in this aproach and compare the speech/non-speech distributions for our decision function based on bispectra and when noise reduction is optionally applied (see section 4). Section 5 describes the experimental framework considered for the evaluation of the proposed end-point detection algorithm. Finally, section 6 summarizes the conclusions of this work.

2   Model Assumptions

Let $\{x(t)\}$ denote the discrete time measurements at the sensor. Consider the set of stochastic variables y_k, $k = 0, \pm 1 \ldots \pm M$ obtained from the shift of the input signal $\{x(t)\}$:

$$\mathbf{y}_k(t) = \mathbf{x}(t + k \cdot \tau) \tag{1}$$

where $k \cdot \tau$ is the differential delay (or advance) between the samples. This provides a new set of $2 \cdot m + 1$ variables by selecting $n = 1 \ldots N$ samples of the input signal. It can be represented using the associated Toeplitz matrix:

$$T_{x(t_0)} = \begin{pmatrix} y_{-M}(t_0) & \cdots & y_{-m}(t_N) \\ y_{-M+1}(t_0) & \cdots & y_{-M+1}(t_N) \\ \cdots & \cdots & \cdots \\ y_M(t_0) & \cdots & y_M(t_N) \end{pmatrix} \tag{2}$$

Using this model the speech-non speech detection can be described by using two essential hypothesis(re-ordering indexes):

$$H_o = \begin{pmatrix} \mathbf{y}_0 = n_0 \\ \mathbf{y}_{\pm 1} = n_{\pm 1} \\ \cdots \\ \mathbf{y}_{\pm M} = n_{\pm M} \end{pmatrix} \tag{3}$$

$$H_1 = \begin{pmatrix} \mathbf{y}_0 = s_0 + n_0 \\ \mathbf{y}_{\pm 1} = s_{\pm 1} + n_{\pm 1} \\ \cdots \\ \mathbf{y}_{\pm M} = s_{\pm M} + n_{\pm M} \end{pmatrix} \tag{4}$$

where s_k's/n_k's are the speech (see section /refsec:speech) /non-speech (any kind of additive background noise i.e. gaussian) signals, related themselves with some differential parameter. All the process involved are assumed to be jointly stationary and zero-mean. Consider the third order cumulant function $C_{y_k y_l}$ defined as:

$$C_{y_k y_l} \equiv E[y_0 y_k y_l] \tag{5}$$

and the two-dimensional discrete Fourier transform (DFT) of $C_{y_k y_l}$, the bispectrum function:

$$C_{y_k y_l}(\omega_1, \omega_2) = \sum_{k=-\infty}^{\infty} \sum_{l=-\infty}^{\infty} C_{y_k y_l} \cdot \exp(-j(\omega_1 k + \omega_2 l))) \tag{6}$$

2.1   Bispectrum Estimators

The set of estimators used in the statistical tests of section 3 are described in the following[1].

Indirect Methods: Sampling the equation 6 and assuming a finite number of samples, the "indirect" bispectrum estimator can be written as:

$$\hat{C}_{y_k y_l}(n, m) = \sum_{k=-M}^{M} \sum_{l=-M}^{M} C_{y_k y_l} \cdot w(k, l) \cdot \exp(-j(\omega_n k + \omega_m l)) \tag{7}$$

where $\omega_{n,m} = \frac{2\pi}{M}(n, m)$ with $n, m = -M, \ldots, M$ are the sampling frequencies, $w(k, l)$ is the window function (to get smooth estimates [15]) and $C_{y_k y_l} = \frac{1}{N} \sum_{i=0}^{N-1} y_0(t_i) y_k(t_i) y_l(t_i) = \frac{1}{N} y_0 y_k y_l|_{t_0}$. Under the assumption that the bispectrum $C_{y_k y_l}$ is sufficiently smooth, the smoothed estimate is known to be consistent, with variance given by:

$$var\left(\hat{C}_{y_k y_l}(n, m)\right) = \frac{1}{N} S_{y_0} S_{y_k} S_{y_l} \int \int w(t, s) dt ds \tag{8}$$

where S is the power spectrum. That is, the data are segmented into possibly overlapping records; biased or unbiased sample estimates of third-order cumulants are computed for each record and then averaged across records; a lag window is applied to the estimated cumulants, and the bispectrum is obtained as the $2 - D$ FFT (fast fourier transform) of the windowed cumulant function. This is the classical method for estimating the Bispectrum function which is known to be consistent. An alternative approach is to perform the smoothing $W(\omega_n, \omega_m)$ in the frequency domain.

Direct Methods: The "direct" class of methods for higher-order spctrum estimation are similar to the "averaged periodogram" or Welch method for power spectrum estimation [16]. In this approach the data are segmented into

---

[1] A deep discussion can be found in the "HOSA" software toolbox (Higher-Order Spectral Analysis Toolbox User's Guide) by Ananthram Swami, Jerry M. Mendel and Chrysostomos L. (Max) Nikias. http://www.mathworks.com/hosa.html

possibly overlapping records; the mean is removed from each record, and the FFT computed; the bispectrum of the Kth record is computed as:

$$\hat{\mathcal{C}}^K_{\mathbf{y}_k \mathbf{y}_l}(n, m) = \mathbf{Y}_K(m)\mathbf{Y}_K(n)\mathbf{Y}_K(m+n) \tag{9}$$

where $\mathbf{Y}_K$ denotes the FFT of the Kth record. The bispectral estimates are averaged across records, and an optional frequency-domain smoother is also applied. Brillinger [17] pointed out that the higher order periodogram is asymptotically "unbiased", and "consistent" if neighbor frequency smoothing is applied.

Integrated Bispectrum: Various VAD algorithms used to date use an averaged Bispectrum function to obtain the decision rule, i.e. in [18, 19]. The following estimation based on Tugnait's work [20] where the connection between the cross-spectrum of a given signal $x(t)$ and its square $y(t) = x^2(t) - E(x^2(t))$ and the integrated bispectrum of the signal is established:

$$S_{yx}(\omega) = \frac{1}{2\pi}\int_{-\pi}^{\pi} C_x(\omega, \widetilde{\omega})d\widetilde{\omega} = \sum_{k=-\infty}^{+\infty} E(y(t)x(t+k))\exp(j\omega)d\omega \tag{10}$$

This implementation improves VAD efficiency and reduces computational effort since just a single FFT has to be computed.

The estimation of the bispectrum is deep discussed in [21] and many others, where conditions for consistency are given. The estimate is said to be (asymptotically) consistent if the squared deviation goes to zero, as the number of samples tends to infinity.

3   Detection Tests for Voice Activity

The decision of our algorithm is based on statistical tests including the Generalized Likelihood ratio tests (GLRT) [22] and the Central χ^2-distributed test statistic under H_O [23]. We will call them GLRT and χ^2 tests. The tests are based on some asymptotic distributions and computer simulations in [24] show that the χ^2 tests require larger data sets to achieve a consistent theoretical asymptotic distribution.

GRLT: Consider the complete domain in bispectrum frequency for $0 \leq \omega_{n,m} \leq 2\pi$ and define P uniformly distributed points in this grid (m, n), called coarse grid. Define the fine grid of L points as the L nearest frequency pairs to coarse grid points. We have that $2M + 1 = P \cdot L$. If we reorder the components of the set of L Bispectrum estimates $\hat{\mathcal{C}}(n_l, m_l)$ where $l = 1, \ldots, L$, on the fine grid around the bifrequency pair into a L vector β_{ml} where $m = 1, \ldots P$ indexes the coarse grid [22] and define P-vectors $\phi_i(\beta_{1i}, \ldots, \beta_{Pi})$, $i = 1, \ldots L$; the generalized likelihood ratio test for the above discussed hypothesis testing problem:

$$H_0 : \mu = \mu_n \quad against \quad H_1 : \eta \equiv \mu^T \sigma^{-1}\mu > \mu_n^T \sigma_n^{-1}\mu_n \tag{11}$$

where $\mu = 1/L \sum_{i=1}^{L} \phi_i$ and $\sigma = 1/L \sum_{i=1}^{L} (\phi_i - \mu)(\phi_i - \mu)^T$ are the maximum likelihood gaussian estimates of vector $C = (C_{\mathbf{y}_k \mathbf{y}_l}(m_1, n_1) \ldots C_{\mathbf{y}_k \mathbf{y}_l}(m_P, n_P))$, leads to the activity voice detection if:

$$\eta > \eta_0 \tag{12}$$

where η_0 is a constant determined by a certain significance level, i.e. the probability of false alarm. Note that:

1. We suppose independence between signal s_k and additive noise n_k bispectrum coeffcients[2] thus:

$$\mu = \mu_n + \mu_s; \quad \sigma = \sigma_n + \sigma_s \tag{13}$$

2. The right hand side of H_1 hypothesis must be estimated in each frame (it's unknown a-priori). In our algorithm the approach is based on the information in the previous non-speech detected intervals.

These assumptions are very restrictive, indeed, somehow the results shown in the experimental section allow them. The statistic considered here η is distributed as a central $F_{2P,2(L-P)}$ under the null hypothesis. Therefore a Neyman-Pearson test can be designed for a significance level α.

χ^2 **Tests:** In this section we consider the χ^2_{2L} distributed test statistic[23]:

$$\eta = \sum_{m,n} 2M^{-1} |\Gamma_{\mathbf{y}_k \mathbf{y}_l}(m,n)|^2 \tag{14}$$

where $\Gamma_{\mathbf{y}_k \mathbf{y}_l}(m,n) = \frac{|\hat{C}_{\mathbf{y}_k \mathbf{y}_l}(n,m)|}{[S_{\mathbf{y}_0}(m) S_{\mathbf{y}_k}(n) S_{\mathbf{y}_l}(m+n)]^{0.5}}$ which is asymptotically distributed as $\chi^2_{2L}(0)$ where L denotes the number of points in interior of the principal domain. The Neyman-Pearson test for a significant level (false-alarm probability) α turns out to be:

$$H_1 \quad if \quad \eta > \eta_\alpha \tag{15}$$

where η_α is determined from tables of the central χ^2 distribution. Note that the denominator of $\Gamma_{\mathbf{y}_k \mathbf{y}_l}(m,n)$ is unknown a priori so they must be estimated as the bispectrum function (that is calculate $\hat{C}_{\mathbf{y}_k \mathbf{y}_l}(n,m)$). This requires a larger data set as we mentioned above in this section.

4   Noise Reduction Block

Almost any VAD can be improved just placing a noise reduction block in the data channel before it. The noise reduction block for high energy noisy peaks, consists of four stages and was first developed in [25]:

$i)$ Spectrum smoothing. The power spectrum is averaged over two consecutive frames and two adjacent spectral bands.

---

[2] Observe that now we do not assume that n_k $k = 0 \ldots \pm M$ are gaussian.

ii) Noise estimation. The noise spectrum $N_e(m, l)$ is updated by means of a 1st order IIR filter on the smoothed spectrum $X_s(m, l)$, that is, $N_e(m, l) = \lambda N_e(m, l-1) + (1 - \lambda)X_s(m, l)$ where $\lambda = 0.99$ and $m = 0, 1, ..., NFFT/2$.

iii) Wiener Filter (WF) design. First, the clean signal $S(m, l)$ is estimated by combining smoothing and spectral subtraction and then, the WF $H(m, l)$ is designed. The filter $H(m, l)$ is smoothed in order to eliminate rapid changes between neighbor frequencies that may often cause musical noise. Thus, the variance of the residual noise is reduced and consequently, the robustness when detecting non-speech is enhanced. The smoothing is performed by truncating the impulse response of the corresponding causal FIR filter to 17 taps using a Hanning window. With this operation performed in the time domain, the frequency response of the Wiener filter is smoothed and the performance of the VAD is improved.

iv) Frequency domain filtering. The smoothed filter H_s is applied in the frequency domain to obtain the de-noised spectrum $Y(m, l) = H_s(m, l)X(m, l)$.

Fig. 1 shows the operation of the proposed VAD on an utterance of the Spanish SpeechDat-Car (SDC) database [26]. The phonetic transcription is: ["siete", "θinko", "dos", "uno", "otSo", "seis"]. Fig 1(b) shows the value of η versus time. Observe how assuming η_0 the initial value of the magnitude η over the first frame (noise), we can achieve a good VAD decision. It is clearly shown how the detection tests yield improved speech/non-speech discrimination of fricative sounds by giving complementary information. The VAD performs an advanced detection of beginnings and delayed detection of word endings which, in part, makes a hang-over unnecessary. In Fig 2 we display the differences between noise and voice in general and in figure we settle these differences in the evaluation of η on speech and non-speech frames.

According to [25], using a noise reduction block previous to endpoint detection together with a long-term measure of the noise parameters, reports important benefits for detecting speech in noise since misclassification errors are significantly reduced.

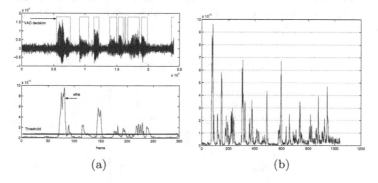

(a)   (b)

Fig. 1. Operation of the VAD on an utterance of Spanish SDC database. (a) Evaluation of η and VAD Decision. (b) Evaluation of the test hypothesis on an example utterance of the Spanish SpeechDat-Car (SDC) database [26].

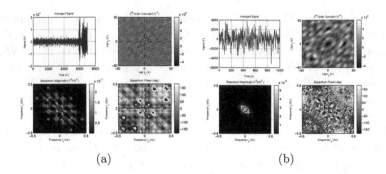

(a)   (b)

Fig. 2. Different Features allowing voice activity detection. (a) Features of Voice Speech Signal. (b) Features of non Speech Signal.

5   Experimental Framework

Several experiments are commonly conducted to evaluate the performance of VAD algorithms. The analysis is mainly focussed on the determination of the error probabilities or classification errors at different SNR levels [11] vs. our VAD operation point, The work about the influence of the VAD decision on the performance of speech processing systems [8] is on the way. Subjective performance tests have also been considered for the evaluation of VADs working in combination with speech coders [27]. The experimental framework and the objective performance tests conducted to evaluate the proposed algorithm are partially showed for space reasons (we only show the results on AURORA-3 database)in this section.

First of all, let's compare the results we obtain using GLRT over the different Bispectrum estimators. The results over the Spanish database shows similar accuracy in voice activity detection depending on the parameters used in each estimator (resolution or number of FFT points "NFTT", different smoothing windows, number of records, etc.) as is shown in 4. Of course the computational effort of the cross-spectrum estimate, essential in on-line applications, is lower than the other approaches.

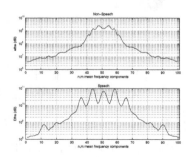

Fig. 3. Speech/non-Speech η values for Speech-Non Speech Frames

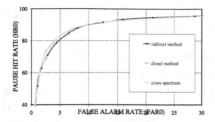

Fig. 4. Receiving Operating Curve in high noisy condition of Aurora 3 the Spanish-Car Database using three bispectrum estimators

The ROC curves are frequently used to completely describe the VAD error rate. The AURORA subset of the original Spanish SpeechDat-Car (SDC) database [26] was used in this analysis. This database contains 4914 recordings using close-talking and distant microphones from more than 160 speakers. The files are categorized into three noisy conditions: quiet, low noisy and highly noisy conditions, which represent different driving conditions with average SNR values between 25dB, and 5dB. The non-speech hit rate (HR0) and the false alarm rate (FAR0= 100-HR1) were determined in each noise condition being the actual speech frames and actual speech pauses determined by hand-labelling the database on the close-talking microphone. These noisy signals represent the most probable application scenarios for telecommunication terminals (suburban train, babble, car, exhibition hall, restaurant, street, airport and train station).

In table 1 shows the averaged ROC curves of the proposed VAD (BiSpectra based-VAD) and other frequently referred algorithms [9, 10, 11, 6] for recordings from the distant microphone in quiet, low and high noisy conditions. The working points of the G.729, AMR and AFE VADs are also included. The results show improvements in detection accuracy over standard VADs and over a representative set VAD algorithms [9, 10, 11, 6]. It can be concluded from these results that:

i) The working point of the G.729 VAD shifts to the right in the ROC space with decreasing SNR.

ii) AMR1 works on a low false alarm rate point of the ROC space but exhibits poor non-speech hit rate.

iii) AMR2 yields clear advantages over G.729 and AMR1 exhibiting important reduction of the false alarm rate when compared to G.729 and increased non-speech hit rate over AMR1.

iv) The VAD used in the AFE for noise estimation yields good non-speech detection accuracy but works on a high false alarm rate point on the ROC space. It suffers from rapid performance degradation when the driving conditions get noisier. On the other hand, the VAD used in the AFE for FD has been planned to be conservative since it is only used in the DSR standard for that purpose. Thus, it exhibits poor non-speech detection accuracy working on a low false alarm rate point of the ROC space.

v) The proposed VAD also works with lower false alarm rate and higher non-speech hit rate when compared to the Sohn's [6], Woo's [9], Li's [10] and

Marzinzik's [11] algorithms in poor SNR scenarios. The BSVAD works robustly as noise level increases.

The benefits are especially important over G.729, which is used along with a speech codec for discontinuous transmission, and over the Li's algorithm, that is based on an optimum linear filter for edge detection. The proposed VAD also improves Marzinzik's VAD that tracks the power spectral envelopes, and the Sohn's VAD, that formulates the decision rule by means of a statistical likelihood ratio test.

It is worthwhile mentioning that the experiments described above yields a first measure of the performance of the VAD. Other measures of VAD performance that have been reported are the clipping errors [27]. These measures provide valuable information about the performance of the VAD and can be used for optimizing its operation. Our analysis does not distinguish between the frames that are being classified and assesses the hit-rates and false alarm rates for a first performance evaluation of the proposed VAD. On the other hand, the speech recognition experiments conducted later on the AURORA databases will be a direct measure of the quality of the VAD and the application it was designed for. Clipping errors are evaluated indirectly by the speech recognition system since there is a high probability of a deletion error to occur when part of the word is lost after frame-dropping.

Performance of ASR systems working over wireless networks and noisy environments normally decreases and non-efficient speech/non-speech detection appears to be an important degradation source [1]. Although the discrimination analysis or the ROC curves are effective to evaluate a given algorithm, this section evaluates the VAD according to the goal for which it was developed by

Table 1. Average speech/non-speech hit rates for SNRs between $25dB$ and $5dB$. Comparison of the proposed BSVAD to standard and recently reported VADs.

(%)	G.729	AMR1	AMR2	AFE (WF)	AFE (FD)
HR0	55.798	51.565	57.627	69.07	33.987
HR1	88.065	98.257	97.618	85.437	99.750
(%)	Woo	Li	Marzinzik	Sohn	χ^2/GLRT
HR0	62.17	57.03	51.21	66.200	66.520/68.048
HR1	94.53	88.323	94.273	88.614	85.192/90.536

Table 2. Average Word Accuracy (%) for the Spanish SDC databases and tasks

		Base	Woo	Li	Marzinzik	Sohn	G.729	AMR1	AMR2	AFE	**GLRT**
	WM	92.94	95.35	91.82	94.29	96.07	88.62	94.65	95.67	95.28	96.28
Sp.	MM	83.31	89.30	77.45	89.81	91.64	72.84	80.59	90.91	90.23	92.41
	HM	51.55	83.64	78.52	79.43	84.03	65.50	62.41	85.77	77.53	86.70
	Ave.	**75.93**	89.43	82.60	87.84	90.58	75.65	74.33	90.78	87.68	**91.80**

assessing the influence of the VAD over the performance of a speech recognition system. The reference framework considered for these experiments was the ETSI AURORA project for DSR [28]. The recognizer is based on the HTK (Hidden Markov Model Toolkit) software package [29].

Table 2 shows the recognition performance for the Spanish SDC databases for the different training/test mismatch conditions (HM, high mismatch, MM: medium mismatch and WM: well matched) when WF and FD are performed on the base system [28]. Again, the VAD outperforms all the algorithms used for reference, yielding relevant improvements in speech recognition. Note that the SDC databases used in the AURORA 3 experiments have longer non-speech periods than the AURORA 2 database and then, the effectiveness of the VAD results more important for the speech recognition system. This fact can be clearly shown when comparing the performance of the proposed VAD to Marzinzik's VAD. The word accuracy of both VADs is quite similar for the AURORA 2 task. However, the proposed VAD yields a significant performance improvement over Marzinzik's VAD for the SDC databases.

6 Conclusions

This paper presented a new VAD for improving speech detection robustness in noisy environments. The approach is based on higher order Spectra Analysis employing noise reduction techniques and order statistic filters for the formulation of the decision rule. The VAD performs an advanced detection of beginnings and delayed detection of word endings which, in part, avoids having to include additional hangover schemes. As a result, it leads to clear improvements in speech/non-speech discrimination especially when the SNR drops. With this and other innovations, the proposed algorithm outperformed G.729, AMR and AFE standard VADs as well as recently reported approaches for endpoint detection. We think that it also will improve the recognition rate when it was considered as part of a complete speech recognition system.

References

1. Karray, L., Martin, A.: Towards improving speech detection robustness for speech recognition in adverse environments. Speech Communitation (2003) 261–276
2. ETSI: Voice activity detector (VAD) for Adaptive Multi-Rate (AMR) speech traffic channels. ETSI EN 301 708 Recommendation (1999)
3. ITU: A silence compression scheme for G.729 optimized for terminals conforming to recommendation V.70. ITU-T Recommendation G.729-Annex B (1996)
4. Sangwan, A., Chiranth, M.C., Jamadagni, H.S., Sah, R., Prasad, R.V., Gaurav, V.: VAD techniques for real-time speech transmission on the Internet. In: IEEE International Conference on High-Speed Networks and Multimedia Communications. (2002) 46–50
5. Gustafsson, S., Martin, R., Jax, P., Vary, P.: A psychoacoustic approach to combined acoustic echo cancellation and noise reduction. IEEE Transactions on Speech and Audio Processing 10 (2002) 245–256

6. Sohn, J., Kim, N.S., Sung, W.: A statistical model-based voice activity detection. IEEE Signal Processing Letters **16** (1999) 1–3
7. Cho, Y.D., Kondoz, A.: Analysis and improvement of a statistical model-based voice activity detector. IEEE Signal Processing Letters **8** (2001) 276–278
8. Bouquin-Jeannes, R.L., Faucon, G.: Study of a voice activity detector and its influence on a noise reduction system. Speech Communication **16** (1995) 245–254
9. Woo, K., Yang, T., Park, K., Lee, C.: Robust voice activity detection algorithm for estimating noise spectrum. Electronics Letters **36** (2000) 180–181
10. Li, Q., Zheng, J., Tsai, A., Zhou, Q.: Robust endpoint detection and energy normalization for real-time speech and speaker recognition. IEEE Transactions on Speech and Audio Processing **10** (2002) 146–157
11. Marzinzik, M., Kollmeier, B.: Speech pause detection for noise spectrum estimation by tracking power envelope dynamics. IEEE Transactions on Speech and Audio Processing **10** (2002) 341–351
12. Chengalvarayan, R.: Robust energy normalization using speech/non-speech discriminator for German connected digit recognition. In: Proc. of EUROSPEECH 1999, Budapest, Hungary (1999) 61–64
13. Tucker, R.: Voice activity detection using a periodicity measure. IEE Proceedings, Communications, Speech and Vision **139** (1992) 377–380
14. Nemer, E., Goubran, R., Mahmoud, S.: Robust voice activity detection using higher-order statistics in the lpc residual domain. IEEE Trans. Speech and Audio Processing **9** (2001) 217–231
15. Nikias, C., Petropulu, A.: Higher Order Spectra Analysis: a Nonlinear Signal Processing Framework. Prentice Hall (1993)
16. Kay, S.: Modern Spectral Esimation. Englewood Cliffs, NJ: Prentice Hall (1988)
17. Brillinger, D.: Analysis of a linear time invariant relation between a stochastic series and several deterministic series. In: Time Series Data Analysis and Theory. Siam in applied Maths (2001)
18. Górriz, J.M., Puntonet, C.G., Ramírez, J., Segura, J.C.: Bispectra analysis-based vad for robust speech recognition. In: IWINAC (2). (2005) 567–576
19. Górriz, J.M., Ramírez, J., Segura, J.C., Puntonet, C.G.: An improved mo-lrt vad based on a bispectra gaussian model. In press IEE Electronic Letters **X** (Jul 2005) XXX–XXX
20. Tugnait, J.: Detection of non-gaussian signals using integrated polyspectrum. IEEE Transaction on signal Processing **42** (1994) 3137–3149
21. Brillinger, D., Rossenblatt, M.: Asymptotic theory of estimates of kth order spectra. In: Spectral Analysis of Time Series. Wiley (1975)
22. Subba-Rao, T.: A test for linearity of stationary time series. Journal of Time Series Analisys **1** (1982) 145–158
23. Hinich, J.: Testing for gaussianity and linearity of a stationary time series. Journal of Time Series Analisys **3** (1982) 169–176
24. Tugnait, J.: Two channel tests fro common non-gaussian signal detection. IEE Proceedings-F **140** (1993) 343–349
25. Ramírez, J., Segura, J., Benítez, C., delaTorre, A., Rubio, A.: An effective subband osf-based vad with noise reduction for robust speech recognition. In press IEEE Transactions on Speech and Audio Processing **X** (2004) X–X
26. Moreno, A., Borge, L., Christoph, D., Gael, R., Khalid, C., Stephan, E., Jeffrey, A.: SpeechDat-Car: A Large Speech Database for Automotive Environments. In: Proceedings of the II LREC Conference. (2000)

27. Benyassine, A., Shlomot, E., Su, H., Massaloux, D., Lamblin, C., Petit, J.: ITU-T Recommendation G.729 Annex B: A silence compression scheme for use with G.729 optimized for V.70 digital simultaneous voice and data applications. IEEE Communications Magazine **35** (1997) 64–73
28. ETSI: Speech processing, transmission and quality aspects (stq); distributed speech recognition; front-end feature extraction algorithm; compression algorithms. ETSI ES 201 108 Recommendation (2000)
29. Young, S., Odell, J., Ollason, D., Valtchev, V., Woodland, P.: The HTK Book. Cambridge University (1997)

On the Acoustic-to-Electropalatographic Mapping

Asterios Toutios and Konstantinos Margaritis

Parallel and Distributed Processing Laboratory,
Department of Applied Informatics,
University of Macedonia, Thessaloniki, Greece
{toutios, kmarg}@uom.gr

Abstract. Electropalatography is a well established technique for recording information on the patterns of contact between the tongue and the hard palate during speech. It leads to a stream of binary vectors, called electropalatograms. We are interested in the mapping from the acoustic signal to electropalatographic information. We present results on experiments using Support Vector Classification and a combination of Principal Component Analysis and Support Vector Regression.

1 Introduction

Electropalatography (EPG) [1] is a widely used technique for recording and analyzing one aspect of tongue activity, namely its contact with the hard palate during continuous speech. It is well established as a relatively non-invasive, conceptually simple and easy-to-use tool for the investigation of lingual activity in both normal and pathological speech. An essential component of EPG is a custom-made artificial palate, which is moulded to fit as unobtrusively as possible against a speaker's hard palate. Embedded in it are a number of electrodes (62 in the Reading EPG system, which is considered herein). When contact occurs between the tongue surface and any of the electrodes a signal is conducted to an external processing unit and recorded. Typically, the sampling rate of such a system is 100-200 Hz. Thus, for a given utterance, the sequence of raw EPG data consists of a stream of binary (1 if there is a contact; -1 if there is not) vectors with both spatial and temporal structure. Figure 1 shows part of such a stream. Observation of both temporal and spatial details of contact across the entire palatal region can be very helpful to identify many phonetically relevant details of lingual activity.

Electropalatography has been succesfully used to study a number of phenomena in phonetic descriptive work, in studies of lingual coarticulation and in the diagnosis and treatment of a variety of speech disorders. It has also been suggested that visual feedback from EPG might be used in the context of second language acquisition.

However, there are difficulties in acquiring EPG data. First, each artificial palate must be individually manufactured from dental moulds of the speaker.

M. Faundez-Zanuy et al. (Eds.): NOLISP 2005, LNAI 3817, pp. 186–195, 2005.

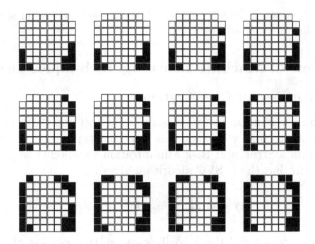

Fig. 1. Typical EPG sequence. Black squares indicate a contact between the tongue and the palate.

Second, the artificial palate in the speaker's mouth may sometimes hinder their ability to produce normal speech.

What is suggested here is that some means of estimating EPG information using only the audio signal (which is far more easier to record and handle) as a source would be beneficial. To this end, we study the mapping from the acoustic signal to the EPG vectors, namely the *acoustic-to-electropalatographic mapping*. We adopt a machine learning point of view, in the sense that we try to infer the mapping only *from the data*, without making a priori use of any kind of speech production related theoretical intuitions.

2   The MOCHA Database

The MOCHA (Multi-Channel Articulatory) [2] database is evolving in a purpose built studio at the Edinburgh Speech Production Facility at Queen Margaret University College.

During speech, four data streams are recorded concurrently straight to a computer: the acoustic waveform, sampled at 16kHz with 16 bit precision, together with laryngograph, electropalatograph and electromagnetic articulograph data. EPG provides tongue-palate contact data at 62 normalised positions on the hard palate, defined by landmarks on maxilla. The EPG data are recorded at 200Hz.

The speakers are recorded reading a set of 460 British TIMIT sentences. These short sentences are designed to provide phonetically diverse material and capture with good coverage the connected speech processes in English. All waveforms are labelled at the phonemic level.

The final release of the MOCHA database will feature up to 40 speakers with a variety of regional accents. At the time of writing this paper three speakers are available. For the experiments herein, the acoustic waveform and EPG data, as well as the phonemic labels for the fsew0 speaker, a female speaker with a Southern English accent, are used.

3　Overview of Machine Learning Techniques Used

3.1　C-Support Vector Classification

Given n training vectors $\mathbf{x_i}$ in two classes and a vector $y \in R^n$ such that $y_i \in \{-1, 1\}$, we want to find a decision function that separates the two classes in an optimal (from a Structural Risk Minimization viewpoint) way [3, 4, 5] . The decision function that the C-SVC algorithm gives is:

$$f(\mathbf{x}) = \text{sgn}\left(\sum_{i=1}^{n} a_i y_i k(\mathbf{x}, \mathbf{x_i}) + b\right), \tag{1}$$

where b is a bias terms and the a coefficients are the solution of the quadratic programming problem:

$$\text{maximize } W(\mathbf{a}) = -\frac{1}{2}\sum_{ij} a_i a_j y_i y_j k(\mathbf{x_i x_j})$$

$$\text{subject to } 0 \le a_i \le C, i = 1, \dots, n, \text{ and } \sum_i a_i y_i = 0. \tag{2}$$

Here C, called the *penalty parameter*, is a parameter defined by the user and $k(\mathbf{x_i x_j})$ is a special function called the *kernel* which serves to convert the data into a higher-dimensional space in order to account for non-linearities in the decision function. A commonly used kernel is the Radial Basis Function (RBF) kernel:

$$k(\mathbf{x}, \mathbf{y}) = \exp(-\gamma \parallel \mathbf{x} - \mathbf{y} \parallel^2), \tag{3}$$

where the γ parameter is selected by the user.

3.2　ϵ-Support Vector Regression

The ϵ-SVR algorithm [6, 5] generalizes the C-SVC algorithm to the regression case. Given n training vectors $\mathbf{x_i}$ and a vector $y \in R^n$ such that $y_i \in R$, we want to find an estimate for the fuction $y = f(\mathbf{x})$. According to ϵ-SVR, this estimate is:

$$f(\mathbf{x}) = \sum_{i=1}^{n}(a_i^* - a_i)k(\mathbf{x_i}, \mathbf{x}) + b, \tag{4}$$

where the coefficients a_i and a_i^* are the solution of the quadratic problem

$$\text{maximize}$$

$$W(\mathbf{a}, \mathbf{a}^*) = -\epsilon\sum_{i=1}^{n}(a_i^* + a_i) + \sum_{i=1}^{n}(a_i^* - a_i)y_i - \frac{1}{2}\sum_{i,j=1}^{n}(a_i^* - a_i)(a_j^* - a_j)k(\mathbf{x_i x_j})$$

$$\text{subject to } 0 \le a_i, a_i^* \le C, i = 1, \dots, n, \text{ and } \sum_{i=1}^{n}(a_i^* - a_i) = 0. \tag{5}$$

$C > 0$ and $\epsilon \geq 0$ are chosen by the user. C may be as high as infinity, while typical values for ϵ are 0.1 or 0.001.

3.3   Principal Component Analysis

PCA [7, 1] is a transform that chooses a new coordinate system for a data set such that the greatest variance by any projection of the data set comes to lie on the first axis, the second greatest variance on the second axis, and so on. The new axes are called the *principal components*. PCA is commonly used for reducing dimensionality in a data set while retaining those characteristics of the data set that contribute most to its variance by eliminating the later principal components.

The direction $\mathbf{w_1}$ of the first principal component is defined by

$$\mathbf{w_1} = \arg \max_{\|w\|=1} E\{(\mathbf{w}^T\mathbf{x})^2\} \tag{6}$$

where $\mathbf{w_1}$ is of the same dimension as the data vectors $\mathbf{x}$. Having determined the direction of the first $k - 1$ principal components, the direction of the kth component is:

$$\mathbf{w_k} = \arg \max_{\|w\|=1} E\left\{\mathbf{w}^T\left(\mathbf{x} - \sum_{i=1}^{k-1} \mathbf{w_i}\mathbf{w_i}^T\mathbf{x}\right)^2\right\}. \tag{7}$$

In practice, the computation of the $\mathbf{w_i}$ can be simply accomplished using the sample covariance matrix $E\{\mathbf{xx}^T\} = \mathbf{C}$. The $\mathbf{w_i}$ are then the eigenvectors of $\mathbf{C}$ that correspond to the largest eigenvalues of $\mathbf{C}$.

4   Data Processing

The MOCHA database includes 460 utterances of the fsew0 speaker. In order to render these data into input-output pairs suitable for our purposes, we proceed as follows.

First, based on the label files we omit silent parts from the beginning and end of the utterances. During silent stretches the tongue can possibly take any configuration, something that could pose serious difficulties to our task.

Next, we perform a standard Mel Frequency Spectral Analysis [8] on the acoustic signal with the VOICEBOX Toolkit [9], using a window of 16ms (256 points) with a shift of 5ms (this is to match the 200Hz sampling rate of the EPG data). We use 30 filterbanks and calculate the first 13 Mel Frequency Cepstral Coefficients. Then, we normalize them in order have zero mean and unity standard deviation.

In order to account for the dynamic properties of the speech signal and cope with the temporal extent of our problem, we just use a commonplace in the speech processing field *spatial metaphor for time*. That is, we construct input vectors spanning over a large number of acoustic frames. Based on some previous

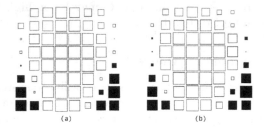

Fig. 2. Distributions of EPG events (a) in tthe training set (b) in the test. The bigger the square, the bigger the difference between positive and negative examples. Black squares indicate excess of positive examples and white squares excess of negative examples.

small-scale experiments of ours, we construct input vectors consisting of the MFCCs of 17 frames: the frame in question, plus the 8 previous ones, plus the 8 next ones.

Thus, we end up with training examples with a 221-dimensional (17×13) real-valued vector as input and a 62-dimensional binary vector as output. We split our data into two big halves: the even-numbered utterances constitute an "extended training set", and the odd-numbered ones an "extended test set". Each one has more than 100.000 examples.

But, since SVR training is a relatively slow process, using the whole "extended training set" for training would merely be out of the question. We would like a reduced training set, that is somehow "representative" of the whole corpus. Knowing (from the label files) the phonemic label of each of our "extended training set" examples, we randomly select 200 training examples corresponding to every one of the 44 distinct phonemic labels. Since some phonemic labels have less than 200 examples in the dataset, we end up with 8686 training examples.

Finally, for our test set, we simply use 10 utterances spanning across our whole "extended test set". This test set consists of 5524 examples.

In both our final training and test sets, the distributions of the output among the EPG points values vary considerably, ranging from EPG points with a nearly equal number of positive (contacts, value 1) and negative (non-contacts, value -1) examples, to points with a 100% of examples belonging to one of the two classes. This fact is depicted graphically in Figure 2.

5   Training and Results

We follow two approaches to the mapping between the MFCCs and the EPG data. For the first one, we make the working assumption that every EPG event (a contact or a non-contact at a certain electrode and point in time) is independent of neighbouring (in space and time) EPG events. Thus, the problem of estimating EPG patterns, becomes a problem of training 62 binary classifiers.

The C-SVC algorithm then offers a straightforward way to independently deal with each one of these classification tasks, where the input is the MFCC vector

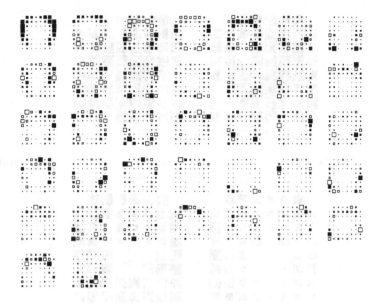

Fig. 3. Principal Components of the EPG data. Each value is represented by a square of size proportional to its absolute value and color black or white whether it is positive or negative.

(constructed as described previously) and the output is a binary value describing the activity of the EPG point in question.

We consider the RBF kernel with $\gamma = 0.0045$ and select $C = 1$, based on heuristics found in [10] The experiments are conducted using the LIBSVM software package [11].

For our second approach to the mapping, we consider accounting for the spatial relationships in the EPG data by applying PCA. We perform PCA on the "extended training set" and keep the 37 first principal components (depicted in Figure 3), which are the ones with eigenvalues larger than the 1/100 of the largest eigenvalue.

PCA transforms the output data by moving them into a new space. In this space the output values are real, so we have to solve 37 regression problems. We use ϵ-SVR for this task.

Just before SVR training we perform two further preprocessing steps on our (PCA transformed) output data. Firstly we *center* them so that the mean value of every channel is zero, and, secondly we scale them by four times their standard deviation, so that they roughly lie in the interval $(-1, 1)$, something crucial for SVR training.

For the actual ϵ-SVR training, we use the RBF kernel with $\gamma = 0.0045$ and select $C = 1$ and $\epsilon = 0.1$. In testing, we need to invert the processes of scaling, centering and PCA.

For assessing the performance of are classifiers (even though we used regression in our second approach, the final outcome is still a set of classifiers) we

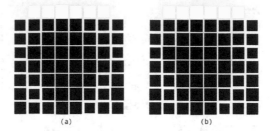

Fig. 4. Classification Rates for (a) the SVC approach (b) the PCA+SVR approach

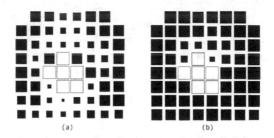

Fig. 5. AUCs for (a) the SVC approach (b) the PCA+SVR approach

use two metrics. The first one is the absolute classification rate (in the second approach by assigning positive output values as contacts, and negative as non-contacts), and the second one is the area under the ROC curve (AUC) [12]. The results are presented in Table 1 and Figures 4, 5. The convention used in the figures is that the size of the black squares is proportional to the value of the metric, while white squares indicate EPG points where the specific metric is meaningless (i.e. there is no AUC when all the examples in the test set belong the same class).

6   Conclusion

We applied two methods to the acoustic-to-electropalatographic mapping task, the first of which (SVC) does not take into account the spatial interrelationships inherent in the EPG data, while the second one (PCA+SVR) does.

The chance level (defined as the average percentage of the class with the most examples among the EPG points) of the data in the test set we used was 85,60%. Both the methods we applied exceed by far this chance level. For the SVC approach the average classification rate is 92,34%, and for the PCA+SVR approach 92,44%.

Between the two approaches, the differences in performance in terms of classification rates is small. The PCA+SVR approach improves upon SVC's classification rate only by 0,1%. Nevertheless, the ROC curves (with the exception of a couple of EPG points) are in general much better for the PCA+SVR, leading to

Table 1. Performances of the sets of classifiers in terms of Classification Rates and AUCs. Also shown the percentages of contacts in the training and test sets.

EPG Point	Training Set % Contacts	Test Set % Contacts	SVC Class. Rate	SVC AUC	PCA+SVR Class. Rate	PCA+SVR AUC
1	16,39	25,53	86,50	0,80	87,56	0,93
2	9,73	16,56	87,65	0,72	88,49	0,93
3	4,02	6,95	93,21	0,54	93,72	0,85
4	8,85	17,20	85,16	0,61	86,12	0,88
5	18,26	30,90	85,63	0,80	86,73	0,94
6	24,41	36,75	86,77	0,85	86,93	0,93
7	26,80	38,78	83,87	0,82	83,74	0,92
8	12,66	15,06	88,38	0,74	88,07	0,88
9	7,52	8,64	92,32	0,59	91,96	0,88
10	3,64	3,86	96,20	0,44	96,18	0,86
11	3,48	4,38	95,49	0,43	95,49	0,74
12	10,17	12,51	89,34	0,64	89,68	0,84
13	24,56	36,35	83,73	0,81	83,80	0,93
14	38,61	49,64	84,29	0,84	83,69	0,92
15	44,02	55,70	87,13	0,87	86,55	0,94
16	9,46	7,46	93,25	0,60	93,12	0,84
17	1,54	1,41	98,57	0,43	98,53	0,59
18	0,46	0,52	99,48	0,41	99,44	0,54
19	0,93	1,19	98,81	0,39	98,75	0,72
20	2,75	3,01	96,92	0,51	96,90	0,70
21	10,98	12,65	89,68	0,59	89,43	0,81
22	50,40	61,62	87,38	0,88	87,64	0,94
23	40,24	47,07	84,43	0,84	84,76	0,92
24	2,60	1,18	98,82	0,43	98,82	0,83
25	0,22	0,24	99,76	0,80	99,76	0,53
26	0,01	0,00	100,00	-	100,00	-
27	0,06	0,18	99,82	0,80	99,82	0,39
28	0,56	0,49	99,51	0,56	99,51	0,69
29	7,03	6,97	93,54	0,51	93,10	0,80
30	39,17	48,21	82,35	0,81	81,88	0,91
31	54,93	59,92	88,00	0,88	88,49	0,95
32	10,47	8,85	93,54	0,62	93,28	0,89
33	0,10	0,00	100,00	-	100,00	-
34	0,00	0,00	100,00	-	100,00	-
35	0,00	0,00	100,00	-	100,00	-
36	0,16	0,18	99,82	0,80	99,82	0,91
37	10,76	9,41	92,98	0,66	92,99	0,92
38	68,77	71,54	91,75	0,92	91,60	0,97
39	79,33	77,34	91,02	0,87	90,41	0,94
40	28,63	24,80	86,01	0,75	85,97	0,91
41	0,67	0,67	99,33	0,29	99,33	0,90
42	0,01	0,00	100,00	-	100,00	-
43	0,00	0,00	100,00	-	100,00	-
44	3,50	2,41	97,52	0,43	97,59	0,86
45	39,90	37,74	84,12	0,80	85,03	0,93
46	90,24	88,90	94,21	0,82	93,72	0,94
47	92,44	90,39	91,67	0,67	91,71	0,85
48	43,33	39,66	80,52	0,76	81,77	0,90
49	4,82	5,38	94,73	0,40	94,73	0,86
50	0,25	0,18	99,82	0,21	99,82	0,88
51	1,66	1,67	98,33	0,49	98,37	0,88
52	9,84	9,49	90,80	0,51	91,13	0,85
53	69,12	69,37	84,90	0,82	84,79	0,92
54	97,54	98,21	98,21	0,90	97,18	0,90
55	93,84	92,85	93,79	0,63	94,41	0,86
56	85,74	84,32	86,08	0,52	86,51	0,81
57	11,19	10,90	90,53	0,50	90,35	0,87
58	1,51	1,39	98,61	0,40	98,48	0,84
59	5,45	6,03	94,77	0,50	94,68	0,92
60	26,62	23,57	80,25	0,61	81,44	0,83
61	87,95	85,48	87,22	0,57	87,93	0,81
62	89,79	87,44	88,50	0,65	89,30	0,79
Overall			92,34	0,64	92,44	0,85

a remarkable increase in the average AUC, as shown in Table 1. Figure 6 shows the ROC curves for some characteristic EPG points.

So, it is mainly the improvement of the ROC curves achieved with the PCA+SVR approach, that makes it a better choice of an approach between the two. This agrees with the intuition that the PCA+SVR approach *should* be better, since it takes into account the spatial structure of the problem at hand.

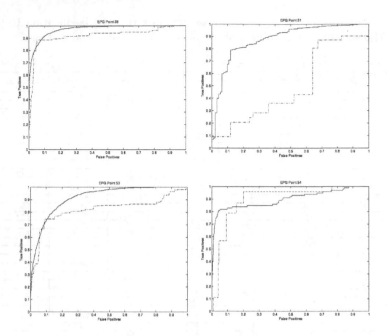

Fig. 6. ROC curves for some EPG points. Dashed-dotted curves correspond to the SVC approach, solid curves to the PCA+SVR approach.

One drawback of our experimental setup was that we trained our machines using only a small set of training examples, selected by a rather ad hoc procedure. As a future work direction, we might employ a more structured approach (i.e. clustering) in order to select training examples. Or, we might directly experiment with more data. Training time is always an issue, but recent findings in the machine learning field, such as Cross-Training [13], seem quite promising in the direction of speeding up things.

As a second future work direction, we could try to account for the temporal structure of our problem, i.e. the fact that the activity of a certain EPG point is depended on its activity at previous time instants. This is a difficult problem, though there are promising proposals from the machine learning field, such as the HMM–SVM method [14].

References

1. Miguel Á. Carreira-Perpiñán. *Continuous Latent Variable Models for Dimensionality Reduction and Sequential Data Reconstruction.* PhD thesis, University of Sheffield, UK, February 2001.
2. Alan A. Wrench and William J. Hardcastle. A multichannel articulatory database and its application for automatic speech recognition. In *5th Seminar on Speech Production: Models and Data*, pages 305–308, Kloster Seeon, Bavaria, 2000.
3. Vladimir Vapnik. *Statistical Learning Theory.* Wiley, New York, 1998.

4. Christopher J. C. Burges. A tutorial on support vector machines for pattern recognition. *Data Mining and Knowledge Discovery*, 2(2):121–167, 1998.
5. Bernhard Schölkopf and Alex Smola. *Learning with Kernels: Support Vector Machines, Optimization, Regularization and Beyond.* MIT Press, 1st edition, 2001.
6. Alex Smola and Bernhard Schölkhopf. A tutorial on support vector regression. *Statistics and Computing*, 14(3):199–222, August 2004.
7. Mike E. Tipping and Christopher M. Bishop. Probabilistic principal component analysis. *Journal of the Royal Statistical Society, Series B*, 61(3):611–622, 1999.
8. Steven B. Davis and Paul Mermelstein. Comparison of parametric representations for monosyllabic word recognition in continuously spoken sentences. In Alexander Waibel and Kai-Fu Lee, editors, *Readings in speech recognition*, pages 65–74. Morgan Kaufmann Publishers Inc., 1990.
9. Mike Brooks. The VOICEBOX toolkit. Software vailable at http://www.ee.ic.ac.uk/hp/staff/dmb/voicebox/voicebox.html.
10. Jason Weston, Arthur Gretton, and Andre Elisseeff. SVM practical session (how to get good results without cheating). Machine Learning Summer School 2003, Tuebingen, Germany.
11. Chih-Chung Chang and Chih-Jen Lin. *LIBSVM: a library for support vector machines*, 2001. Software available at http://www.csie.ntu.edu.tw/~cjlin/libsvm.
12. Tom Fawcett. ROC graphs: Notes and practical considerations for researchers. Technical report, HP Laboratories, Palo Alto, April 2004.
13. Gökhan Bakir, Léon Bottou, and Jason Weston. Breaking SVM complexity with cross training. In *18th Annual Conference on Neural Information Processing Systems, NIPS-2004*, 2004.
14. Yasemin Altun, Ioannis Tsochantaridis, and Thomas Hofmann. Hidden markov support vector machines. In *20th International Conference on Machine Learning ICML-2004*, Washington DC, 2003.

Issues in Clinical Applications of Bilateral Multi-step Predictive Analysis of Speech

J. Schoentgen, E. Dessalle, A. Kacha, and F. Grenez

National Fund for Scientific Research, Belgium
Department "Signals and Waves", Faculty of Applied Sciences,
Université Libre de Bruxelles, Brussel, Belgium

Abstract. The article concerns methodological problems posed by multi-step predictive analysis of speech, carried out with a view to estimating vocal dysperiodicities. Problems that are discussed are the following. First, the stability of the multi-step predictive synthesis filter; second, the decrease of quantization noise by means of multiple prediction coefficients; third, the implementation of multi-step predictive analyses via lattice filters; fourth, the adequacy per se of the multi-step predictive analysis paradigm for estimating vocal dysperiodicities. Results suggest that implementations of multi-step predictive analyses that are considered to be optimal for speech coding are sub-optimal for clinical applications and vice versa. Also, multi-step predictive analyses as such do not appear to be under all circumstances a paradigm adequate for analysing vocal dysperiodicities unambiguously. An alternative is discussed, which is based on a generalized variogram of the speech signal.

1 Introduction

The presentation concerns issues in clinical applications of bilateral multi-step predictive analysis of speech. Multi-step analysis designates the linear prediction of the present speech sample by means of samples that are distant. Because the purpose is the estimation of dysperiodicities in speech, the prediction distance is assigned to the lag for which the correlation between the present and a distant speech frame is maximal. This lag is indeed expected to agree with an integer multiple of the vocal cycle lengths of voiced speech sounds. In the case of unvoiced sounds or highly irregularly voiced sounds, this lag remains mathematically meaningful but is not interpreted in terms of the glottal cycle length. Bilateral means that predictive analyses are performed to the right and left of the current speech frame and that the minimal prediction error is kept and assigned to the vocal dysperiodicity trace.

Voice disorders, or dysphonias, are common consequences of disease, injury or faulty use of the larynx. A frequent symptom of dysphonia is increased noise in the speech signal or lack of regularity of the vocal cycles. Speech analyses are therefore carried out routinely in the context of the functional assessment of voice disorders.

At present, these analyses are most often carried out on steady fragments of sustained vowels. The reason is that the signal processing is often based on the

M. Faundez-Zanuy et al. (Eds.): NOLISP 2005, LNAI 3817, pp. 196–205, 2005.
© Springer-Verlag Berlin Heidelberg 2005

assumption that the speech cycles are locally quasi-identical in length and amplitude. Therefore, such analyses may fail on sustained vowels or connected speech produced by severely hoarse speakers. Studies devoted to vocal dysperiodicities in connected speech or vowels including onsets and offsets are therefore comparatively rare. An overview of published research is given in [2].

Clinicians have, however, expressed the wish to be able to analyze any speech fragment produced by any speaker, including vowel onsets and offsets as well as connected speech. Arguments in favour are that, compared to stationary speech fragments, connected speech is more difficult to produce because of more frequent voice onsets and offsets, the voicing of obstruents, the maintaining of voicing while the larynx continually ascends and descends in the neck, as well as because of intonation and accentuation.

Qi et al. [1] and Bettens et al. [2] have presented methods that enable estimating vocal dysperiodicities without making any strong assumptions with regard to the regularity of the vibrations of the vocal folds or recorded speech sounds. These methods have been inspired by speech coding based on multi-step linear predictive analysis. The method presented by Qi et al. [1] involves a conventional single-step predictive analysis followed by a multi-step analysis of the residual error of the single-step prediction. In a clinical context, the multi-step prediction error is construed as the vocal dysperiodicity trace.

The method presented by Bettens et al. [2] involves a bilateral multi-step predictive analysis. It may be carried out on the speech signal directly or on any other signal considered to be clinically apposite, because the method omits the single-step analysis and avoids predicting across phonetic boundaries.

The topic of this article is an examination of methodological problems posed by bilateral multi-step predictive analyses when applied clinically.

2 Models

Formally, bilateral multi-step prediction is based on models (1). In [2], bilateral prediction is called bidirectional. In the present text, the term bilateral is preferred because it stresses the distinction between multi-step predictive analyses that are carried out to the left and right of the current speech frame, on the one hand, and the forward and backward errors involved in the lattice filter implementation of unilateral multi-step analyses, on the other.

$$e(n) = \begin{cases} e_{right}(n) = s(n) + \sum\limits_{i=0}^{M} a_{rigthi} s(n + T_{right} - i), \\ e_{left}(n) = s(n) + \sum\limits_{i=0}^{M} a_{left,i} s(n - T_{left} + i). \end{cases} \tag{1}$$

Symbol $s(n)$ is the current speech sample; $e(n)$ is the bilateral multi-step prediction error; weights a are the prediction coefficients. For each analysis frame, the multi-step prediction error, the energy of which is smallest, is assigned to the dysperiodicity trace. The comparison of the present speech frame to frames to the left and right guarantees that it is compared at least once to a frame that belongs to the same phonetic segment, provided that the segment is at least two vocal cycles long. The selection of the minimum prediction error to the left or right so removes predictions that are performed across phonetic boundaries. Cross-boundary prediction errors must be discarded, because cycle-to-cycle differences owing to the evolving phonetic identity dwarf cycle discrepancies that are due to vocal noise.

Order M is typically equal to 1 or 2. The purpose of including more than one prediction coefficient is the expected reduction of quantization noise. Indeed, lag T is an integer, whereas the vocal cycle lengths are likely to be equal to a non-integer number of sampling steps. Lags T in relations (1) are determined for each analysis frame either by an exhaustive search for the minimum error or by means of the empirical inter-correlation between present and lagged frames. In the case of the latter, lag T is assigned to the position, within the open lag interval, for which the inter-correlation function is a maximum.

3 Problems and Solutions

Results show that methods proposed in [1] as well as [2] enable computing markers of vocal noise that are plausible and that co-vary with the degree of perceived hoarseness of sustained vowels or connected speech. This article is devoted to methodological issues that are raised by these proposals, as well as to their solutions.

3.1 Burg's Rule

Multi-step predictive analyses have been implemented by means of lattice filters, the coefficients of which obey Burg's rule [3]. That is, the filter coefficients are determined by means of the harmonic mean of unilateral forward and backward prediction errors, a choice that guarantees filter stability. A consequence is that the filter may be unable to track rapid signal onsets faithfully. Transients may therefore give rise to prediction errors that are higher than the prediction errors that one would obtain by means of unstable filters.

Owing to the bilateral analysis, however, this is likely to be a problem only when a rapid signal boost ends or a rapid signal drop starts at a phonetic boundary. When no risk of cross-boundary prediction is involved, the bilateral analysis turns the prediction of onsets into the retro-diction of offsets and vice versa.

Be that as it may, in the framework of clinical applications linear multi-step prediction is carried out for analysis purposes only. Filter stability is therefore not an issue and can be omitted in favour of a direct form implementation the coefficients of which are determined by means of the conventional covariance method, for instance.

3.2 Lattice Filter Implementation

When more than one multi-step prediction coefficient is involved, the prediction error obtained by a lattice filter comprises several recent as well as several distant speech samples. For instance, when order M is equal to 1, the lattice filter output is the following [3].

$$e_{left}(n) = s(n) + c_T c_{T-1} s(n-1) + c_{T-1} s(n-T+1) + c_T s(n-T).$$

(2)

Symbol T is the prediction distance in number of samples, $s(n)$ is the nth speech sample and c_j are the lattice-filter coefficients. Sample $s(n-1)$ in error (2) obscures the conceptual simplicity of relations (1) and upsets the straightforward interpretation of the multi-step prediction error as a measure of vocal dysperiodicity. The intercalation of additional recent samples is typical of the lattice filter implementation and can be avoided in the framework of implementations that are direct or involve single coefficients only.

3.3   Multiple Prediction Coefficients

Relations (1) may involve multiple prediction coefficients. A consequence is that the present speech sample is compared to a weighted sum of distant speech samples. The goal is to decrease quantization noise. A sample-by-sample comparison by means of a single-coefficient multi-step prediction would be easier to interpret, however, given the overall objective, which is to estimate vocal dysperiodicities.

A solution consists in decreasing quantization noise by over-sampling first and replacing the multiple coefficients by a single one. This removes the risk of decreasing genuine vocal noise via the weighted sum that is involved in the distant prediction.

3.4   Multi-step Linear Predictive Analysis as a Paradigm for the Analysis of Vocal Dysperiodicities

This section addresses a basic issue, which is the adequacy per se of the multi-step prediction paradigm as a framework for analyzing vocal dysperiodicities. Hereafter, one assumes that the multi-step prediction involves a single coefficient the value of which is determined by means of the conventional covariance method. The conclusions are valid, however, for any implementation of the multi-step predictive analysis filter.

The covariance method consists in minimizing the energy of the prediction error cumulated over a rectangular frame of length N. When a single coefficient is involved, one easily shows that the (unilateral) multi-step prediction error is equal to the following.

$$E = \sum_{n=1}^{N} \left[s(n) - s(n-T) \frac{\sum_{i=1}^{N} s(i)s(i-T)}{\sum_{i=1}^{N} s^2(i-T)} \right]^2. \tag{3}$$

From (3) follow solutions (4). Parameter b is a positive gain that is constant over the analysis frame. It demonstrates that the prediction coefficient in (3) automatically compensates for slow variations of the vocal amplitude.

$$E = 0 \quad \begin{cases} s(n) = +bs(n-T), \\ s(n) = -bs(n-T). \end{cases} \tag{4}$$

Solutions (4) show that, formally, the multi-step prediction error is not a measure of vocal dysperiodicity. The reason is parasitic solution $s(n) = -bs(n-T)$. For a sinusoid of period T, for instance, solutions (4) suggest that the multi-step prediction error is a minimum for shifts $T/2$ and T, of which only the latter has an interpretation in terms of the period of the sinusoid. In practice, this means that an exhaustive search for optimal shift T is likely to produce erroneous measures of dysperiodicity for phonetic segments that are quasi-sinusoidal, i.e. voiced plosives, for example.

Determining optimal shift T by means of the empirical inter-correlation between present and lagged frames is less likely to give rise to parasitic solutions. The reason is that the optimal shift is assigned to the lag for which the inter-correlation is a maximum. Formally, the removal of parasitic solutions is not guaranteed, however.

Moreover, the interpretation of error E remains ambiguous even when parasitic solutions are discarded. Because of the inter-correlation that is involved in (3), error E is a measure of signal dysperiodicity only when the vocal noise is feeble. The prediction error turns into a measure of signal energy when the vocal noise is strong (Table 1).

3.5 Generalized Variogram

A possible alternative is based on the observation that for a periodic signal $s(n)$, the following expression is expected to be true for any shift T that is an integer multiple of the signal period, assuming that the quantization noise can be neglected.

$$\sum_{n=-\infty}^{n=+\infty}[s(n) - s(n-T)] = 0. \tag{5}$$

In practice, voiced speech segments are locally-periodic at best, speech cycle amplitudes are expected to evolve slowly and the glottal cycle length is not known a priori. This suggests analyzing the signal frame by frame, squaring expression (5), and inserting a positive gain g that is constant over the analysis frame.

$$V(T) = \sum_{n=0}^{N}[s(n) - gs(n-T)]^2. \tag{6}$$

When gain $g = 1$, cumulated difference (6) is known as the empirical variogram of signal $s(n)$. Length N fixes the frame length. The squaring guarantees that difference (6) is a minimum for lags that are integer multiples of the period of the signal.

Gain g enables neutralizing drifts of the signal amplitude that are due to onsets, offsets or prosody. Gain g is chosen so that it is always positive and the interpretation of generalized variogram $V(T)$ is the same whatever the strength of the vocal noise. A definition of g that satisfies these criteria equalizes the signal energies in the present and lagged analysis frames.

$$V(T) = \sum_{n=0}^{N}\left[s(n) - s(n-T)\sqrt{\frac{\sum_{i=0}^{N}s^2(i)}{\sum_{i=0}^{N}s^2(i-T)}}\right]^2. \tag{7}$$

Table 1. Variogram (7) and multi-step prediction error (3) for periodic, odd-periodic and white noise signals

	white noise	$s(n) = -bs(n-T), b > 0$	$s(n) = bs(n-T), b > 0$
V	$\Sigma[s(n)-s(n-T)]^2$	$4\Sigma s^2(n)$	0
E	$\Sigma s^2(n)$	0	0

Inspecting multi-step prediction error (3) and generalized variogram (7) suggests that they are proportional when $s(n)$ is approximately equal to $s(n-T)$. Otherwise, they are different. Table 1 summarizes the values of expressions (3) and (7) when, for example, $s(n) = bs(n-T)$, $s(n) = -bs(n-T)$, $b > 0$, as well as when $s(n)$ is white noise.

One sees that generalized variogram V is different from zero when the signal is odd-periodic and lag T equal to the odd-period. Also, expression V is the cumulated squared difference between the present and lagged signal samples, whether the signal is deterministic or stochastic. The minimum of V is therefore a measure of signal dysperiodicity in the analysis frame.

On the contrary, the multi-step prediction error E is zero when the signal is periodic or odd-periodic and lag T equal to the period or odd-period. Also, error E is the cumulated squared difference between the present and lagged signal samples only when they are (strongly) correlated. When they are uncorrelated, error E is the signal energy. Error E is therefore a measure of signal (un)-predictability. Because predictability is a more general property than periodicity, variogram V and error E only agree for special instances of signals and lags.

4 Methods

The experimental part of the study involves seven analysis methods, which are listed in Table 2. The objective is to investigate whether issues that are discussed above give rise to statistically significant differences in the vocal dysperiodicity traces. For each method, the length of the rectangular analysis frame was equal to 2.5 milliseconds [2]. The analysis frames were non-overlapping, but contiguous. Prediction lag T was assigned to the position of the maximum of the inter-correlation between present and lagged frames or, when appropriate, to the position of the minimum of the variogram. The prediction lag was requested to be within an interval between 2.5 and 20 milliseconds. This interval includes the phonatory cycle lengths that are typical of male and female speakers. Per frame, each analysis method was applied twice, once for positive and once for negative lag values, and the minimum prediction error or variogram-determined signal difference was kept and assigned to the vocal dysperiodicity trace.

For several analyses, the speech signals, inter-correlation function or variogram were interpolated linearly or parabolically. The purpose was to test the use of non-integer prediction lags.

4.1 Analysis Methods

Table 2. Characteristics of analysis methods

Label	Analysis method	Nber of coefficients	Interpolation
1	Burg, covariance-lattice	3	no
2	covariance	1	no
3	covariance	3	no
4	covariance	1	linear
5	covariance	1	parabolic
6	variogram	n.a.	no
7	variogram	n.a.	linear

4.2 Corpora

The corpora have been sinusoids; as well as vowels and short sentences produced by normophonic or dysphonic speakers. Sinusoids as well as speech signals have been sampled at 20 kHz. The sinusoids have been contaminated by additive or frequency modulation noise. The purpose was to test interpolation with a view to reducing quantization noise.

The speech corpus comprised sustained vowels [a] and two French sentences spoken affirmatively by 22 normophonic or dysphonic, male or female speakers. The sentences were "le garde a endigué l'abbé" (S1) and "une poule a picoré ton cake" (S2). All phonetic segments in sentence S1 are voiced by default, whereas sentence S2 comprises voiced as well as unvoiced phonetic segments. The sentences are matched grammatically and comprise the same number of syllables. Strident fricatives were omitted on purpose.

4.3 Noise Marker

The vocal dysperiodicity trace $e(n)$ is summarized by means of a signal-to-dysperiodicity ratio (SDR) that is defined as follows [1]. Symbol I is the number of samples in the total analysis interval.

$$SDR = 10 \log \frac{\sum_{i=1}^{I} s^2(i)}{\sum_{i=1}^{I} e^2(i)}. \tag{7}$$

Table 1 shows that $SDR \rightarrow 0$ when the signal is white noise and analyzed by means of multi-step prediction. On the contrary, $SDR \rightarrow -3$ dB when the signal is white noise and analyzed by means of the generalized variogram. The reason is that variogram (6) is the cumulated squared difference between present and lagged samples. Prediction error (3) is, on the contrary, equal to a cumulated squared difference between present and lagged samples only when the signal is periodic or pseudo-periodic.

5 Results and Discussion

5.1 Sinusoidal Signals

Analyses of sinusoids confirm that dysperiodicity traces obtained by single-coefficient multi-step predictive or variogram analyses may be altered by quantization noise. *SDR* values of clean sinusoids sampled at 20 kHz were typically comprised in the interval 30 – 40 dB when the sampling frequency was not an integer multiple of the frequency of the sinusoid.

Non-integer lags, determined via interpolation, have been shown to increase the distance between vocal and quantization noise. Simulations suggest that interpolation moves the *SDR* values of sampled clean sinusoids to values greater than 65 dB.

5.2 Sustained Vowels and Running Speech

Table 3 summarizes the quartiles of the *SDR* values (in dB) obtained for a corpus of sustained vowels [a], including onsets and offsets, and sentences *S1* and *S2* spoken by 22 speakers. The labels of the analysis methods agree with the labels given in Table 2. The *SDR* values have been rounded to the nearest decimal after the comma.

For each speech corpus, a single-factor repeated measures analysis of variance of the *SDR* values has been carried out to check whether differences between methods 1 to 7 are statistically significant. Subsequently, methods have been compared pair-wise by means of paired *t*-tests. The levels of significance of the individual tests have been adjusted by means of Bonferroni's correction to fix to *0.05* the overall level of significance of a total of *21* pair-wise comparisons [5]. Statistical analyses of the data show the following.

   a) For vowel [a], the analysis of variance shows that the inter-method differences are statistically significant ($F = 249$, $p < 0.001$). Out of the *21* pair-wise comparisons, *17* are statistically significant. Of these, all involve differences between analysis methods (covariance lattice, covariance of order 0 or 2, variogram).

Table 3. Quartiles of the *SDR* values (in dB) obtained for a corpus of sustained vowels [a], including onsets and offsets, and sentences *S1* and *S2* spoken by 22 speakers

	Method label	1	2	3	4	5	6	7
	First quartile	23.5	16.8	17.0	16.7	16.7	16.7	16.7
[a]	Median	26.7	20.2	20.6	20.3	20.4	20.1	20.2
	Third quartile	28.7	22.4	22.9	22.6	22.8	22.4	22.6
	First quartile	19.5	14.4	14.6	14.4	14.4	14.2	14.2
S1	Median	22.3	17.2	17.4	17.2	17.1	17.2	17.1
	Third quartile	24.6	18.1	18.4	18.1	17.9	18.0	18.0
	First quartile	19.0	16.7	17.2	16.6	15.7	16.5	16.8
S2	Median	22.6	18.3	18.5	18.2	17.7	18.1	18.1
	Third quartile	24.4	19.7	20.0	19.4	19.2	19.6	19.4

b) For sentence *S1*, the analysis of variance shows that the inter-method differences are statistically significant ($F = 129, p < 0.001$). Of the *21* pair-wise comparisons, *16* are statistically significant. Of these, *15* pairs involve differences between analysis methods (covariance-lattice, covariance of order 0 or 2, variogram). One pair differs by the interpolation method (linear versus parabolic).

c) For sentence *S2*, the analysis of variance shows that the inter-method differences are statistically significant ($F = 67, p < 0.001$). Of the *21* pair-wise comparisons, *15* are statistically significant. Of these, all involve differences between analysis methods (covariance-lattice, covariance of order 0 or 2, variogram).

Results therefore suggest that different analysis methods cause *SDR* values to differ statistically significantly. Possible explanations are the following.

a) The covariance-lattice implementation (Table 3, column 1) implicates running averages of the recent as well as distant samples. The original purpose of involving several prediction coefficients has been the decrease of quantization noise. Results suggest that multiple prediction coefficients decrease genuine vocal noise as well as quantization noise.

   Also, the lattice filter is stable. Stability would let one expect a boost of the prediction error because of an increased difficulty in tracking rapid transients. This is not observed. This would suggest that either the corresponding error increase is masked by the decrease of genuine vocal noise owing to local averaging (2), or by the bilateral analysis (1) that turns onsets into offsets.

b) The 3-coefficient covariance method (Table 3, column 3) involves a running average of the distant samples only. The original purpose has been the decrease of quantization noise. Single-coefficient covariance analyses omit this local smoothing. As a consequence, single-coefficient (column 2) and 3-coefficient (column 3) covariance analyses give rise to *SDR* values that differ statistically significantly. Inspecting data averages suggests that the corresponding *SDR* values typically differ by less than 1 dB. The difference is due to a decrease of the genuine vocal noise by local averaging rather than to a decrease of the quantization noise.

c) The variogram (Table 3, columns 6 and 7) involves an energy-normalisation coefficient the mathematical properties of which differ from those of the prediction coefficients implicated in methods labelled 1 to 5. Consequently, *SDR* values obtained by variogram and linear predictive analyses differ statistically significantly. Inspection of the data averages suggests, however, that *SDR* values obtained via 1-coefficient covariance and variogram analyses typically differ by less than 1 dB. Simulations indeed show that variogram and 1-coefficient linear predictive analyses give comparable *SDR* values as long as these are greater than roughly 10 dB [4].

Statistical analyses show that interpolation does not cause the *SDR* values to increase statistically significantly for a same analysis method. The purpose of interpolation is to decrease quantization noise. Inspecting data averages suggests that *SDR* differences owing to interpolation are typically less than 0.1 dB. A possible explanation is that, in the absence of interpolation, the *SDR* ceiling owing to quantization noise is in the vicinity of 30 dB. Therefore, quantization noise is negligible compared to vocal noise in signals the *SDR* value of which is typically 17 dB.

6 Conclusion

Implementations of linear predictive analyses that are considered to be optimal for speech coding are sub-optimal for clinical applications and vice versa. For clinical applications, the recommended implementation would involve a single prediction coefficient the value of which is fixed by means of a conventional covariance method. Interpolation or over-sampling would be the preferred method for decreasing quantization noise. Moreover, the presentation shows that multi-step prediction is not a paradigm that would enable interpreting under all circumstances the prediction error as a trace of the vocal dysperiodicity. The generalized variogram of the speech signal is an alternative that does not admit any ambiguity in interpretation.

References

[1] Qi Y., Hillman R. E., and Milstein C. (1999) "The estimation of signal to-noise ratio in continuous speech for disordered voices,"J.Acoust. Soc. Am. 105, 4, 2532–2535.
[2] Bettens F., Grenez F. and Schoentgen J. (2005) "Estimation of vocal dysperiodicities in disordered connected speech by means of distant-sample bidirectional linear predictive analysis", J. Acoust. Soc Am., 117, 1, 10 pp.
[3] Ramachandran R., and Kabal P. (1989) "Pitch prediction filters in speech coding,"IEEETrans. Acoust., Speech, Signal Process. 37, 4, 467–478.
[4] Dessalle, E. (2004) "Estimation en ligne des dispériodicités vocals dans la parole connectée", unpublished Master Thesis, Université Libre de Bruxelles, Bruxelles.
[5] Moore D., McCabe G. (1999) "Introduction to the practice of statistics", Freeman, New York.

Optimal Size of Time Window in Nonlinear Features for Voice Quality Measurement

Jesús B. Alonso[1], Fernando Díaz-de-María[2], Carlos M. Travieso[1],
and Miguel A. Ferrer[1]

[1] Dpto. de Señales y Comunicaciones, Universidad de Las Palmas de Gran Canaria,
Campus de Tafira, 35017 - Las Palmas de Gran Canaria, Spain
{jalonso, ctravieso, mferrer}@dsc.ulpgc.es
http://www.gpds.ulpgc.es/index.htm
[2] Dpto. de Teoría de la Señal y Comunicaciones, Universidad Carlos III de Madrid,
Avda. de la Universidad, 30, 28911 Leganes (Madrid), Spain
fdiaz@tsc.uc3m.es

Abstract. In this paper we propose the use of nonlinear speech features to improve the voice quality measurement. We have tested a couple of features from the Dynamical System Theory, namely: the Correlation Dimension and the largest Lyapunov Exponent. In particular, we have studied the optimal size of time window for this type of analysis in the field of the characterization of the voice quality. Two systems of automatic detection of laryngeal pathologies, one of them including these features, have been implemented with the purpose of validating the usefulness of the suggested nonlinear features. We obtain slight improvements with respect to a classical system.

1 Introduction

The medical community uses subjective techniques (evaluation of the voice quality by the specialist doctor's direct audition) or invasive methods (which allow the direct inspection of vocal folds thanks to the use of laryngoscopical techniques) for the evaluation and the diagnostic of voice pathologies. The voice quality measurement has received much attention during the last decade ([2] [3] [4] [5] are good examples). These systems allow us to quantify the voice quality effectively and to document the patient's evolution using objective measures. These techniques provide the ability to detect quickly and simply laryngeal pathologies; thus they can be applied in preventive medicine and telemedicine environments.

On the other hand, automatic laryngeal pathologies detection systems have been developed [6] [7] [8] [9]. In these works, different success rates are obtained in the classification between healthy voices and pathological voices, being evaluated each system with different data bases, since a data base of reference does not exist.

In [1], the authors proposed a classification system to distinguish healthy from pathologic voices using a Neuronal Networks (NN). In the feature extraction phase, diverse measures based on the High Order Statistics (HOS) were used in addition to a

M. Faundez-Zanuy et al. (Eds.): NOLISP 2005, LNAI 3817, pp. 206–218, 2005.
© Springer-Verlag Berlin Heidelberg 2005

selection of classical voice quality measurements present in the current literature. These measurements of the voice quality based on the HOS achieve good results, but in exchange for a high computational cost.

In this work, the viability of the nonlinear dynamic-based speech analysis has been studied with the purpose of obtaining information on the voice signal nonlinear behavior. The tested nonlinear features are less computationally demanding than HOS-based ones. The viability of characterizing the voice signal by means of the Lyapunov Exponents has been already suggested in other works [10] [11] . In another paper [12] , the utility of the correlation dimension to detect the presence of laryngeal pathologies has also been proposed. However, different aspects of these measurements are explored, for example, the optimal size of the time window. Some preliminary results on this topic are presented in this work.

2   Nonlinear Dynamical System: The Embedding Theorem

The Chaos Theory can be used to gain a better understanding and interpretation of observed complex dynamical behaviour. Besides, It can give some advantages in predicting or controlling such time evolution 13.

Deterministic dynamical systems describe the time evolution of a system in some state space $\Gamma \subset R^d$. Such an evolution can be described case by ordinary differential equations:

$$\dot{x}(t) = F(x(t)) \tag{1}$$

or in discrete time $t = n\Delta t$ by maps of the form:

$$x_{n+1} = F(x_n) \tag{2}$$

Unfortunately, the actual state-vector only can be inferred for quite simple systems, and as anyone can imagine, the dynamical system underlying the speech production process is very complex. Nevertheless, as established by the "embedding theorem" 14, it is possible to reconstruct a state space equivalent to the original one. Furthermore, a state-space vector formed by time-delayed samples of the observation (in our case, the speech samples) could be an appropriate choice:

$$\mathbf{s_n} = [s(n), s(n-T), \ldots, s(n-(d-1)T)]^t \tag{3}$$

where $s(n)$ is the speech signal, d is the dimension of the state-space vector, T is a time delay and t means transpose.

Finally, the reconstructed state-space vector dynamic, $\mathbf{s_{n+1}} = F(\mathbf{s_n})$, can be learned through either local or global models, which in turn will be polynomial mappings, neural networks, etc.

2.1   Correlation Dimension

The correlation dimension D_2 gives an idea of the complexity of the dynamics. A more complex system has a higher dimension, which means that more state variables are needed to describe its dynamics. The correlation dimension of a random noise is not bounded while the correlation dimension of a deterministic system yields a finite value. The correlation dimension can be obtained as follows:

$$D_2 = \lim_{r \to 0} \lim_{N \to \infty} \frac{\log C(N,r)}{\log r} \tag{4}$$

with C(N,r) being,

$$C(N,r) = \frac{1}{N(N-1)} \sum_{i=1}^{N} \sum_{\substack{j=1 \\ i \neq j}}^{N} \theta(r - \|X_i - X_j\|) \tag{5}$$

where r is the radius around X_i and $\theta(x)$ is the step function. Equation [4] converges very slowly as r tends to zero. To circumvent this problem, the local slope can be estimated:

$$D_2 = \frac{d \log C(N,r)}{d \log r} \cong \lim_{\Delta \log r \to 0} \frac{\Delta \log C(N,r)}{\Delta \log r} \tag{6}$$

When the length N is significantly large, D_2 will converge with the increase of the embedding dimension, m.

2.2 The Largest Lyapunov Exponent

Chaotic behaviour arises from the exponential growth of infinitesimal perturbations. This exponential instability is characterized by the Lyapunov exponents. Lyapunov exponents are invariant under smooth transformations and are thus independent of the measurement function or the embedding procedure.

The largest Lyapunov exponent can be determined without the explicit construction of a model for the time series. It considers the representation of the time series as a trajectory in the embedding space, and assume that you observe a very close return $s_{n'}$ to a previously visited point s_n. Then one can consider the distance $\Delta_0 = s_n - s_{n'}$ as an small perturbation, which should grow exponentially in time. Its evolution can be followed from the time series: $\Delta_l = s_{n+l} - s_{n'+l}$. If one finds that $|\Delta_l| \approx \Delta_o e^{\lambda l}$, λ is the largest Lyapunov exponent.

3 New Voice Disorder Parameterisation

In the current literature, some works suggest the viability of characterizing the voice signal by means of the Lyapunov Exponents (for example in synthesis of phonemes 10 and 11), and characterizing the voice disorder signal by means of the correlation dimension 12.

For the study of the presence of laryngeal pathologies based on the voice recording, it is very common to use recordings of sustained vowels.

In this work, we have studied which is the optimal size of the time window for the nonlinear analysis (Correlation Dimension and the Largest Lyapunov Exponent), with the purpose of deciding whether a vowel utterance comes from a healthy or a pathological voice. Different sizes of time window has been studied: 10, 30, 50, 100, 150, 300, 500 ms or the whole vowel utterance and pitch-synchronous segments of 3, 5, 7, 10 pitch periods.

In the case of obtaining multiple frames for each vowel, the following parameters have been extracted for each feature (Correlation Dimension and the Largest Lyapunov Exponent):

- The mean value of the feature P for the different frames $\{T_i\}_N$:

$$M_P = \frac{1}{N}\sum_{i=1}^{N} P_i \tag{7}$$

- Variation of the value of the feature along the time:

$$V_P = \frac{1}{(N-1)}\frac{1}{\max\{P_i\}}\sum_{i=2}^{N}|P_i - P_{i-1}| \tag{8}$$

3.1 Voice Database

The voice signals used in this study were digitalized with a sample frequency of 22050 Hz and 16 bits per sample. The speaker's voice was recorded with a conventional sound card and a basic microphone. The database consists of 100 voices of healthy speakers and 68 voices of pathological speakers. Each sample of the database is composed by the five Spanish vowels ('a', 'e', 'i', 'o' and 'u') pronounced in a sustained way by the speakers during approximately two seconds for each vowel. In case of pathological speakers there are situations of vocal folds without lesion (hypofunction, hyperfunction, vocal fold paralysis,...) and vocal folds with lesion (carcinoma, vocal folds nodule, sessile polyp, pedunculated polyp, Reninke's edema, adult papiloma,...). The database has been created contemplating different disphonia levels: "light pathological voice", "moderate pathological voice" and "severe pathological voice."

3.2 Evaluation of the Parameterization

The attractor dimension has been fixed to 2 since the result obtained does not justify the increment of the time consuming, and the delay, T, has been estimated to 8 samples.

A one-second interval, located in the centre of the utterance, has been studied. This alteration has been carried out with the purpose of eliminating the beginning and end of the phonation, because it presents a transitory character. This modification has been implemented except when the whole vowel is used.

Four different attributes have been studied:

- *Atrib1*: Mean value of the Correlation Dimension.
- *Atrib2*: Time Variation of the Correlation Dimension values.
- *Atrib3*: Mean value of the Largest Lyapunov Exponent.
- *Atrib4*: Time Variation of the Largest Lyapunov Exponent values.

A neural network has been used to evaluate the benefits of the different attributes in the environment of the automatic pathologies detection. Each attribute has separately been evaluated using neural network Multilayer feedforward with 2 hidden layers, with Backpropagation train algorithm. Different sizes of asynchronous time window have been evaluated of using like evaluation function the success rate in the classification. Each attribute has been evaluated separately, differentiating between the five vowels. The different sizes of asynchronous time window are: 10, 30, 50, 100, 150, 300, 500 milliseconds and the whole vowel utterance ('full' in the figures). The result is showed in the Figures 1, 2, 3 and 4.

The asynchronous time window has a disadvantage: because the vibration frequency of the vocal folds (picth) of the women is greater than the men, for a certain temporal window is obtained different number of periods between men and women. In order to be able to make the parameterization process independent of the pitch frequency, it is possible dividing the vowel in pitch-synchronous segments of 3, 5, 7, 10 pitch periods (To). The result is showed in the Figures 5, 6, 7 and 8.

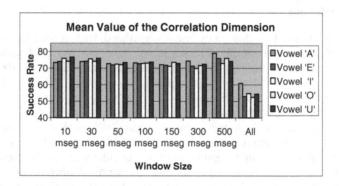

Fig. 1. Results of the study about size of asynchronous time window for "mean value of the Correlation Dimension"

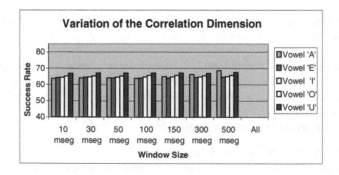

Fig. 2. Results of the study about size of asynchronous time window for "Time Variation of the Correlation Dimension values"

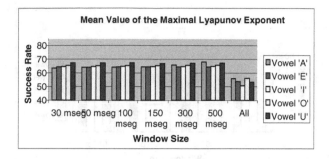

Fig. 3. Results of the study about size of asynchronous time window for "mean value of the Maximal Lyapunov Exponent"

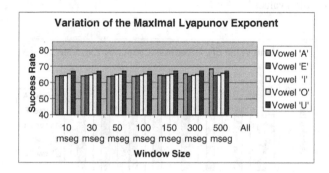

Fig. 4. Results of the study about size of asynchronous time window for "Time Variation of the Maximal Lyapunov Exponent values"

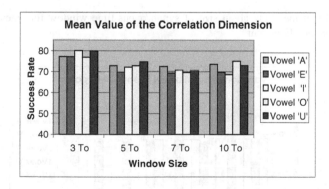

Fig. 5. Results of the study about size of synchronous time window for "mean value of the Correlation Dimension"

To sum up, it is observed better results dividing the vowel in pitch-synchronous segments of 3 pitch periods. It is also observed better results for the attribute "mean value of the Correlation Dimension", during the individual evaluation of the parameter.

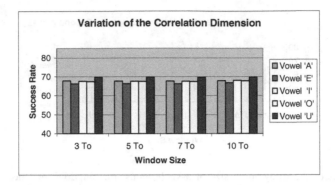

Fig. 6. Results of the study about size of synchronous time window for "Time Variation of the Correlation Dimension values"

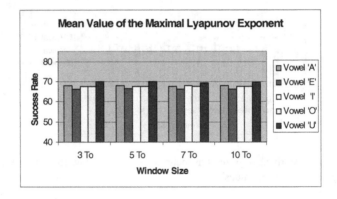

Fig. 7. Results of the study about size of synchronous time window for "mean value of the Maximal Lyapunov Exponent"

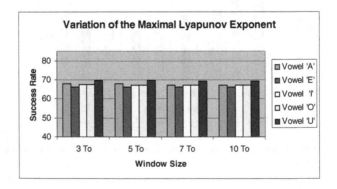

Fig. 8. Results of the study about size of synchronous time window for "Time Variation of the Maximal Lyapunov Exponent values"

4 Detector Model

The voice automatic classification system allows us to discriminate healthy voices from pathological ones. It is based on a pattern recognition model.

These systems are typically structured in three steps, namely: "Voice Acquisition", "Parameterization" and "Classification". The proposed automatic laryngeal pathologies recognition system follows this structure, illustrated in figure 9. Firstly, it captures the speaker's voice using a sound card and a microphone. The parameterization step uses parameters presented in [1], where a combination of a selection of parameters exposed in the current literature with new parameters based on Higher Order Statistics (HOS) was exposed. Finally, a net of classifiers based on Neural Networks (NN) is used to classify between healthy and pathological voices.

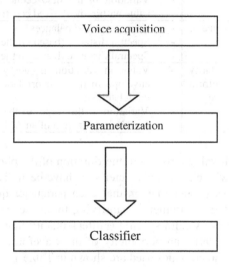

Fig. 9. Pattern recognition model

4.1 Parameterization

The parameterization step uses the same parameters that has been used in [1], where the authors made a selection of 17 parameters for the laryngeal pathologies classification (in the rest of the paper "classic parameters"), among multiple voice quality characterization parameters well-known in the literature.

4.1.1 Classic Parameters
There is no parameter which is completely conclusive in the detection of laryngeal pathologies, because each pathology affects the voice in a different way. For example, there are pathologies that present a great content of non-stationary noise in the high frequency components. On the other hand, other pathologies are characterised by the

Table 1. Classic characteristics

Group of characteristics	Name of the attributes
Quantifying the variation in amplitude (shimmer)	- Variation in the mean quadratic value of each voice frame - Variation in the highest value of the short time cross correlation function of each voice frame
Quantifying the presence of unvoiced frames	- Relationship between the number of unvoiced frames and the total number of frames of the sample voice - The unvoiced periodicity index of a sample voice
Quantifying the absence of wealth spectral (Hitter)	- Variation of pitch energy cepstral measure - Variation in the first harmonic value in the derived cepstrum domain - Variation in the first/second harmonic relationship value within the derived cepstrum domain
Quantifying the presence of noise	- Energy spectral balances - Spectral distance (based on the spectral module) - Spectral distance (based on the spectral phase)
Quantifying the regularity an periodicity of the waveform of a sustained voiced voice	- Value an variation in energy of the slope of the envelope in the autocorrelation function of an AM modulated signal - Variation of the slope of the envelope in the auto-correlation function of an AM modulated signal

uncertainty of the pitch value throughout the duration of the phonation of a sustained voiced sound. This is why classical characteristics have been divided into five groups depending on the physical phenomenon that each parameter quantifies: quantifying the variation in amplitude (shimmer), quantifying the presence of unvoiced frames, quantifying the absence of wealth spectral (Hitter), quantifying the presence of noise and quantifying the regularity an periodicity of the waveform of a sustained voiced voice. All the classic characteristics used are shown in Table 1.

4.1.2 New Nonlinear Parameters
In this work the possibility of using nonlinear features with the purpose of detecting the presence of laryngeal pathologies has been explored. The four measures proposed will be used: mean value and time variation of the Correlation Dimension and mean value and time variation of the Maximal Lyapunov Exponent values.

4.2 Classifier

The proposed system is based on the use of a net of classifiers, where each one discriminates frames of a certain vowel. Combinational logic has been added to evaluate the success rate of each classifier.

The structure of the proposed classifier is similar to the one proposed in [1], and represented in figure 10. Five NN–based vowel classifiers have been used to discriminate between healthy and pathological vowels, one for each vowel. The inputs of each vowel classifier are the feature vectors of the sequence of frames in which the

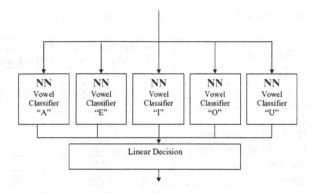

Fig. 10. Classification System Scheme

vowel corresponding to this classifier has been divided. The length of each frame is a three pitch period, for a voiced sound, or 30 ms if it is an unvoiced sound (typical of the pathological voices). Only the 500 central milliseconds of the vowel have been considered to avoid considering the frames that exhibit non-stationary behaviour at the beginning and end of each vowel.

The dependence of the parameters on the analysed vowel has been taken into account, as pointed out by Jacques Koreman and Manfred Pützer [15]. Consequently, a "vowel classifier" has been used for each vowel, such as is shown in figure 10.

First of all, each "vowel classifier" emits an estimation dependent on whether the analysed vowel is related to a "healthy vowel" or to "pathological vowel". Secondly, the results of the different vowel classifiers are evaluated by means of an "output logic".

In each "vowel classifier", the different voice frame are evaluated in two neural networks, and an assessment is emitted: "healthy frame" or "pathological frame". If 70% or more of the frames correspond to healthy frame, the analysed vowel will be labelled as a "healthy vowel", otherwise it will be labelled as a "pathological vowel." The scheme of a vowel classifier is shown in figure 11. In this study, normalized data (zero-mean and variance one) have been used.

The characteristics of the Neural Network are described in the table 2.

The output logic will indicate that the voice sample corresponds to a "pathological voice" if two or more vowels are classified as "pathological vowels", whereas the voice sample will be classified as a "healthy voice" if only one vowel or none of them are classified as "pathological vowels".

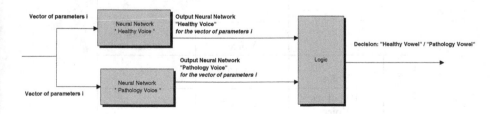

Fig. 11. Classification System Scheme

Table 2. Characteristics of the Neural Network

	Characteristic	Neural Network "Healthy Voice"	Neural Network "Pathological Voice"
Topology	Number of Layer	2	2
	Number of inputs	Number of parameters	Number of parameters
	Number of neurons in the first layer	40	40
	Number of outputs	50	50
Training	Maximum Threshold of absolute error	0.01	0.01
	Maximum Threshold of relative error	0.015	0.015
	Maximum number of epochs	10000	10000
	Training method	Back-propagation	Back-propagation
	No linear function	Hidden layer: "tansig"	Hidden layer: "tansig"
		Output layer: "purelin"	Output layer: "purelin"

5 Results

Two systems have been compared using the same data base. The first one, only works with the "classic parameters", while the second one uses both "classic" and "nonlinear" parameters, obtaining the results displayed in Tables 2 and 3.

A global success rate of 91,77% is obtained using "classic parameters", whereas a global success rate of 92,76% using "classic parameters" and "nonlinear" parameters. These results show the utility of new parameters.

Table 3. Success Rate using "Classic characteristics"

		Input	
		Healthy Voice	Pathological Voice
Output	Healthy Voice	95.65 %	12.10 %
	Pathological Voice	4.35 %	87.90 %

Table 4. Success Rate using "Classic characteristics + New parameters"

		Input	
		Healthy Voice	Pathological Voice
Output	Healthy Voice	96.12 %	10.60 %
	Pathological Voice	3.88 %	89.40 %

6 Conclusions

In this work, the possibility of using nonlinear features to improve the performance of an automatic detector of laryngeal pathologies has been explored.

Two features have been tested: Correlation Dimension and the Largest Lyapunov Exponent. In particular, the system works with their mean value and variation.

An experimental study aiming at selecting the best size for the time window of the nonlinear analysis has been conducted, concluding that the best option is using a pitch-synchronous window containing three periods.

Finally, the results of the classification system including the mean value and variation of the correlation dimension are slightly better than those achieved by the system using only the classic parameters.

Though the improvement is slight, we consider it an encouraging result, since the research is currently in the first stages. Further work is necessary in diverse directions.

Acknowledgement

This work was partially supported by the Spanish government (TIC2003-08956-C02-02).

References

1. J.B Alonso, J. de Leon, I. Alonso, M.A. Ferrer: Automatic detection of pathologies in the voice by HOS based parameters, *Proc. EUROASIP Journal on Applied Signal Processing*, (2001), vol.1, 275-284.
2. M. Fröhlich, D.Michaelis, H.W. Srube: Acoustic 'Breathiness Measures' in the description of Pathologic Voices, *Porc. ICASSP-98*, Seattle, WA, (1998), vol.2, 937-940.
3. L.Gavidia, J. Hansen: Direct Speech Feature Estimation Using an Iterative EM Algorithm for Vocal Fold Pathology Detection, *Porc. IEEE Transactions on Biomedical Engineering,* (1996), vol.43, no.4, 373-383.
4. S. Feijoo, C. Hernandez, A. Carollo, R.C Hermida, E.Moldes: Acoustic Evaluation of glottal cancer based on short-term stability measures, *Porc. IEEE Engineering in Medicine & Biology Society 11th Annual Internatioonal Conference*, (1989), vol. 2, 675-676.
5. B.Boyanov, S.Hadjitodorov Ivanov: Analysis of voiced speech by means of bispectrum, *Electronics Letters*, (1991), vol. 27, no. 24, 2267-2268.
6. B. Boyanov, S.Hadjitodorov: Acoustic analysis of pathological voices: a voice analysis system for screening of laryngeal diseases, *Proc. IEEE Engineering in Medical and Biology*, (1997), vol. 16, no. 4, 74-82.
7. J.H.L Hansen, L. Gavidia, F. James: A nonlinear operator-based speech feature analysis method with application to vocal fold pathology assessment, *Proc. IEEE Transactions on Biomedical Engineering* (1998), vol. 45, no. 3, 300-313.
8. M.O. Rosa, J.C. Pereria, C.P.L.F. Carvalho: Evaluation of neural classifier using statistics methods for identification of laryngeal pathologies, (1998), *Proc..Neural Networks*, vol.1, 220-225.
9. E. J. Wallen, J.H.L Hansen: A Screening test for speech pathology assessment using objective quality measures, *Proc. International Conference on Spoken Language Processing*, (1996), Philadelphia, PA, vol. 2, 776-779.

10. Michael Banbrook, Stephen McLaughlin, Iain Mann: Speech Characterization and Synthesis by Nonlinear Methods, *Proc.IEEE Transactions on Speech and Audio Processing*, January (1999), vol. 7, no.1.
11. V Pitsikalis, I Kokkinos, P Maragos: Nonlinear Analysis of Speech Signals: Generalized Dimensions and Lyapunov Exponents, *Proc EUROSPEECH-2003*,Geneva,(2003), 817-820.
12. J.J. Jiang,Yu Zhang: Nonlinear dynamic analysis of speech from pathological subjects, *Proc IEEE Electronics Letters*, March (2002), vol.38, no.6.
13. R. Hegger, H. Kantz, T. Schreiber: *Practical implementation of nonlinear time series methods: The TISEAN package*, CHAOS 9, 413, (1999)
14. E. Ott: Chaos in Dynamical Systems: Cambridge: Cambridge University Press, (1993).
15. Jacques Koreman and Manfred Pützer: Finding Correlates of Vocal Fold Adduction Deficiencies, *Phonus 3*, Institute of Phonetics, University of the Saarland, (1997), pp. 155-178

Support Vector Machines Applied to the Detection of Voice Disorders

Juan Ignacio Godino-Llorente[1], Pedro Gómez-Vilda[1],
Nicolás Sáenz-Lechón[1], Manuel Blanco-Velasco[2], Fernando Cruz-Roldán[2],
and Miguel Angel Ferrer-Ballester[3]

[1] Universidad Politécnica de Madrid, EUIT de Telecomunicación,
Ctra. de Valencia km. 7, 28031, Madrid, Spain
igodino@ics.upm.es
[2] Universidad de Alcalá, Escuela Politécnica, Ctra. de Madrid-Barcelona, km. 33,6,
28871, Alcalá de Henares, Madrid, Spain
[3] Universidad de Las Palmas de Gran Canaria, ETSI de Telecomunicación,
Campus de Tarifa, 35017, Las Palmas de Gran Canaria, Spain

Abstract. Support Vector Machines (SVMs) have become a popular tool for discriminative classification. An exciting area of recent application of SVMs is in speech processing. In this paper discriminatively trained SVMs have been introduced as a novel approach for the automatic detection of voice impairments. SVMs have a distinctly different modelling strategy in the detection of voice impairments problem, compared to other methods found in the literature (such a Gaussian Mixture or Hidden Markov Models): the SVM models the boundary between the classes instead of modelling the probability density of each class. In this paper it is shown that the scheme proposed fed with short-term cepstral and noise parameters can be applied for the detection of voice impairments with a good performance.

1 Introduction

Voice diseases are increasing dramatically nowadays due mainly to unhealthy social habits and voice abuse. These diseases have to be diagnosed and treated at an early stage, especially larynx cancer. Acoustic analysis is a useful tool to diagnose such diseases; furthermore, it presents two main advantages: it is a non-invasive tool, and provides an objective diagnosis, being a complementary tool to those methods based on the direct observation of the vocal folds using laryngoscopy.

The state of the art in acoustic analysis allows to estimate a large amount of long-term acoustic parameters such the pitch, jitter, shimmer, Amplitude Perturbation Quotient (APQ), Pitch Perturbation Quotient (PPQ), Harmonics to Noise Ratio (HNR), Normalized Noise energy (NNE), Voice Turbulence Index (VTI), Soft Phonation Index (SPI), Frequency Amplitude Tremor (FATR), Glottal to Noise Excitation (GNE), and many others [1-8], conceived to measure the quality and "degree of normality" of voice records. Former studies [9;10] show that the detection of voice alterations can be carried out by means of the before mentioned long-term estimated

M. Faundez-Zanuy et al. (Eds.): NOLISP 2005, LNAI 3817, pp. 219–230, 2005.

acoustic parameters, so each voice frame is quantified by a single vector. However, their reliable estimation is based on an accurate measurement of the fundamental frequency: a difficult task, especially in the presence of certain pathologies.

In the last recent years newer approaches are found using short-time analysis of the speech or electroglottographic (EGG) signal. Some of them, address the automatic detection of voice impairments from the excitation waveform collected with a laryngograph [11] or extracted from the acoustic data by inverse filtering [12]. However, due to the fact that inverse filtering is based on the assumption of a linear model, such methods do not behave well when pathology is present due to non-linearities introduced by pathology in itself.

On the other hand, it is well known that the acoustic signal itself contains information about the vocal tract and the excitation waveform as well. The basic idea for this research is to use a non-parametric approach able of modeling the effects of pathologies on both the excitation (vocal folds) and the system (vocal tract), although through the present research emphasis has been placed in pathologies affecting mainly to the vocal folds.

In this study, a novel approach to detect the presence of pathology from voice records is proposed and discussed by means of short-time parameterization of the speech signal. The automatic detection of voice alterations is addressed by means of Support Vector Machines (SVM) using non-parametric short-term Mel Frequency Cepstral Coefficients (MFCC) [13] complemented with short-term noise measurements. Each voice record is characterized with as many vectors as time frames are produced from each speech sample. The detection is carried out for each frame, and the final decision is taken establishing a threshold over the frame account classified as normal or pathological.

The present study is focused on those organic pathologies resulting in an affection of the vocal folds, which are due most of the times to vocal misuse, and reveal themselves as a modification of the excitation organ morphology (i.e. vocal folds), which may result in the increment of mass or rigidity of certain organs, thus resulting in a different pattern of vibration altering the periodicity (bimodal vibration), reducing higher modes of vibration (mucosal wave), and introducing more turbulent components in the voice record. Within this group the following pathologies can be enumerated among others: polyps, nodules, paralysis, cysts, sulcus, edemas, carcinomas, etc. [14]

2 Database

This study has been carried out using the database developed by the Massachusetts Eye and Ear Infirmary Voice and Speech Lab [15], due mainly to its availability. The speech samples were collected in a controlled environment and sampled with 16-bit resolution. A downsampling with a previous half band filtering has been done to adjust every utterance to the sampling rate of 25 kHz. The acoustic samples correspond to sustained phonations of vowel /ah/ from patients (males and females) with normal voices and a wide variety of organic, neurological, traumatic, and psychogenic voice disorders.

We have considered only a subset of all the possible files, 53 normal and 173 pathological voices, according to [16]. This decision was adopted to avoid recordings without a diagnosis and because features sex and age are uniformly distributed between the two classes. The length of normal files is around 3 seconds, whereas pathological files are around 1 second, because people with voice disorders have more difficulties to sustain a vowel during 2 or 3 s.

3 Methodology

Fig. 1 shows a block diagram describing the process set up for the detection of voice alterations. A short description of each step is presented in the following sections.

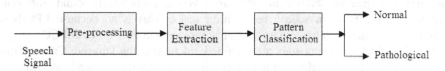

Fig. 1. Block diagram of the speech pathology detector: preprocessing front-end, feature extraction and detection module

3.1 Pre-processing

Each recording from the database contains one "target value" (class label) and several "attributes" (features). These features are calculated from short-time windows extracted from the speech utterances. The window length was selected to contain at least two consecutive pitch periods $(2 \cdot T_0)$ [17] even for the lowest fundamental frequency, so feature extraction was performed using a 40 ms Hamming windows, with an overlap of 50% between adjacent frames. The frame rate thus obtained is 50 frames/s.

As the pre-processing front-end divides the speech signals into overlapping frames, the feature extraction block produces the input vectors that will be used to train and test the classifier. The total amount of vectors generated by the system is around 16700, each one corresponding to a framed window. Around 48% of them correspond to normal voices, and the remaining 52% to pathological ones.

3.2 Parameterization

Through this approach, the detection of voice disorders is carried out by means of short-time features. For each frame the following features were extracted: a) 11 MFCCs; b) 3 noise measurements: Harmonics to Noise Ratio (HNR), Normalized Noise Energy (NNE), and Glottal to Noise Excitation Ratio (GNE); c) the energy of the frame; d) and the first temporal derivatives (Δ) extracted from each enumerated parameter. The final feature vector had dimension 30 (11 MFCCs, 3 Noise features, Energy, and 15 Δ). A brief description of these parameters is given next.

Calculation of the MFCC parameters: MFCCs have been calculated following a non-parametric modeling method, which is based on the human auditory perception system. The term mel refers to a kind of estimate related to the perceived frequency. The mapping between the real frequency scale (Hz) and the perceived frequency scale (mels) is approximately linear below 1 kHz and logarithmic for higher frequencies. The bandwidth of the critical band varies accordingly to the perceived frequency [13]. Such mapping converts real into perceived frequency and matches with the idea that a well trained speech therapist is able, most of the times, to detect the presence of a disorder just listening the speech.

MFCCs can be estimated using a parametric approach derived from Linear Prediction Coefficients (LPC), or using a non-parametric FFT-based approach. However, FFT-based MFCCs typically encode more information from excitation, while LPC-based MFCCs remove the excitation. Such an idea is demonstrated in [18], were FFT-based MFCCs are found to be more dependent on high-pitched speech resulting from loud or angry speaking styles than LPC-based MFCCs, witch were found more sensitive to additive noise in speech recognition tasks. This is so because LPC-based MFCCs ignore the pitch-based harmonic structure seen in FFT-based MFCCs.

FFT-based MFCC parameters are obtained calculating the Discrete Cosine Transform (DCT) over the logarithm of the energy in several frequency bands as in ec. 1:

$$c_m = \sum_{k=1}^{M} \log(S_k) \cdot \cos\left[m \cdot (k - 0.5) \cdot \frac{\pi}{M} \right] \qquad (1)$$

where $m=(1:L)$; L being the order, and S_k given by ec. 2.

$$S_k = \sum_{j=1}^{NFFT} W_k(j) \cdot X(j) \qquad (2)$$

where $k=(1:M)$; M being the band number in mel scale; $W_k(j)$ is the triangular weighting function associated with the kth mel band in mel scale, , and $X(j)$ is the $NFFT$-point magnitude spectrum $(j=1:NFTT)$.

Each band in the frequency domain is bandwidth dependant of the filter central frequency. The higher the frequency is, the wider the bandwidth is.

The alterations related with the mucosal waveform due to an increase of mass are reflected in the low bands of the MFCC, whereas the higher bands are able to model the noisy components due to a lack of closure. Both alterations are reflected as noisy components with poor outstanding components and wide band spectra. The spectral detail given by the MFCC can be considered good enough for our purpose.

Noise features. MFCCs have been complemented with three classical short-term measurements, specifically developed to measure the degree of noise present due to disorders: Harmonics to Noise Ratio (HNR), Normalized Noise Energy (NNE), and Glottal to Noise Excitation Ratio (GNE). The aim of these features is to separate the contribution of the excitation and the noise present, that is much higher in pathological conditions.

Harmonics to Noise Ratio (HNR). This parameter [3] is a measurement of the voice pureness. It is based on calculating the ratio of the energy of the harmonics related to the noise energy present in the voice (both measured in dB). Such measurement is carried out from the speech cepstrum, removing the energy present at the rahmonics by liftering. The resulting liftered cepstrum provides a noise spectrum which is subtracted from the original cepstrum. The result is a spectrum that contains only the harmonic components. After performing a baseline correction, the modified noise spectrum is subtracted from the original log spectrum in order to provide the HNR ratio estimation.

Normalized Noise Energy (NNE). This parameter [4] is a measurement of the noise present in the voice respect to the total energy (i.e. NNE is the ratio between the energy of noise and total energy of the signal -both measured in dB). Such measurement is carried out from the speech spectrum, separating by comb filtering the contribution of the harmonics in the frequency domain, from the valleys (noise). Between the harmonics, the noise energy is directly obtained from the spectrum. Within a harmonic, the noise energy is assumed to be the mean value of both adjacent minima in the spectrum.

Glottal to Noise Excitation Ratio (GNE). This parameter [8] is based on the correlation between Hilbert envelopes of different frequency channels extracted from the inverse filtering of the speech signal. The bandwidth of envelopes is 1 kHz, and frequency bands are separated 500 Hz. Triggered by a single glottis closure, all the frequency channels are simultaneously excited, so that the envelopes in all channels share the same shape, leading to high correlation between the envelopes. The shape of each excitation pulse is practically independent of preceding or following pulses. In case of turbulent signals (noise, whisper) a narrowband noise is excited in each frequency channel. These narrow band noises are uncorrelated (if the windows that define adjacent frequency channels do not overlap too much). The GNE is calculated picking the maximum of each correlation functions between adjacent frequency bands. The parameter indicates whether a given voice signal originates from vibrations of the vocal folds or from turbulent noise generated in the vocal tract.

Temporal derivatives. A representation better showing the dynamic behavior of speech can be obtained by extending the analysis to include the temporal derivatives of the parameters among neighbor frames. First (Δ) derivative has been used in the present study. To introduce temporal order into the parameter representation, let's denote the m_{th} coefficient at time p by $c_m[p]$ [13]:

$$\Delta c_m[p] \approx \mu \cdot \sum_{k=-K}^{K} k \cdot c_m[p+k]$$ (3)

where μ is an appropriate normalization constant and $(2K+1)$ is the number of frames over which the computation is performed.

For each frame p, the result of the analysis is a vector of L coefficients, to which another L-dimensional vector giving the first time derivative is appended; that is:

$$o[p] = \{c_1[p], c_2[p], ..., c_L[p], \Delta c_1[p], \Delta c_2[p], ..., \Delta c_L[p]\}$$ (4)

where $o[p]$ is a feature vector with $2 \cdot L$ elements.

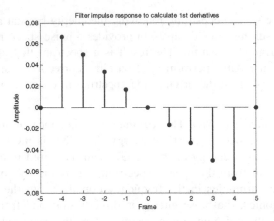

Fig. 2. Filter impulse response to calculate the Δs of the temporal sequence of parameters. The filter coefficients are $a_k=\{0.067, 0.05, 0.033, 0.017, 0, -0.017, -0.033, -0.05, -0.067\}$.

The derivatives provide relevant information about the dynamics of the short-time variation in the parameters. A priori, these features have been considered significant because, due to the presence of disorders a lower degree of stationarity may be expected in the speech signal [11]. Therefore, larger temporal variations of the parameters may be expected. Another reason to complement the feature vectors with speed is that SVMs do not consider any temporal dependence by themselves as Hidden Markov Models (HMM) do [19]. The calculation of Δ has been achieved by means of anti-symmetric Finite Impulse Response (FIR) filters to avoid phase distortion of the temporal sequence (Fig. 2).

3.3 Pattern Classification: The SVM Detector

A support vector machine (SVM) [20] is a two-class classifier constructed from sums of a kernel function $K(\cdot,\cdot)$:

$$f(x) = \sum_{i=1}^{N} \alpha_i t_i K(x, x_i) + b \tag{5}$$

where the t_i are the target values, $\sum_{i=1}^{N} \alpha_i t_i = 0$, $\alpha_i > 0$, and b is the bias term. The vectors x_i are support vectors and obtained from the training set by an optimization process [20]. The target values are either 1 or -1 depending upon whether the corresponding support vector is in class 0 or class 1. For classification, a class decision is based upon whether the value, $f(x)$, is above or below a threshold.

The kernel $K(\cdot,\cdot)$ is constrained to have certain properties (the Mercer condition), so that $K(\cdot,\cdot)$ can be expressed as:

$$K(x, y) = b(x)^t b(y) \tag{6}$$

where $b(x)$ is a mapping from the input space to a possibly infinite dimensional space. In this paper, a Radial Basis Function (RBF) kernel (ec. 7) has been used.

$$K(x, y) = e^{-\gamma\|x-y\|^2}, \quad \gamma > 0 \tag{7}$$

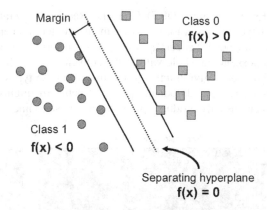

Fig. 3. Basis of the Support Vector Machine

The optimization condition relies upon a maximum margin concept (Fig. 3). For a separable data set, the system places a hyperplane in a high dimensional space so that the hyperplane has maximum margin. The data points from the training set lying on the boundaries are the support vectors in ec. 5. For the RBF kernel, the number of centers, the centers themselves x_i, the weights α_i, and the threshold b are all calculated automatically by the SVM training by an optimization procedure. The training imply adjusting the parameter of the kernel, γ, and a penalty parameter, C, of the error term (a larger C value corresponds to assign a higher penalty to errors). The goal is to identify good (C, γ) pairs, so that the classifier can accurately predict unknown data.

3.4 Evaluation Procedure

Results are obtained following a cross-validation strategy, according to which the training and validation process of the model is repeated 10 times.

In each repetition, the database is split randomly into two subsets: first one, with the 70% of the files, is used for training the model, and the second one, with the remaining 30%, is used to evaluate the generalization of the model. Each subset keeps the same proportion of normal and pathological files as the original data set. The division of the database into training and evaluation sets was carried out in a file basis (not in a frame basis) in order to prevent the system from learning speaker-related features and so to favor the generalization. Both male and female voices have been mixed indistinctly in the two sets.

The final estimations of the model's performance are achieved averaging the 10 partial results. These values are presented by means of a confusion matrix (as shown in Figure I), based on the number of frames correctly or incorrectly classified.

For computing the results on a file basis, all the scores corresponding to the frames of a given file are averaged, thus obtaining a single score per file. With these file scores, a new threshold value that maximizes the correct classification rate within the training set is established. Then, the threshold is employed to compute the system's performance with the validation set. Finally, the confusion matrix file-based is filled, averaging the 10 trials.

The Detection Error Tradeoff (DET) [21] and Receiver Operating Characteristic (ROC) [22] curves have been used for the assessment of detection performance (Fig. 6). ROC displays the diagnostic accuracy expressed in terms of sensitivity against (1-specificity) at all possible threshold values in a convenient way. In the DET curve we plot error rates on both axes, giving uniform treatment to both types of error, and use a scale for both axes which spreads out the plot and better distinguishes different well performing systems and usually produces plots that are close to linear.

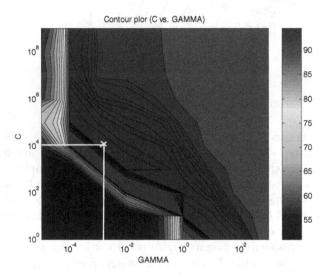

Fig. 4. Contour plot (penalty parameter C vs. γ) to show the cell where the detector performs better. The grid selected is $(C, \gamma) = (10^4, 10^{-3})$.

Table 1. Confusion matrix to show the performance of the classifier in terms of frame and file accuracy; a) True negative (TN) or correct rejection: the detector found no event (normal voice) when none was present; b) True positive (TP) or correct detection: the detector found an event (pathological voice) when one was present; c) False negative (FN) or false rejection: the detector found no event (normal) when present (pathological); d) False positive (FP) or false alarm: the detector found an event (pathological) when none was present (normal); e) Sensitivity: probability for an event to be detected given that it is present; f) Specificity: probability for the absence of an event to be detected given that it is absent; g) Efficiency: probability for the correct detection and rejection. Results are given for the Equal Error Rate point

			Event		
			Present	**Absent**	*Efficiency (%):* **94,1±2**
Decision	Frame accuracy	**Present**	TP: 90.3 (%)	FP: 2.1 (%)	*Sensitivity: 0,97*
		Absent	FN: 9.7 (%)	TN: 97.9 (%)	*Specificity: 0,90*
	File Accuracy		**Present**	**Absent**	*Efficiency (%):* **95±2**
		Present	TP: 94,9 (%)	FP: 5,1 (%)	*Sensitivity: 0,95*
		Absent	FN: 5,1 (%)	TN: 94.9 (%)	*Specificity: 0,95*

4 Results

Data were normalized into the interval [-1, 1] before feeding the net. The parameters (C, γ), were chosen to find the average optimum accuracy. At each (C, γ) grid, ten sets were sequentially used following the previously commented cross-validation scheme. The grid finally selected is $(C, \gamma) = (10^4, 10^{-3})$ (Fig. 4).

Table 1 shows the performance of the detector in terms of frame and file accuracy. The specificity of the system may be increased at the expense of sensitivity shifting the threshold used for the detection (Fig. 5). Table 1, Fig. 6a and Fig. 6b reveals that the Equal Error Rate (EER) point is around 6% and 5% for the frame and file accuracy respectively. The results shown are a little bit worse (in terms of accuracy around the

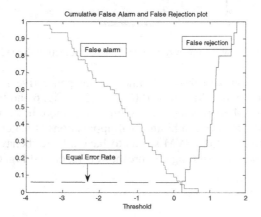

Fig. 5. Cumulative false alarm and false rejection plot. The Equal Error Rate point is shown.

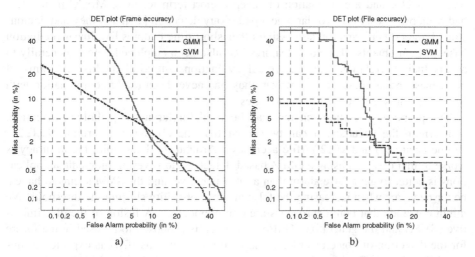

Fig. 6. a) DET plot to show the False Alarm vs. Miss Probability (False Negative) for the frame accuracy; b) DET for the file accuracy

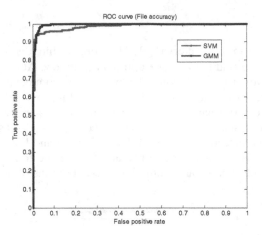

Fig. 7. ROC plot to show the False Alarm vs. Correct Detection (True Positive) probability

Equal Error Rate point) to those obtained using generative methods such as the classical Gaussian Mixture Models (GMM) [23;24] (Fig. 6 and Fig. 7).

The classical GMM approach [23;24] seems to be more linear in terms of the DET plot than the SVM. However, in biomedical applications is better to get a false alarm than a false negative, and the SVM seems to have a better behavior when increasing the false alarm probability shifting the threshold (Fig. 6a and Fig. 6b).

5 Conclusions

The paper has shown a system to automatically detect voice pathologies, based on a SVM detector and a combination of several short term features. MFCC parameters had been previously used for laryngeal pathology detection [25], and they had demonstrated a good performance, surpassing other short time features like linear prediction based measurements. The noise features had also been employed before, generally to assess their correlation with perceptual evaluations of vocal quality (in terms of hoarseness, breathiness, etc). However, they had never been used in combination with MFCC features to detect voice pathologies.

The proposed detection scheme may be used for laryngeal pathology detection. Concerning the accuracy, it can be shown that the efficiency is around 94,1±2% (frame accuracy) and 95±2% (file accuracy). Around the EER point the accuracy is worse to that obtained using a GMM based detector. However, in biomedical applications is better to get a false alarm than a false negative, and the SVM seems to have a better behavior when increasing the false alarm probability shifting the threshold. We might conclude that both approaches are complementary. A fusion of the discriminative (SVM) and the generative (GMM) approach is proposed to be used in the future for the detection of voice disorders. The fusion of both classifiers is expected to improve the overall efficiency of the system keeping in mind that the decision criteria of both systems are rather different.

The current study opens up the way to extend this method for classification tasks between different disorders (polyps, nodules, cists, etc) or perceptual vocal qualities: hoarseness, breathiness, etc.

Acknowledgments

This research was carried out under grants: TIC2003-08956-C02-00 and TIC-2002-0273 from Ministry of Science and Technology of Spain; and AP2001-1278 from the Ministry of Education of Spain.

References

[1] Baken, R. J. and Orlikoff, R., *Clinical measurement of speech and voice*, 2 ed. Singular Publishing Group, 2000.

[2] Feijoo, S. and Hernández, C., "Short-term stability measures for the evaluation of vocal quality", *Journal of Speech and Hearing Research*, vol. 33 pp. 324-334, June1990.

[3] de Krom, G., "A cepstrum-based technique for determining a harmonics-to-noise ratio in speech signals", *Journal of Speech and Hearing Research*, vol. 36, no. 2, pp. 254-266, Apr.1993.

[4] Kasuya, H., Ogawa, S., Mashima, K., and Ebihara, S., "Normalized noise energy as an acoustic measure to evaluate pathologic voice", *Journal of the Acoustical Society of America*, vol. 80, no. 5, pp. 1329-1334, Nov.1986.

[5] Winholtz, W., "Vocal tremor analysis with the vocal demodulator", *Journal of Speech and Hearing Research*, no. 35, pp. 562-563, 1992.

[6] Boyanov, B. and Hadjitodorov, S., "Acoustic analysis of pathological voices. A voice analysis system for the screening of laryngeal diseases", *IEEE Engineering in Medicine & Biology Magazine*, vol. 16, no. 4, July/August, pp. 74-82, 1997.

[7] Deliyski D. Acoustic model and evaluation of pathological voice production. 1969-1972. 1993. Berlin, Germany. Proceedings of Eurospeech '93.

[8] Michaelis, D., Gramss, T., and Strube, H. W., "Glottal-to-Noise Excitation ratio - a new measure for describing pathological voices", *Acustica/Acta acustica*, vol. 83 pp. 700-706, 1997.

[9] Yumoto, E., Sasaki, Y., and Okamura, H., "Harmonics-to-noise ratio and psychophysical measurement of the degree of hoarseness", *Journal of Speech and Hearing Research*, vol. 27, no. 1, pp. 2-6, Mar.1984.

[10] Hadjitodorov, S., Boyanov, B., and Teston, B., "Laryngeal pathology detection by means of class-specific neural maps", *IEEE Transactions on Information Technology in Biomedicine*, vol. 4, no. 1, pp. 68-73, Mar.2000.

[11] Childers, D. G. and Sung-Bae, K., "Detection of laryngeal function using speech and electroglottographic data", *IEEE Transactions on Biomedical Engineering*, vol. 39, no. 1, pp. 19-25, Jan.1992.

[12] Gavidia-Ceballos, L. and Hansen, J. H. L., "Direct speech feature estimation using an iterative EM algorithm for vocal fold pathology detection", *IEEE Transactions on Biomedical Engineering*, vol. 43, no. 4, pp. 373-383, Apr.1996.

[13] Deller, J. R., Proakis, J. G., and Hansen, J. H. L., *Discrete-time processing of speech signals* New York: Macmillan Series for Prentice Hall, 1993.

[14] Aronson, A. E., *Clinical voice disorders. An interdisciplinary approach*, 3 ed. New York: Theme publishers, 1990.

[15] Kay Elemetrics Corp. Disordered Voice Database. Version 1.03. 1994. Lincoln Park, NJ, Kay Elemetrics Corp.

[16] Parsa, V. and Jamieson, D. G., "Identification of pathological voices using glottal noise measures", *Journal of Speech, Language and Hearing Research*, vol. 43, no. 2, pp. 469-485, Apr.2000.

[17] Manfredi, C., D'Aniello, M., Bruscaglioni, P., and Ismaelli, A., "A comparative analysis of fundamental frequency estimation methods with application to pathological voices", *Medical Engineering & Physics*, vol. 22, no. 2, pp. 135-147, Mar.2000.

[18] Bou-Ghazale, S. E. and Hansen, J. H. L., "A comparative study of traditional and newly proposed features for recognition of speech under stress", *IEEE Transactions on Speech and Audio Processing*, vol. 8, no. 4, pp. 429-442, July2000.

[19] Rabiner, L. R. and Juang, B. H., *Fundamentals of speech recognition* Englewood Cliffs, NJ: Prentice Hall, 1993.

[20] Vapnik, V., "An overview of statistical learning theory", *IEEE Transactions on Neural Networks*, vol. 10, no. 5, pp. 988-1000, Sept.1999.

[21] Martin A, Doddington GR, Kamm T, Ordowski M, Przybocki M. The DET curve in assessment of detection task performance. IV, 1895-1898. 1997. Rhodes, Crete. Proceedings of Eurospeech '97.

[22] Hanley, J. A. and McNeil, B. J., "The meaning and use of the area under a receiver operating characteristic (ROC) curve", *Radiology*, vol. 143, no. 1, pp. 29-36, Apr.1982.

[23] Reynolds, D. A. and Rose, R. C., "Robust text-independent speaker identification using Gaussian mixture speaker models", *IEEE Transactions on Speech and Audio Processing*, vol. 3, no. 1, pp. 72-83, Jan.1995.

[24] Reynolds, D. A., "Speaker identification using Gaussian mixture speaker models", *Speech Communication*, vol. 17 pp. 91-108, 1995.

[25] Godino-Llorente, J. I. and Gómez-Vilda, P., "Automatic detection of voice impairments by means of short-term cepstral parameters and neural network based detectors", *IEEE Transactions on Biomedical Engineering*, vol. 51, no. 2, pp. 380-384, Feb.2004.

Synthesis of Disordered Voices

Julien Hanquinet, Francis Grenez , and Jean Schoentgen

Department Signals and Waves, Université libre de Bruxelles,
50, Avenue F.-D. Roosevelt, 1050 Brussels, Belgium
jhanquin@ulb.ac.be
National Fund for Scientific Research, Belgium

Abstract. The presentation concerns the simulation of disordered voices. The phonatory excitation model is based on shaping functions, which are nonlinear memoryless input-output characteristics that transform a trigonometric driving function into a synthetic phonatory excitation signal. The shaping function model enables controlling the instantaneous frequency and spectral brilliance of the phonatory excitation via two separate parameters. The presentation demonstrates the synthesis of different types of dysperiodicities via a modulation of the amplitude and instantaneous frequency of the harmonic driving function. The voice disorders that are simulated are short- and long-term perturbations of the vocal frequency and cycle amplitude, biphonation, diplophonia and raucity. Acoustic noise due to turbulent airflow is modeled by means of additive white noise.

1 Introduction

Often, human speech production is mimicked by means of a linear model of speech production, which represents voiced speech by means of a periodic pulse sequence that is filtered by the vocal tract transfer function [1]. Natural voices, especially disordered voices, are not simply periodic, however.

Vocal dysperiodicities may have several causes. One distinguishes dysperiodicities that are due to (i) the vocal fold dynamics, e.g. diplophonia, biphonation and random cycle lengths; (ii) external perturbations, such as phonatory jitter and frequency tremor; (iii) turbulence noise, i.e. additive noise.

Motivations for developing synthesizers able to simulate vocal dysperiodicities are the following. First, the simulation of vocal dysperiodicities may enable discovering acoustic features of perceptual traits of hoarse voices, as well as facilitate the training of clinicians evaluating disordered voices perceptually. Second, a better understanding of the perceptual effects of vocal dysperiodicities may enable improving the naturalness of synthetic speech. Third, synthetic test signals may be used to calibrate analysis methods of vocal dysperiodicity. Fourth, synthetic reference signals may mark perceptual boundaries within which disordered voices are perceptually classified by clinicians.

Existing synthesizers of disordered voices, based on the Liljencrants-Fant model, for instance, do not enable controlling the instantaneous glottal cycle length [2]. This

M. Faundez-Zanuy et al. (Eds.): NOLISP 2005, LNAI 3817, pp. 231 – 241, 2005.

means that any change of the vocal cycle length must be synchronized with the onset or offset of the glottal excitation pulse. Also, the Liljencrants-Fant model requests resetting all other model parameters to guarantee area balance. Finally, the bandwidth evolves unpredictably while modulating the model parameters to simulate vocal dysperiodicities. The risk of aliasing is not nil, because the bandwidth of concatenated-curve models is not known a priori.

We therefore propose to synthesize the phonatory excitation, which is the acoustic signal that is generated at the glottis by the vibrating vocal folds and pulsatile airflow, by means of shaping functions. A shaping function is an operator that transforms a trigonometric driving function into any desired waveform.

The shaping function-based model has the following desirable properties. First, the instantaneous frequency and harmonic richness of the signal are controlled by two separate parameters, which are the instantaneous frequency and amplitude of the driving function. These parameters can be varied continuously and asynchronously. Second, the instantaneous amplitude, frequency, and harmonic richness of the synthetic vocal excitation, and the shaping function coefficients that encode speaker identity, can be set independently.

2 Model

A shaping function is a memoryless input-output characteristic that transforms a cycle of a harmonic into any desired shape [3]. The shaping function used here consist in an equivalent polynomial formulation of the Fourier series. The relation between the Fourier series of a signal and its shaping function model is demonstrated hereafter.

2.1 Model I: Fourier Series Representation

Under some mild assumptions, a discrete signal $y(n)$ of cycle length N can be approximated by a Fourier series truncated at the Mth harmonic.

$$y(n) \approx \frac{1}{2}a_0 + \sum_{k=1}^{M} a_k \cos(k\frac{2\pi}{N}n) + b_k \sin(k\frac{2\pi}{N}n). \tag{1}$$

In series (1), coefficients a_k and b_k encode the shape of a cycle of signal y, and constant N is the cycle length. Changing N generates signals with the same cycle shape, but with different cycle lengths.

If N is a real number, series (1) can be rewritten as follows.

$$y(n) = \frac{1}{2}a_0 + \sum_{k=1}^{M} a_k \cos(k\theta_n) + b_k \sin(k\theta_n), \tag{2}$$

with $\theta_{n+1} = \theta_n + 2\pi f\Delta$. Symbol f is the instantaneous frequency and Δ the sampling step. Letting f assume real values introduces quantization errors of one sample at most in the cycle length. This error depends on the sampling frequency, which must be chosen accordingly.

A signal with a fixed cycle shape, but with arbitrary instantaneous frequency, can therefore be synthesized by means of series (2). The default cycle shape is set by means of a glottal cycle template.

The harmonic richness or spectral balance (i.e. brightness of timbre) of the synthetic signal may be fixed by a parameter that modifies the Fourier coefficients a_k and b_k. Indeed, non-linear oscillators are known to output signals the shape of which coevolves with signal amplitude. Often, the lower the amplitude is the more sinus-like is the shape. This suggest letting evolve Fourier coefficients as follows.

$$a_k -> a'_k = A^k a_k,$$
$$b_k -> b'_k = A^k b_k. \qquad \text{with } 0 < A < 1. \tag{3}$$

Transform (3) is loosely inspired by [3]. Series (2) then becomes the following.

$$y(n) = \frac{1}{2} a_0 + \sum_{k=1}^{M} a_k A^k \cos(k\theta_n) + b_k A^k \sin(k\theta_n). \tag{4}$$

Series (4) demonstrates that the harmonics decrease the faster the higher their order, when parameter A is less than one.

By means of series (4), several types of dysperiodicity may be simulated by modulating parameter A and instantaneous frequency f. These modulations may introduce aliasing. To facilitate the control of the bandwidth of series (4), it is reformulated polynomially.

2.2 Model II: Distortion Function Representation

It has been shown elsewhere that series (2) can be rewritten as a sum of powers of cosines (5) [3][4].

$$y(n) = F[\cos(\theta_n)] + \sin(\theta_n) G[\cos(\theta_n)], \tag{5}$$

with

$$F[\cos(\theta_n)] = \sum_{k=0}^{M} c_k \cos^k(\theta_n),$$
$$G[\cos(\theta_n)] = \sum_{k=0}^{M} d_k \cos^k(\theta_n). \tag{6}$$

The relation between polynomial coefficients c_k and d_k and Fourier coefficients a_k and b_k is given by expressions (7) and (9) [3][4].

$$\begin{pmatrix} a_0 \\ a_1 \\ \dots \\ a_M \end{pmatrix} = 2M_e \begin{pmatrix} c_0 \\ c_1/2 \\ \dots \\ c_M/2^M \end{pmatrix}, \tag{7}$$

with

$$M_e = \begin{pmatrix} 1 & 0 & 2 & 0 & 6 & 0 & 20 & \dots \\ 0 & 1 & 0 & 3 & 0 & 10 & 0 & \dots \\ 0 & 0 & 1 & 0 & 4 & 0 & 15 & \dots \\ 0 & 0 & 0 & 1 & 0 & 5 & 0 & \dots \\ 0 & 0 & 0 & 0 & 1 & 0 & 6 & \dots \\ 0 & 0 & 0 & 0 & 0 & 1 & 0 & \dots \\ 0 & 0 & 0 & 0 & 0 & 0 & 1 & \dots \\ \dots & \dots & \dots & \dots & \dots & \dots & \dots & \dots \end{pmatrix}, \tag{8}$$

and,

$$\begin{pmatrix} b_1 \\ b2 \\ \dots \\ b_{M+1} \end{pmatrix} = M_o \begin{pmatrix} d_0 \\ d_1/2 \\ \dots \\ d_M/2^{M+1} \end{pmatrix}, \tag{9}$$

with

$$M_o = \begin{pmatrix} 1 & 0 & 1 & 0 & 2 & 0 & 5 & \dots \\ 0 & 1 & 0 & 2 & 0 & 5 & 0 & \dots \\ 0 & 0 & 1 & 0 & 3 & 0 & 9 & \dots \\ 0 & 0 & 0 & 1 & 0 & 4 & 0 & \dots \\ 0 & 0 & 0 & 0 & 1 & 0 & 5 & \dots \\ 0 & 0 & 0 & 0 & 0 & 1 & 0 & \dots \\ 0 & 0 & 0 & 0 & 0 & 0 & 1 & \dots \\ \dots & \dots & \dots & \dots & \dots & \dots & \dots & \dots \end{pmatrix}. \tag{10}$$

To set the spectral brilliance of the synthetic excitation, polynomials F and G may be driven by trigonometric functions, the amplitude of which differ from unity. Representation (5) then generalizes to the following.

$$y(n) = F[A\cos(\theta_n)] + A\sin(\theta_n)G[A\cos(\theta_n)]. \tag{11}$$

The harmonic richness decreases and increases with amplitude A. Increasing A above 1 is not recommended, however. Indeed, increasing $A > 1$ is equivalent to extrapolating the cycle shape, the outcome of which lacks plausibility as soon as A exceeds unity by more than a few percent.

By means of model (11), several types of dysperiodicities may be simulated by modulating the amplitude and frequency of the trigonometric driving sine and cosine. Aliasing is avoided on the condition that the upper bound of the effective bandwidth of the modulated driving functions times the order of the shaping polynomials is less than half the sampling frequency [5]. The respect of this condition can be guaranteed by low-pass filtering the modulated harmonic excitation before insertion into polynomials F and G. Aliasing is indeed easier to avoid while modulating a harmonic, than while modulating a full-bandwidth Fourier series (4).

2.3 Phonatory Excitation

Parameter A does not influence the mean value of Fourier series (4). Indeed, if the average of the glottal cycle template equals zero, series (4) is rewritten as follows.

$$y(n) = \sum_{k=1}^{M} a_k A^k \cos(k\theta_n) + b_k A^k \sin(k\theta_n). \tag{12}$$

One sees that the average of each cycle remains zero, whatever the value of A.

This is not true of trigonometric polynomial (11). Indeed, if A is zero, polynomial (11) is equal to c_0 which involves a linear combination of Fourier coefficients a_k that are different from zero.

Also, the average of a cycle must be zero, whatever its amplitude, because the phonatory excitation is an acoustic signal. To warrant this property mathematically, the derivative of polynomial (11) is taken. Indeed, in the framework of the modeling of the phonatory excitation signal, the phonatory excitation is conventionally considered to be the derivative of the glottal airflow rate [2]. The derivative mathematically guarantees that the average of a glottal excitation pulse is zero as well as that its amplitude goes to zero with driving function amplitude A [3].

In practice, the coefficients of polynomials F and G are computed on the base of a template flow rate and the derivative of polynomials (11) is taken. To avoid the dependency of the amplitude of the phonatory excitation on phonatory frequency, this derivative is taken with respect to phase instead of time. Consequently, the phonatory excitation is written as follows, assuming that phase θ is continuous and derivable. Discretization of phase θ is carried out after the derivation.

$$e(\theta) = \frac{d}{d\theta}\{F[A\cos(\theta)] + A\sin(\theta)G[A\cos(\theta)]\}, \text{ with } 0 \le A \le 1. \tag{13}$$

Figure 1 summarises the simulation of the phonatory excitation. First, a template cycle of the glottal flow rate is selected, e,g, the glottal flow rate that is given by the Fant model. The Fourier coefficients are computed numerically and the polynomial coefficients are obtained by means of matrixes (7) and (9). The polynomial coefficients are inserted into derivative (13). The derivative of the flow rate is then obtained by driving (13) with a sine and cosine. This signal is construed as the phonatory excitation. To avoid aliasing, the modulated sine and cosine are low-pass filtered so that the product of the upper bound of their effective bandwidth with the order of (13) is less than half the sampling frequency.

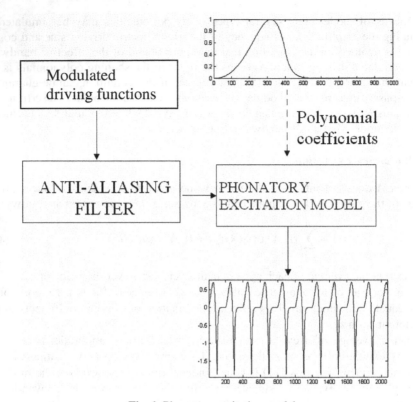

Fig. 1. Phonatory excitation model

Instantaneous frequency, spectral balance and speaker identity are controlled via dinstinct parameters. Instantaneous frequency and harmonic richness are controlled via the instantaneous frequency and amplitude of the trigonometric driving functions. Speaker identity (i.e. the default cycle shape) is encoded via the polynomial coefficients. The instantaneous control of frequency and spectral balance is illustrated in Figures 2 and 3.

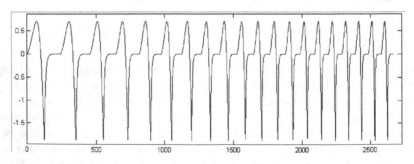

Fig. 2. Synthetic phonatory excitation, the instantaneous frequency of which evolves linearly from 75 to 200 Hz. The vertical axis is in arbitrary units and the horizontal axis is labeled in number of samples.

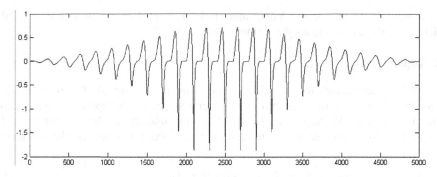

Fig. 3. Synthetic phonatory excitation for which driving amplitude A evolves from zero to one and from one to zero. The vertical axis is in arbitrary units and the horizontal axis is labeled in number of samples.

2.4 Vocal Tract Model

To simulate the propagation of the phonatory excitation through the vocal tract, a concatenation of cylindrical tubelets is used. Each tubelet has the same length. For each tubelet, viscous, thermal and wall vibrations losses are modeled by means of filters [8]. A conical tubelet the opening of which is controlled, is added at the lip-end of the vocal tract model to simulate the transition at the lips from one-dimensional to three-dimensional wave propagation. In practice, this conical tubelet is quasi-cylindrical, because the high-pass filtering involved in the 3-D radiation has already been taken into account via the modeling of the derivative of the glottal air flow rate.

3 Synthesis of Disordered Voices

3.1 Vocal Jitter and Microtremor

Jitter and microtremor designate small stochastic cycle-to-cycle perturbations and low-frequency modulations of the glottal cycle lengths respectively. Jitter and microtremor are therefore inserted into the phonatory excitation model by perturbing the instantaneous frequency of the driving function by two random components [6]. Consequently, the discrete-time evolution of the phase of the sinusoidal driving function is written as follows.

$$\theta_{n+1} = \theta_n + 2\pi(f_0\Delta + j_n + m_n) \ . \tag{14}$$

Symbol f_0 is the unperturbed instantaneous vocal frequency; Δ is the time step; j_n is uniformly distributed white noise that simulates intra-cycle frequency perturbations that give rise to jitter; m_n is uniformly distributed white noise filtered by a linear second order filter, which sets the microtremor frequency and bandwidth, which is typically 3 Hz and 4 Hz respectively.

Investigating relations between perturbations of the phonatory excitation and the speech signal demonstrates that vocal shimmy and amplitude tremor are due to modulation distortion by the vocal tract. Modulation distortion converts frequency into amplitude perturbations. Vocal amplitude tremor also involves tremor of the articulators, which modulates the formant frequencies [7].

3.2 Diplophonia

Diplophonia here refers to periodic phonatory excitation signals the mathematical period of which comprises several unequal glottal cycles. A repetitive sequence of different glottal cycle shapes can be simulated by modulating the amplitude of the driving function (15) because it influences the spectral balance of the phonatory excitation.

$$A_n = A_0 + A_1 \sin(\theta_n / Q). \tag{15}$$

Parameter A_0 is the mean value of driving amplitude A. Symbol A_1 is the amplitude of the modulation of driving amplitude A. Parameter Q sets the number of different glottal cycle shapes within the mathematical period of the vocal excitation. In practice, parameter Q is a ratio of two small integers.

Similarly, a modulation of the instantaneous frequency of the driving function may simulate a repetitive sequence of glottal cycles of unequal lengths. The temporal evolution of the amplitude and phase of the driving function are then written as follows.

$$\theta_{n+1} = \theta_n + 2\pi\Delta[f_0 + f_1 \sin(\theta_n / Q)]. \tag{16}$$

Parameter f_0 is the mean value of the instantaneous frequency of the synthetic phonatory signal and f_1 is the amplitude of the modulation. Parameter Q sets the number of different glottal cycle length within the mathematical period of the vocal excitation. In practice, parameter Q is a ratio of two small integers.

3.3 Biphonation

Biphonation is characterized by discrete spectra with irrational quotients of the frequencies of some of the partials. As a consequence, biphonation is also described by a sequence of glottal cycles of different shapes and lengths. But, in this case, two glottal cycles are never exactly identical. Consequently, biphonation is simulated by means of expressions similar to (15) and (16). The difference is that parameter Q is equal to an irrational number. Physiologically speaking, biphonation implies quasi-independence of the vibrations of at least two structures (e.g. left and right vocal folds).

3.4 Random Cycle Lengths

Contrary to jitter, which is due to external perturbations of a dynamic glottal regime that is periodic, random vocal cycle lengths are the consequence of a random vibratory regime of the vocal folds. The relevant model parameter is therefore the total cycle length. In the framework of model (14), the selection of a new instantaneous frequency is synchronized with the onset of the excitation cycle. The statistical distribution of the cycle lengths is requested to be a gamma distribution. The gamma distribution is the simplest distribution that enables setting independently the variance of positive cycle lengths and their average. Speech characterized by random vocal cycles is perceived as rough.

3.5 Turbulence Noise

Turbulence noise is taken into account by means of additive noise, which simulates the acoustic effect of excessive turbulent airflow through the glottis. These turbulences are expected to occur when the glottis closes. Uniformly distributed white noise is therefore added to the phonatory excitation signal when it is negative. No noise is added when the signal is positive or zero.

4 Methods

The template flow rate used to compute the coefficients of the shaping polynomials is a synthetic cycle mimicked by means of the Liljencrants-Fant model. Its Fourier coefficients are computed numerically, 40 cosine coefficients and 40 sine coefficients. The polynomial coefficients are obtained from the Fourier coefficients by means of linear transformations (7) and (9). The modulated sinusoidal driving function is sampled at 100 kHz. The anti-aliasing filter is chosen so that its attenuation is −5dB at $1/M+1$ times half the sampling frequency, with $M+1$ equal to the order of polynomials (13).

The simulated vocal tract shape is fixed on the base of published data. The number of cylindrical tubelets used in the vocal tract model is comprised between 20 and 30.

5 Results

Preliminary tests show that the model that is presented here enables synthesizing vocal timbres that are perceived as plausible exemplars of disordered voices. We here illustrate graphically the ability of the synthesizer to simulate different vocal timbres by means of examples of synthetic diplophonic, biphonic and rough voices.

Figure 4 shows an example of diplophonia obtained by modulating the driving functions following expression (15) and (16) with Q set to two.

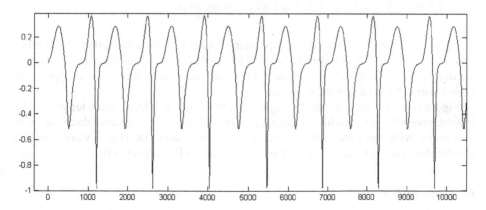

Fig. 4. Synthetic excitation signal simulating diplophonia. The horizontal axis is labeled in number of samples and the vertical axis is in arbitrary units.

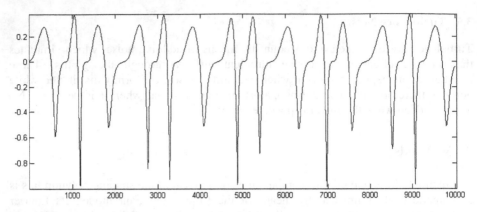

Fig. 5. Synthetic excitation signal simulating biphonation. The horizontal axis is labeled in number of samples and the vertical axis is in arbitrary units.

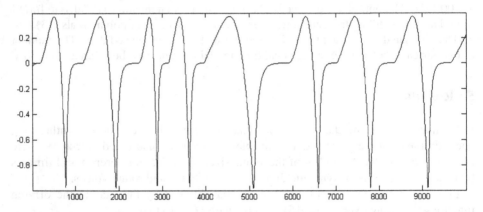

Fig. 6. Synthetic excitation signal simulating random vibrations. The horizontal axis is labeled in number of samples and the vertical axis is in arbitrary units.

Figure 5 shows an example of biphonation obtained by modulating the driving functions following expression (15) and (16) with Q set to Euler's number e. This setting is for demonstration purposes only and does not claim to be typical of biphonation observed in human speakers.

Figure 6 shows an example of rough voice (i.e. random glottal cycle lengths). The instantaneous frequencies have been randomly selected from a gamma distribution in synchrony with the onsets of the cycles. The mean and standard deviation of the gamma distribution have been equal to 100 Hz and 25 Hz respectively.

6 Conclusion

The presentation concerns a simulation of disordered voices. The model is based on a nonlinear memoryless input-output characteristic that transforms a trigonometric driving function into a synthetic phonatory excitation signal. Dysperiodicities are simulated by modulating the amplitude and/or frequency of the trigonometric driving function. The model enables synthesizing a wide range of vocal phenomena, such as jitter, microtremor, diplophonia, biphonation, raucity and breathiness. Preliminary tests show that voices may be simulated that are perceived as plausible exemplars of voice disorders.

References

1. Fant, G., "Acoustic theory of Speech Production". The Hague: Mouton 1960.
2. Fant G., Liljencrants J., Lin Q., "A four-parameter model of glottal flow ", STL-QSPR, 4: 1-13, 1985.
3. Schoentgen, J., "Shaping function models of the phonatory excitation signal", J. Acoust. Soc. Am. 114 (5): 2906-2912, 2003.
4. Schoentgen J., "Nonlinear signal representation and its application to the modelling of the glottal waveform", Speech Comm. 9, 189-201, 1990.
5. Schoentgen, J., "On the bandwidth of a shaping function model of the phonatory excitation signal", Proceedings *NOLISP-03*, 2003.
6. Schoentgen, J., "Stochastic models of jitter", J. Acoust. Soc. Am. 109 (4): 1631-1650, 2001.
7. Schoentgen, J., "Spectral models of additive and modulation noise in speech and phonatory excitation signals", J Acoust Soc Am. 113(1):553-62, 2003.
8. Hanquinet J., Grenez F., Schoentgen J., "Synthesis of disordered speech", Proceedings Interspeech 2005, Lisboa (accepted for presentation).

Voice Pathology Detection by Vocal Cord Biomechanical Parameter Estimation

Pedro Gómez, Rafael Martínez, Francisco Díaz, Carlos Lázaro, Agustín Álvarez, Victoria Rodellar, and Víctor Nieto

Facultad de Informática, Universidad Politécnica de Madrid,
Campus de Montegancedo, s/n28660 Boadilla del Monte, Madrid, Spain
pedro@pino.datsi.fi.upm.es

Abstract. Voice pathologies have become a social concern, as voice and speech play an important role in certain professions, and in the general population quality of life. In these last years emphasis has been placed in early pathology detection, for which classical perturbation measurements (*jitter, shimmer, HNR*, etc.) have been used. Going one step ahead the present work is aimed to estimate the values of the biomechanical parameters of the vocal fold system, as mass, stiffness and losses by the inversion of the vocal fold structure, which could help non only in pathology detection, but in classifying the specific patient's pathology as well. The model structure of the vocal cord will be presented, and a method to estimate the biomechanical parameters of the cord body structure will be described. From these, deviations from normophonic cases, and unbalance between cords may be extracted to serve as pathology correlates. The relevance of deviations and unbalance in Pathology Detection is shown through Principal Component Analysis. Results for normal and pathological cases will be presented and discussed.

1 Introduction

Voice pathology detection is a field of important research area in voice and speech processing as it may affect the quality of life of the population, especially in people who use voice extensively in their professional activity, as speakers, singers, actors, lawyers, broadcasters, priests, teachers, call center workers, etc [13][16][22]. The success in treating voice pathologies depend on their early detection, and as such simple yet powerful inspection procedures are desirable. Among those procedures patient's voice inspection is a simple, low cost and fast method to obtain an estimation of the presence of pathology, which can be used as a screening routine to decide if other specialized inspection methods –as videoendoscopy- are to be used, as these being more precise in pathology classification, are at the same time less comfortable, more expensive and complicate, and their use should be obliviated if a simple inspection could help in screening patients before being subject to full inspection procedures. The estimation of biomechanical parameters associated to the structure of the phonation organs would suppose an important improvement in the use of voice for pathology screening. Up-to-date techniques use time- and frequency-domain estimation of *perturbation parameters*, which measure the deviation of the specific patient's

M. Faundez-Zanuy et al. (Eds.): NOLISP 2005, LNAI 3817, pp. 242–256, 2005.
© Springer-Verlag Berlin Heidelberg 2005

voice from certain normal standards [6][20][15][7]. These techniques have revealed efficient themselves in the detection of pathology, but supply little information on the nature of the pathology. Trying to go one step ahead a study has been initiated to estimate the values of the biomechanical parameters of the vocal fold system (mass, stiffness and losses) from the glottal source obtained from voice after removing the vocal tract transfer function. This procedure is well documented in the literature (see for example [2] [5] [1]), and produces a trace which can be shown to be directly related with the glottal source (the average aperture measured between vocal cords during the phonatory cycle) [8]. The use of *k-mass* vocal fold models [18][3] help in determining that there are two main components in the movement of the vocal cord, contributed by the structure of the cord: the movement of the bulk muscular tissue of the cord body (see **Figure 1.a**) and a traveling wave known as the *mucosal wave* [21][17], which is contributed by the epithelial tissue of the cord cover (see **Figure 1.b**).

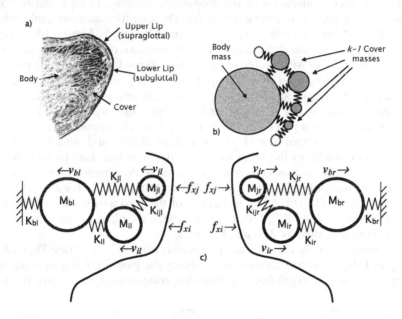

Fig. 1. a) Cross-section of the left vocal cord showing the body and cover structures (taken from [19]). **b)** *k-mass* model of the body and cover. **c)** *3-mass* model used to establish the dynamics of the body-cover system.

In previous research [8][9] it has been shown that both contributions present in the glottal source can be separated to produce two traces, known as the *avrage acoustic wave* or *average glottal source (AGS)* and the *mucosal wave correlate (MWC)*. Their relative energy ratio may be used as a clue for the presence of certain pathologies which induce the reduction or complete disappearance of the *mucosal wave* [14]. In the present study the emphasis will be placed in using the *average glottal source* to measure the main biomechanical parameters involved in the dynamics of the cord body. For such the model structure of the vocal cord will be presented, and a method to estimate the biomechanical parameters of the cord body structure will be described. This method is based on hypothesizing that the fingerprint of the cord body dynamics is responsible for the *power spectral density (psd)* of the *AGS*, thus allowing the identification of the biomechanical parameters of the cord body from the theoretical dynamical transfer function between forces and speeds in the cord body model. In this way a first estimation of the biomechanical parameters is obtained, which can be later refined adaptively.

2 Estimating Cord Dynamics

The vocal cords are two folds which can be found in the phonatory system located in the larynx supported by a complex structure of cartilage and muscles. These folds can be brought to a close contact to stop the flow of air through the respiratory system, and under convenient lung pressure can produce a vibration which is the basics of the phonation. A good explanation of the phonatory function can be found in [20]. A cross section of a vocal cord can be seen in 0.a, showing its tissular structure, which is composed by the body and the cover as mentioned before. In **Figure 1.b** an equivalent *k-mass* model is presented, where the main structure (the body) has been represented by a large lump which referred to as M_{lb} (left cord) and M_{rb} (right cord). The cover is represented by a set of *k-1* lumped masses M_{li} and M_{ri}, $1 \leq i \leq k-1$, linked by springs among themselves and to the body mass. Each spring is represented by a stiffness parameter given by K_{lij} (left cord) and K_{rij} (right cord) where i and j refer to the masses linked ($i,j=0$ will point to the body mass). It will be assumed that a loss factor $R_{l,rij}$ will be also associated to each spring to have viscous and other losses into account. A representation of the vocal fold dynamical relations may be seen in 0.c including a body mass and two cover masses. This is the simplest model which can grant a proper study of the *mucosal wave* phenomenon, and has been widely studied in the literature ([3][17][18]). The estimation of the cord movement is based on the pioneering work by Alku ([2][23]), which has been modified for an iterative implementation in several steps as shown in **Figure 2**.

Step 1 consists in removing the radiation effects from voice $s(n)$ (see **Figure3.a**) by filtering with $H_r(z)$. Step 2 consists in removing the glottal pulse generating model $F_g(z)$ by by its inverse $H_g(z)$ from the radiation compensated voice $s_l(n)$. In the first

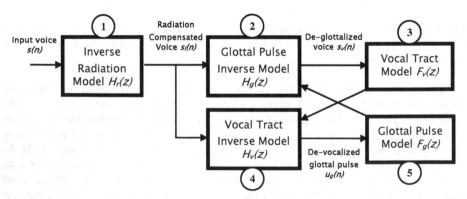

Fig. 2. Estimation of the *glottal pulse* $u_g(n)$ by coupled model estimation and inversion

iteration $H_g(z)$ need not be a very precise estimation, as it will be refined by successive iterations. In step 3 the vocal tract model $F_v(z)$ is estimated from the deglottalized voice $s_v(n)$. Step 4 will consist in removing the vocal tract model by filtering $s_v(n)$ with the vocal tract inverse function $H_v(z)$ to obtain a better estimation of the glottal source $u_g(n)$. Step 5 produces a more precise model of the glottal source $F_g(z)$, which could be used to refine $H_g(z)$. The procedure will repeat steps 2-5 to a desired end. The whole process is described in more detail in previous work [8]. The *glottal source* $u_g(n)$ as shown in **Figure 3.c** is composed by the body mass movement (cord body dynamics) and by the *mucosal wave oscillation* produced by the cover masses (cord cover dynamics). The *mucosal wave correlate* will be defined as:

$$y_m(n)=u_g(n)-\overline{u}_g(n); \quad n\in W_k \tag{1}$$

where $y_m(n)$ is the *mucosal wave correlate*, and W_k is the *k-th period window* on $u_g(n)$.

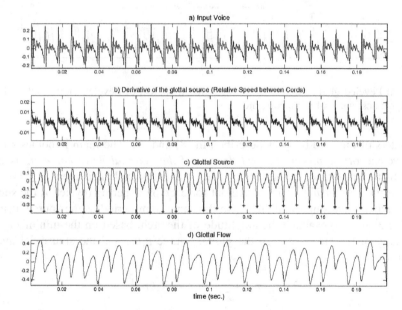

Fig. 3. a) Input voice s(n). b) Glottal source derivative. c) Glottal source $u_g(n)$ unfolding points (*). d) Unleveled glottal pulse.

The effects of vocal tract coupling have been neglected.The traces shown in **Figure 4.a**, **b** and **c** are respectively the ground leveled version of the *glottal source* $u_g(n)$, the *average glottal wave* $\overline{u}_g(n)$, and the leveled *glottal pulse*. As the *average glottal source* may be associated with the simplest cord body dynamics, the difference between the *glottal source* and the *average glottal source* may be considered as contributed by the cord cover dynamics, and can be seen as the *mucosal wave correlate*, as shown in **Figure 4.d**. In the present study emphasis will be placed in adjusting the spectral behavior of the *mucosal wave correlate* to estimate the biomechanical

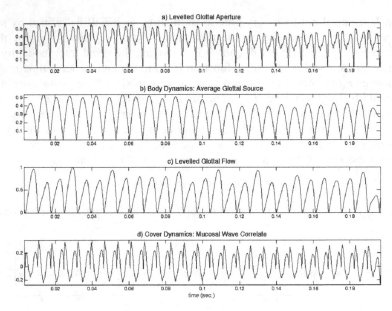

Fig. 4. a) Leveled glottal source, b) average glottal source, c) Leveled glottal pulse, d) cord body dynamics: mucosal wave correlate

parameters of the cord body as described in [8] and [9]. The main hypothesis is that the envelope of the *power spectral density of the mucosal wave correlate* is determined by the admittance of the cord body dynamics, as explained later.

The estimation of $\bar{u}_g(n)$ is carried out for each pitch period (cycle) as the subtraction of a half-wave sinusoidal arch with the same semi-period as the source, using an adaptive method to evaluate the amplitude of the arch, based on the minimization of the energy of the error between the glottal source $u_g(n)$ and the sinusoidal average $\bar{u}_g(n)$ as

$$L = \sum_{n \in W_k} y_m^2(n) = \sum_{n \in W_k} \left(u_g(n) - \bar{u}_g(n)\right)^2 \tag{2}$$

with

$$\bar{u}_g(n) = u_{ok} \, sin(\omega_k n \tau); \quad n \in W_k \tag{3}$$

ω_k being the angular frequency associated to the k-*th* cycle semi-period and τ being the sampling period. The optimization of the amplitude of each sinusoidal arch will be derived minimizing the cost function L in terms of u_{0k} as

$$\frac{\partial L}{\partial u_{0k}} = 0 \quad \Rightarrow \quad u_{0k} = \frac{\displaystyle\sum_{n \in W_k} y_{gk}(n) \, sin(\omega_k n \tau)}{\displaystyle\sum_{n \in W_k} sin^2(\omega_k n \tau)}. \tag{4}$$

3   Estimation of the Body Biomechanical Parameters

Detecting the cord body mass, stiffness and damping is based on the inversion of the integro-differential equation of the one-mass cord model, which for the left vocal cord would be

$$f_{xl} = v_{lb}R_{lb} + M_{lb}\frac{dv_{xl}}{dt} + K_{lb}\int_{-\infty}^{t}v_{xl}dt \tag{5}$$

where the biomechanical parameters involved are the lumped masses M_{lb}, the stiffness K_{lb} and the losses R_{lb}. The equivalent model is shown in **Figure 5**. The estimation of the body biomechanical parameters is related to the inversion of this model, associating the force f_{xl} on the body with the velocity of the cord centre of masses v_{xl} in the frequency domain.

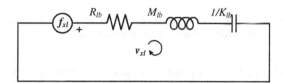

Fig. 5. Electromechanical equivalent of a cord body

The relationship between velocity and force in the frequency domain is expressed as the cord body admittance. The working hypothesis for the process of biomechanical parameter estimation will be based on the assumption that the envelope of the power spectral distribution of the *mucosal wave correlate* (cover dynamic component) is directly related with the square modulus of the input admittance to the electromechanical equivalent $Y_{bl}(s)$ given as

$$T_b(\omega) = |Y_{bl}|^2 = \left|\frac{V_{xl}(\omega)}{F_{xl}(\omega)}\right|^2 = \left[\left(\omega M_{lb} - \omega^{-1}K_{lb}\right)^2 + R_{lb}^2\right]^{-1} \tag{6}$$

The robust estimation of the model parameters is based in the determination of two points on the power spectral density of the cover dynamic component [10] $\{T_{b1}, \omega_1\}$ and $\{T_{b2}, \omega_2\}$, from which the lumped body mass (BM) may be estimated as

$$M_{lb} = \frac{\omega_2}{\omega_2^2 - \omega_1^2}\sqrt{\frac{T_{b1} - T_{b2}}{T_{b1}T_{b2}}} \tag{7}$$

On its turn the elastic parameter (body stiffness: BS) K_{lb} may be estimated from the precise determination of the position of the resonant peak, this being $\{T_r, \omega_r\}$

$$K_{lb} = M_{lb}\omega_r^2 \tag{8}$$

whereas the of body losses (BL) may be estimated (but for a scale factor G_b) as

$$R_{lb} = \frac{G_b}{\sqrt{T_r}} \tag{9}$$

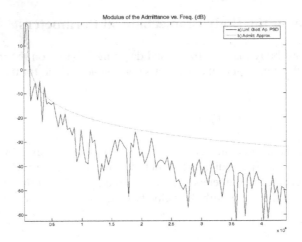

Fig. 6. Parametric fitting of the *mucosal wave power spectral density* for a cycle of the sample trace (full line) against the admittance approximation (dot line)

The estimations obtained from a phonation cycle of a normophonic voice trace have been used to reconstruct the approximated square modulus of the admittance, which is presented in **Figure 6** against the power spectral density of the cover dynamics component for comparison.

Table 1 illustrates the values obtained for the biomechanical parameters of the cord body accordingly with two estimation algorithms (direct and adaptative) from a 2-mass model synthetic voice trace.

Table 1. Comparison between the biomechanical parameters obtained from (6-9) by direct and adaptive estimations

Estimation method	Body Mass (M_{lb})	Losses (R_{lb})	Elasticity (K_{lb})
Direct (3rd harmonic)	2,1500e-004	5,5331e-005	138.500
Adaptive	2,6710e-004	5,5331e-005	171.900

It may be seen that the divergence between both methods is on the order of a 24%. The fact that the mass of the cord body seems to be clearly related to the ratio between the values of the *mucosal wave correlate power spectral density* for the first and third harmonics if $\omega_1 = \omega_t$ and $\omega_2 = 3\omega_t$ gives substantial support to the use of this parameter as an important distortion measure as certain studies on pathological voice suggest [14][4].

4   Results for Synthetic Voice

At this point what seems most crucial is to evaluate the accuracy in the exposed method. Obtaining direct *in vivo* estimations of the biomechanical parameters and voice records from normophonic and pathological cases to establish the accuracy of the method seems to be rather difficult. Another more practical approach is to use a *k-mass* model of the vocal folds to produce voice traces, assigning *a priori* known values for the biomechanical parameters, and use the estimation methods proposed in the present study to infer the values of the parameters, comparing the estimates obtained against the values introduced in the model. For such 16 voice traces where synthesized using a *2-mass* model of the vocal folds. The value of the subglottal mass $(M_{ll}=M_{rl})$ was fixed to *0.2 g*. The supraglottal mass was varied from *0.005 to 0.05 g*. (see **Figure 7.1.b** and **c**).

On its turn the springs linking both masses to the reference wall $(K_{ll}=K_{rl})$ were set to *110* $g.sec^{-2}$ whereas the stiffness linking subglottal and supraglottal masses $(K_{ll2}=K_{rl2})$ was varied from *5* to *255* $g.sec^{-2}$ in alternating steps as shown in Figure **7.3.b** and **c**. The value for the theoretical pitch generated by the model values was fixed to *120 Hz* for all cases. The value of the losses was fixed to 4.10^{-2} $g.sec^{-1}$ for the whole set of traces. A model of the acoustic tube (vocal tract) with *64 sections* for the vowel /a/ was chained to the vocal fold model to generate vowel-like voice. Traces lasting *0.5 sec.* were generated at a sampling frequency of *48.000 Hz*. These were treated as described in **section 2** to obtain the *mucosal wave correlate*, and used in determining its power spectral density and the body biomechanical parameters as

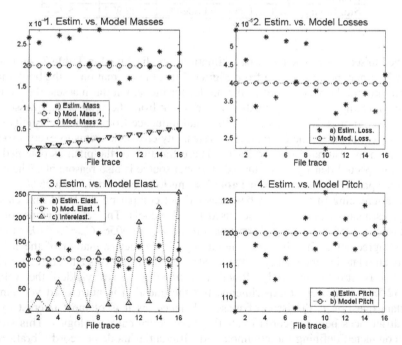

Fig. 7. 1.a) Estimated values for cord body masses. 1.b) Model values for subglottal masses. 1.c) Model values for supraglottal masses. 2.a) Estimated values for cord body losses. 2.b) Model values for subglottal and supraglottal losses. 3.a) Estimated values for cord body elasticity. 3.b) Model values for subglottal and supraglottal elasticity. 3.c) Model values for interelasticity. 4.a) Estimated values for the pitch. 4.b) Model values for pitch.

Table 2. Values of the estimated body parameters for the set of synthetic voice traces plotted in Figure 7

File No.	Mb	Rb	Kb	fp
1	2.660e-004	5.456e-005	122.255	107.900
2	2.539e-004	4.638e-005	126.641	112.400
3	1.793e-004	3.369e-005	98.919	118.201
4	2.714e-004	5.262e-005	145.814	116.665
5	2.643e-004	3.607e-005	132.950	112.873
6	2.838e-004	5.156e-005	151.128	116.149
7	2.046e-004	4.041e-005	94.956	108.414
8	2.850e-004	5.090e-005	168.515	122.384
9	2.064e-004	3.790e-005	112.653	117.584
10	1.591e-004	2.221e-005	100.238	126.342
11	1.700e-004	3.184e-005	94.090	118.407
12	2.385e-004	3.424e-005	140.773	122.277
13	1.971e-004	3.562e-005	107.553	117.576
14	2.334e-004	3.723e-005	142.443	124.347
15	1.726e-004	3.239e-005	100.078	121.194
16	2.300e-004	4.238e-005	134.663	121.771
Means:	2.260e-004	4.000e-005	123.354	117.780
Std. Dev.:	4.264e-005	9.060e-006	22.972	5.344

described in **section 3**. The resulting estimations are displayed in **Table 2** and listed in **Figure 7**. It may be appreciated from **Figure 7** that the estimation of the body mass is centered around the value fixed in the model for the subglottal masses, the estimates showing slight apparent contamination by crosstalk from the supraglottal masses. This is also the case of body stiffness, where a small influence from the interelasticity seems to slightly contaminate the estimates. Interelasticity crosstalk seems to exert also some influence in the estimation of the losses. The estimation of the pitch as obtained from the power spectral density of the unfolded glottal source is also reasonable. The dispersion of the parameters as seen from **Table 2** seems to be in the order of a 25 %.

The referencing of traces has been carried out comparing the mass and elasticity average estimates against the values used in the models. The relative gains for mass and elasticity coefficients have been found to be $G_{ma}=0.0056$, $G_{ka}=0.0040$, which are in good agreement. The absolute referencing for the determination of the losses is very much related to the energy of the trace as obtained from its autocorrelation function, and is still under study. Practical estimations have yielded the value of $G_{ra}= 32.53$ for this set of experiments, but the question is not closed yet. Another important question is the issue of mass unbalance, as it is of most interest to infer mass differences between cords related to several critical pathologies. This study is being conducted defining the common and differential modes of cord vibration, and from these a contribution associated to each cord body could be established. The same may be said for cord stiffness. A slight unbalance between waveform cycles may be observed in **Figure 4.a)** and **c)**. As estimations of mass, stiffness and losses will be available by cycles, the unbalance of these parameters (**BMU** – Body Mass

Unbalance, **BLU** – Body Losses Unbalance and **BSU** – Body Stiffness Unbalance) may be defined as

$$\mu_{uk} = \left(\hat{M}_{bk} - \hat{M}_{bk-1}\right)/\left(\hat{M}_{bk} + \hat{M}_{bk-1}\right)$$
$$\rho_{uk} = \left(\hat{R}_{bk} - \hat{R}_{bk-1}\right)/\left(\hat{R}_{bk} + \hat{R}_{bk-1}\right) \tag{10}$$
$$\gamma_{uk} = \left(\hat{K}_{bk} - \hat{K}_{bk-1}\right)/\left(\hat{K}_{bk} + \hat{K}_{bk-1}\right)$$

where $1 \leq k \leq K$ is the *cycle window* index and $\hat{M}_{bk}, \hat{R}_{bk}, and \hat{K}_{bk}$ are the *k-th cycle* estimates of mass, losses and stiffness on a given voice sample (intra-speaker). Other parameters of interest are the deviations of the average values of mass, losses and compliance for the *j-th* sample $\overline{M}_{bj}, \overline{R}_{bj}, and \hat{K}_{bj}$ relative to average estimates from a normophonic set of speakers (inter-speaker) as

$$\mu_{dj} = \left(\overline{M}_{bj} - \overline{M}_{bs}\right)/\overline{M}_{bs}$$
$$\rho_{dj} = \left(\overline{R}_{bj} - \overline{R}_{bs}\right)/\overline{R}_{bs} \tag{11}$$
$$\gamma_{dj} = \left(\overline{K}_{bj} - \overline{K}_{bs}\right)/\overline{C}_{bs}$$

these parameters are known as **BMD** (Body Mass Deviation), **BLD** (Body Losses Deviation) and **BSD** (Body Stiffness Deviation).

5   Results from Natural Voice

A variant of Principal Component Analysis (PCA) known as *multivariate measurements analysis* (see [12], pp. 429-30) Hierarchical Clustering and have been used with the distortion parameters given in **Table 3** [11].

PCA is conceived as the optimal solution to find the minimum order of a linear combination of random variables x_j showing the same variance as the original set, where the components of x_j correspond to different observations (samples) of a given input parameter (*j-th* parameter) for a set of 20 normophonic and 20 pathologic samples (4 samples with polyps, 6 samples with bilateral nodules, 5 samples with Reinke's Edema, and 5 samples with reflux inflammation) as listed in **Table 4**.

Table 3. List of parameters estimated from voice

Coeff.	Description
x_1	pitch
x_2	jitter
x_{3-5}	shimmer-related
x_{6-7}	glottal closure-related
x_{8-10}	HNR-related
x_{11-14}	mucosal wave psd in energy bins
x_{15-23}	mucosal wave psd singular point values
x_{24-32}	mucosal wave psd singular point positions
x_{33-34}	mucosal wave psd singularity profiles
x_{35-37}	biomechanical parameter deviations (11)
x_{38-40}	biomechanical parameter unbalance (10)

Table 4. Values of x_{35-39} for the samples studied. Sample conditions are: N – *Normophonic;
BP* – *Bilateral Polyp; LVCP* – *Left Vocal Cord Polyp; BRE* – *Bilateral Reinke's Edema; BN* –
Bilateral Noduli; LR – *Larynx Reflux; RE* – *Reinke's Edema; RVCP* – *Right Vocal Cord Polyp.*

Trace	Condit.	BMD	BLD	BSD	BMU	BLU
001	N	-0.632	-0.136	-0.540	0.027	0.039
003	N	-0.154	-0.145	-0.137	0.079	0.056
005	N	-0.039	-0.299	-0.213	0.078	0.044
007	N	-0.492	-0.461	-0.573	0.036	0.046
00A	N	-0.542	-0.207	-0.567	0.065	0.064
00B	N?	1.320	0.642	1.250	0.149	0.191
00E	N	-0.054	0.012	-0.128	0.159	0.098
010	N	-0.408	0.164	-0.491	0.115	0.103
018	N	-0.031	-0.205	-0.167	0.078	0.076
01C	N	-0.557	-0.315	-0.581	0.058	0.052
024	N?	0.631	1.330	1.200	0.120	0.124
029	N	0.101	-0.111	0.416	0.057	0.048
02C	N	-0.329	-0.253	-0.079	0.035	0.040
02D	N	-0.227	-0.193	0.022	0.116	0.053
032	N	-0.507	-0.019	-0.367	0.038	0.071
035	N	0.424	-0.302	-0.021	0.099	0.065
043	N	0.219	0.156	0.466	0.059	0.030
047	N	-0.497	1.070	-0.180	0.076	0.052
049	N	-0.157	0.160	0.029	0.113	0.079
04A	N	-0.005	1.770	0.073	0.098	0.075
065	BP	0.240	7.490	3.220	0.835	0.712
069	LVCP	0.560	3.490	2.460	0.408	0.318
06A	BRE	0.142	2.860	1.760	0.300	0.331
06B	BN	0.427	3.860	2.150	0.339	0.326
06D	BN	0.573	3.540	2.160	0.338	0.339
071	BRE	0.417	3.210	1.870	0.306	0.348
077	LR	2.000	3.170	3.660	0.460	0.320
079	RE	0.658	2.860	2.170	0.396	0.333
07E	BN	0.843	2.990	2.340	0.328	0.303
07F	LR	0.420	2.850	1.950	0.332	0.309
083	LR	0.253	2.880	1.900	0.391	0.333
092	BRE	0.216	2.750	1.720	0.469	0.353
098	RE	0.187	2.830	1.720	0.360	0.339
09E	BN	1.400	11.700	5.510	0.637	0.518
09F	LR	0.062	2.920	1.660	0.309	0.334
0A0	RVCP	0.156	3.020	1.720	0.333	0.338
0A9	LVCP	0.012	3.600	1.660	0.293	0.311
0AA	LR	-0.091	2.970	1.600	0.268	0.315
0B4	BN	0.154	4.280	1.870	0.305	0.338
0CA	BN	-0.057	3.040	1.630	0.310	0.361

These samples were processed to extract the set of *40* parameters listed in **Table 3**,
of which two subsets were defined for classification: $S_1=\{x_{2-39}\}$, including most of the
parameters available, and $S_2=\{x_2, x_3, x_8, x_{35-39}\}$ including *jitter, shimmer, HNR*, deviations (**BMD, BLD** and **BSD**), and unbalances (**BMU** and **BLU**). The results of the

clustering process are shown in **Figure 8** as biplots against the two first principal components from PCA analysis. It may be seen that the clustering process assigned most of normophonic samples to one cluster (with the exception of *00B* and *024*) both for S_1 as well as for S_2. The results using S_2 are given in **Table 5**.

Table 5. Clustering results for S_2

Cluster	Samples
c_{21} (o)	001, 003, 005, 007, 00A, 00E, 010, 018, 01C, 029, 02C, 02D, 032, 035, 043, 047, 049, 04A
c_{22} (◊)	00B, 024, 065, 069, 06A, 06B, 06D, 071, 077, 079, 07E, 07F, 083, 092, 098, 09E, 09F, 0A0, 0A9, 0AA, 0B4, 0CA

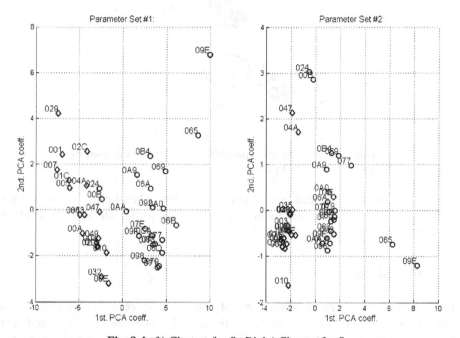

Fig. 8. Left) Clusters for S_1. Right) Clusters for S_2.

To further clarify the analysis a 3D plot of the results vs the three most relevant input parameters in S_2 as established by PCA is presented in **Figure 9**. The most relevant parameter according to this combination seems to be **BSD** (x_{37}). The larger x_{37}, the stiffer the cord and the less normophonic the production. The second most relevant parameter seems to be **jitter** (x_2). The third most relevant parameter is **BLD** (x_{36}) associated to the profile of the spectral profile peak (Q factor).

The behaviour of cases *00B* and *024*, classified as pathological by PCA analysis deserves a brief comment. These appear in **Figure 9** (encircled) not quite far from normal cases 001-04A, but showing a stiffness that doubles those of normophonic

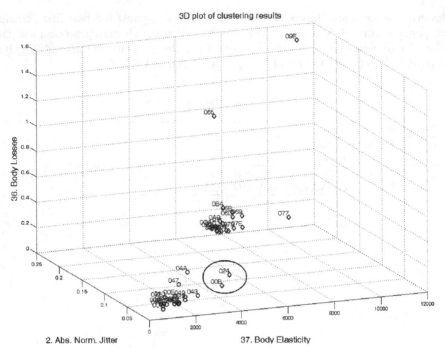

Fig. 9. 3D Clustering Plot showing the separation in the manifold defined by the parameter subset {x37, x2 and x36} – ordered by relevance

samples. Apparently this detail was determinant in their classification as not normophonic by PCA. This fact was confirmed by their values for the **BSD** in **Table 4**, being 1.25 and 1.2 respectively, or 225% and 220%.

6 Conclusions

Through the present paper the possibility of obtaining indirect estimates of the vocal cord biomechanical parameters from the voice trace has been shown. This could open new possibilities for the non-invasive distant exploration of patients both for pathology detection and classification by analysis of the voice trace. The method is still subject to revision to take into account the influence of second-order biomechanical parameters. Its possible extension to unbalanced parameter estimation is also under study. The methodology presented detects biomechanical unbalance from voice records for pathology detection by common pattern recognition techniques. Normophonic samples show small unbalance indices, as opposed to pathologic ones. There is not a specific pattern of unbalance related to a given pathology (although more cases need to be studied). Biomechanical parameter unbalance is a correlate to pathology quantity rather than quality. Although mild pathologies may appear as normophonic from subjective analysis the use of the proposed methods may spot them and help in keeping trace of their evolution in time. Adequately combining classical distortion parameters with deviation parameters renders fairly good results in pathology detection. These conclusions need to be confirmed by more experiments.

Acknowledgments

This research is being carried out under grants TIC2002-02273 and TIC2003-08756 from the *Programa Nacional de las Tecnologías de la Información y las Comunicaciones (Spain).*

References

[1] Akande, O. O. and Murphy, P. J., "Estimation of the vocal tract transfer function with application to glottal wave analysis, Speech Communication, Vol. 46, No. 1, May 2005, pp. 1-13.

[2] Alku, P., "An Automatic Method to Estimate the Time-Based Parameters of the Glottal Pulseform", ICASSP'92, pp. II/29-32.

[3] Berry, D. A.: Mechanisms of modal and nonmodal phonation. Journal of Phonetics, Vol. 29, (2001), pp. 431-450.

[4] Cannito, M. P., Buder, E. H., Chorna, L. B., "Spectral Amplitude Measures of Adductor Spasmodic Dysphonic Speech", Journal of Voice, Vol. 19, No. 3 (2005), pp. 391-410

[5] Childers, D. G.: Speech Processing and Synthesis Toolboxes. John Wiley & Sons, New York (2000).

[6] Dibazar, A. A., Narayanan, S.: A System for Automatic Detection of Pathological Speech. Proc. of the 36th Asilomar Conf. on Signals, Systems and Computers (2002).

[7] Godino, J. I., Gómez, P.: Automatic Detection of Voice Impairments by means of Short Term Cepstral Parameters and Neural Network based Detectors. IEEE Trans. on Biomed. Eng. Vol. 51, No. 2 (2004), pp. 380-384.

[8] Gómez, P., Díaz, F., Martínez, R., Godino, J. I., Álvarez, A., Rodríguez, F. , Rodellar, V.: Precise Reconstruction of the Mucosal Wave for Voice Pathology Detection and Classification. Proceedings of the EUSIPCO'04 (2004) 297-300.

[9] Gómez, P., Godino, J. I., Díaz, F., Álvarez, A., Martínez, R., Rodellar, V.: Biomechanical Parameter Fingerprint in the Mucosal Wave Power Spectral Density, Proc. of the ICSLP'04 (2004) 842-845.

[10] Gómez, P., Godino, J. I., Álvarez, A., Martínez, R., Nieto, V., Rodellar, V., "Evidence of Glottal Source Spectral Features found in Vocal Fold Dynamics", Proceedings of the 2004 Int. Conf. on Acoustics, Speech and Signal Processing, ICASSP'05 (2005), pp. 441-444.

[11] Gómez, P., Díaz, F., Álvarez, A., Martínez, R., Rodellar, V., Fernández, R., Nieto, A., Fernández, F. J.: PCA of Perturbation Parameters in Voice Pathology Detection. Proc. of EUROSPEECH'05. (2005), pp. 645-648.

[12] Johnson, R. A., Wichern, D. W.: Applied Multivariate Statistical Analysis. Prentice-Hall, Upper Saddle River, NJ, 2002.

[13] Growing Problem for Call Centre Workers Suffering from Repetitive Voice Injury. http://www.healthandsafetytips.co.uk/News.htm

[14] Kuo, J., Holmberg, E. B., Hillman, R. E.: Discriminating Speakers with Vocal Nodules Using Aerodynamic and Acoustic Features. Proc. of the ICASSP'99, Vol. 1, 15-19 March (1999), pp. 77-80.

[15] Ritchings, T.; McGillion, M.; Moore, C.: Pathological voice quality assessment using artificial neural networks. Medical Engineering and Physics, vol. 24 (2002), pp. 561-564.

[16] Simberg, S.: Prevalence of Vocal Symptoms and Voice Disorders among Teacher Students and Teachers and a Model of Early Intervention, Ph.D. Thesis, Department of Speech Sciences, University of Helsinki, http://ethesis.helsinki.fi/.

[17] Story, B. H.: An overview of the physiology, physics and modeling of the sound source for vowels. Acoustic Sci. & Tech., Vol. 23, 4 (2002), pp. 195-206.

[18] Story, B. H., and Titze, I. R.: Voice simulation with a bodycover model of the vocal folds. J. Acoust. Soc. Am., Vol. 97 (1995), pp. 1249–1260.

[19] The Voice Center of Eastern Virginia Medical School. http://www.voice-center.com/larynx_ca.html.

[20] Titze, I. R.: Principles of Voice Production. Prentice-Hall, Englewood Cliffs (1994).

[21] Titze, I. R.: The physics of small-amplitude oscillation of the vocal folds. Journal of the Acoustical Society of America, Vol. 83, No. 4 (1988) 1536-1552.

[22] Titze, I.R., Lemke, J., & Montequin, D.: Populations in the U.S. workforce who rely on voice as a primary tool of trade: A preliminary report. *Journal of Voice 11,* (1997) 254–259.

[23] Varho, S., Alku, P., "A new predictive method for all-pole modelling of speech spectra with a compressed set of parameters", *Proc. ICASSP99*, 1999, pp. III/126-129.

A Speech Recognizer Based on Multiclass SVMs with HMM-Guided Segmentation

D. Martín-Iglesias, J. Bernal-Chaves, C. Peláez-Moreno,
A. Gallardo-Antolín, and F. Díaz-de-María

Signal Theory and Communications Department,
EPS-Universidad Carlos III de Madrid,
Avda. de la Universidad, 30, 28911-Leganés (Madrid), Spain

Abstract. Automatic Speech Recognition (ASR) is essentially a problem of pattern classification, however, the time dimension of the speech signal has prevented to pose ASR as a simple static classification problem. Support Vector Machine (SVM) classifiers could provide an appropriate solution, since they are very well adapted to high-dimensional classification problems. Nevertheless, the use of SVMs for ASR is by no means straightforward, mainly because SVM classifiers require an input of fixed-dimension. In this paper we study the use of a HMM-based segmentation as a mean to get the fixed-dimension input vectors required by SVMs, in a problem of isolated-digit recognition. Different configurations for all the parameters involved have been tested. Also, we deal with the problem of multi-class classification (as SVMs are initially binary classifers), studying two of the most popular approaches: 1-vs-all and 1-vs-1.

1 Introduction

Hidden Markov Models (HMMs) are, undoubtedly, the most employed core technique for Automatic Speech Recognition (ASR). During the last decades, research in HMMs for ASR has brought about significant advances and, consequently, the HMMs are currently accurately tuned for this application. Nevertheless, we are still far from achieving high-performance ASR systems. Some alternative approaches, most of them based on Artificial Neural Networks (ANNs), were proposed during the last decade ([1], [2], [3], [4] and [5] are some examples). Some of them tackled the ASR problem using predictive ANNs, while others proposed hybrid (HMM-ANN) approaches. Nowadays, however, the preponderance of HMMs in practical ASR systems is a fact.

Speech recognition is essentially a problem of pattern classification, but the high dimensionality of the sequences of speech feature vectors has prevented researchers to propose a straightforward classification scheme for ASR. Support Vector Machines (SVMs) are state-of-the-art tools for linear and nonlinear knowledge discovery [6], [7]. Being based on the maximum margin classifier, SVMs are able to outperform classical classifiers in the presence of high dimensional data even when working with nonlinear machines.

M. Faundez-Zanuy et al. (Eds.): NOLISP 2005, LNAI 3817, pp. 257–266, 2005.

Some researchers have already proposed different approaches to speech recognition aiming at taking advantage of this type of classifiers. Among them, [8], [9] and [10] use different approaches to perform the recognition of short duration units, like isolated phoneme or letter classification. In [8], the authors carry out a length adaptation based on the triphone model approach. In [9] and [10], a normalizing kernel is used to achieve the adaptation. Both cases show the superior discrimination ability of SVMs. Moreover, in [9], a hybrid approach based on HMMs has been proposed and tested in a CSR (Continuous Speech Recognition) task.

Nevertheless, the use of SVMs for ASR is by no means straightforward. The main problem is the required dimensional normalization, due to the fact that the usual kernels can only deal with vectors of fixed size. However, speech analysis generates sequences of feature vectors of variable lengths (due to the different durations of the acoustic units and the constant frame rate commonly employed). A possible solution is that showed in [11], where the non-uniform distribution of analysis instants provided by the internal states of an HMM with a fixed number of states and a Viterbi decoder is used for dimensional normalization.

Another difficulty is that speech recognition is a problem of multi-class classification, while in the original formulation, an SVM is a binary classifier. Although some versions of multi-class SVMs have been proposed, they are computationally expensive. A more usual approach to cope with this limitation is combining a number of binary SVMs to construct a multi-class classifier. In this paper we have studied two of the most popular approaches (1-vs-1 and 1-vs-all), testing it in a specific ASR task.

This paper is organized as follows. In next section, we describe the fundamentals of SVMs and we describe the procedures for multiclass implementation. Afterwards, we make a review of the HMM-guided segmentation method to produce input vectors with fixed dimension. Then, in Section 4, we present the experimental framework and the results obtained, explaining the different criterions followed to chose the several parameters of the system. Finally, some conclusions and further work close the paper.

2   SVM Fundamentals

2.1   SVM Formulation

An SVM is essentially a binary classifier capable of guessing whether an input vector $\mathbf{x}$ belongs to a class $y_1 = +1$ or to a class $y_2 = -1$. The decision is made according to the following expression:

$$g(\mathbf{x}) = \mathbf{w} \cdot \phi(\mathbf{x}) + b, \tag{1}$$

where $\phi(\mathbf{x})$: $\Re^n \mapsto \Re^{n'}$, ($n << n'$), is a nonlinear function which maps the vector $\mathbf{x}$ to a *feature space* with higher dimensionality (possibly infinite) where classes are linearly separable, and $\mathbf{w}$ defines the separating hyper-plane in such a space.

What makes SVMs more effective than other methods based on linear discriminants is the learning criterion, because instead of minimizing only the empirical risk, they also try to minimize the structural risk, being the solution found a compromise between the empirical error and the generalization capability.

The solution is given by the following minimization problem:

$$\min_{\mathbf{w},b,\xi_i} \frac{1}{2}\mathbf{w}\cdot\mathbf{w} + C\sum_{i=1}^{N}\xi_i, \tag{2}$$

$$[3pt]\text{subject to} \quad y_i(\mathbf{w}\cdot\phi(\mathbf{x}_i)+b) \geq 1 - \xi_i, \tag{3}$$

$$[3pt]\xi_i \geq 0, \text{ for } i = 1,\cdots,N, \tag{4}$$

where $\mathbf{x}_i \in \Re^n$, $i = 1,\ldots,N$ are the training vectors corresponding to the labels $y_i \in \{\pm 1\}$, and the parameter C establishes the compromise between error minimization and generalization capability.

The SVM is usually solved introducing the restrictions in the minimizing function using Lagrange multipliers, leading to the maximization of the Wolfe dual:

$$L_d = \sum_{i=1}^{n}\alpha_i - \sum_{i=1}^{n}\sum_{j=1}^{n}y_iy_{ij}\alpha_i\alpha_j\phi^T(\mathbf{x}_i)\phi(\mathbf{x}_j) \tag{5}$$

with respect to α_i and subject to $\sum_{i=1}^{n}\alpha_i = 0$ and $0 \leq \alpha_i \leq C$. This problem is quadratic and convex, so its convergence to a global minimum is guaranteed using quadratic programming (QP) schemes. The value of $\mathbf{w}$ and b can be recovered from the Lagrange multipliers α_i, that are associated with the first linear restriction in the SVM formulation:

$$\mathbf{w} = \sum_{i=1}^{N}\alpha_iy_i\phi(\mathbf{x}_i),$$

$$[3pt]b = \sum_{j}\alpha_jy_j\phi(\mathbf{x}_i)\cdot\phi(\mathbf{x}_j) + y_i, \forall i. \tag{6}$$

According to (6), only vectors with an associated $\alpha_i \neq 0$ will contribute to determine the weight vector $\mathbf{w}$ and, therefore, the separating boundary, and they receive the name of *support vectors*.

Generally, function $\phi(\mathbf{x})$ is not explicitly known (in fact, in most of the cases its evaluation would be impossible as long as the feature space dimensionality can be infinite). However, we don't actually need to know it, since the only we need to evaluate are the dot products $\phi(\mathbf{x}_i)\cdot\phi(\mathbf{x}_j)$ that, by using what has been called the *kernel trick*, can be evaluated using a kernel function $K(\mathbf{x}_i,\mathbf{x}_j)$.

By this way, the form that finally adopts an SVM is:

$$g(\mathbf{x}) = \sum_{i=1}^{N}\lambda_iy_iK(\mathbf{x}_i,\mathbf{x}) + b. \tag{7}$$

The most widely used kernel functions are the *gaussian radial basis function*,

$$K(\mathbf{x}_i, \mathbf{x}_j) = \exp\left(-\frac{\|\mathbf{x}_i - \mathbf{x}_j\|^2}{2\gamma^2}\right), \tag{8}$$

with an associated feature space of infinite dimensionality, and the polynomial kernel

$$K(\mathbf{x}_i, \mathbf{x}_j) = (1 + \mathbf{x}_i \cdot \mathbf{x}_j)^p, \tag{9}$$

which associated feature space are the polynomials up to grade p.

2.2   Multiclass SVM

Besides of the necessity of using input vectors with fixed-length, there is another important issue that must be solved whenever we work with SVMs for ASR. While in speech recognition we have to make a decision among several classes, support vector machines were originally designed for binary classification and their generalization to the multi-class case is still an on-going research field.

Although some of the proposed approaches make a reformulation of the SVM equations to consider all classes at once, this option is very expensive computationally and, therefore, we haven't consider them in this work.

Another different approach is combining results of several binary classifiers to construct a multi-class classifier. We have experimented with two different versions of this method. The former consists in comparing each class against all the rest (1-vs-all) while in the latter each class is confronted against all the other classes separately (1-vs-1) [12].

In the 1-vs-all method we have to construct k SVMs (with k the number of classes) and in the 1-vs-1 $k(k-1)/2$, but, since in the second approach the number of training vectors for each class is smaller, the necessary computational effort can be ever lower in the latter case, as shown in [13].

For the 1-vs-1 alternative we have used the implementation described in [14], where error correcting codes are used to compare the outputs of the classifiers, and, for the 1-vs-all approach, we obtained the probability-like outputs using the implementation in [15]. Afterwards, the outputs of the binary 1-vs-all classifiers were compared, and the most probable class among the ones showing a positive output was chosen (positive meaning that the binary classifier had selected the 'one' against the 'rest').

3   Feature Extraction and Dimensional Normalization

Since the speech signal is quasi-stationary, speech analysis must be performed on a short-term basis. Typically, the speech signal is divided into a number of overlapping time windows and a speech feature vector is computed to represent each of these frames. The size of the analysis window, w_a, is usually of 20-30 ms. The frame period, T_f, (the time interval between two consecutive analysis windows) is set to a value between 10 and 15 ms. Habitually, $w_a = KT_f$, where K is called the overlapping factor.

With respect to the feature vectors themselves, for each analysis window, twelve Mel-Frequency Cepstral Coefficients (MFCC) are obtained using a mel-scaled filter-bank with 40 channels. Then, the log-energy, the twelve delta-cepstral coefficients and the delta-log energy are appended, making a total vector dimension of 26.

Typically, the values of w_a and T_f are kept constant for every utterance that, on the other hand, exhibits a different time duration. Consequently, the speech analysis generates sequences of feature vectors of variable length. As we have already mentioned, a normalization of these lengths is required to use SVM classifiers.

In a previous work [11], three procedures to perform this dimensional normalization are proposed. Two of them were very straightforward approaches consisting on adjusting either the analysis window size or the frame period to obtain a fixed number of time analysis instants. The third one, more sophisticated, used and HMM-based segmentation to select the time analysis instants. The next subsection describes this last method, that is the one selected for the experiments conducted in this work.

3.1   Non-uniform Distribution of Analysis Instants

An appropriate selection of the time instants at which the speech signal is analysed can presumably improve the classification results.

To determine the appropriate analysis instants, we propose to use the implicit information in the segmentation made by HMM, i.e., to consider those instants at which state transitions occur (very likely related to those at which the changes of the speech spectra happen).

This HMM-guided parameterization procedure consists of two main stages. The first stage is a HMM classifier (a Viterbi decoder) that yields the best sequence of states for each utterance and also provides a set of state boundary time marks. The second stage extracts the speech feature vectors at the time instants previously marked. For the first stage, we have used left-to-right continuous density HMMs with three Gaussian mixtures per state. Each HMM represents a whole-word and consists of Ns states with the topology shown in Figure 1. These models have been trained using only the training set of the speech database and the conventional parameterization module used for the baseline experiments. In particular, the speech parameters con-sists of 12 MFCC, the log-energy, 12 delta-MFCC and the delta-log energy, extracted using a frame period of 10 ms and an analysis Hamming window of 25 ms.

As mentioned before, these acoustic models are used to generate alignments at state level for each utterance in the speech database. In this process, each utterance is compared to each of the HMMs and only the segmentation produced by the acoustic model yielding the best score is saved for the next stage. Note that the obtained seg-mentation may not be correct, even when the utterance is properly recognized by the HMM-based system. Segmentation errors may

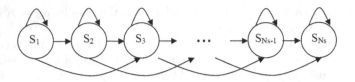

Fig. 1. HMM topology

produce some degradation in the per-formance of the whole system, however, for our task, the results obtained show that the segmentation is accurate enough. Anyway, it is necessary to consider this issue for further research.

In the second stage, the feature vectors are extracted at the time instants derived from the HMM-guided segmentation. In particular, a 25 ms analysis window is sub-sequently located at these time instants. In this way, the number of feature vectors per utterance used as the SVM input turns out to be equal to the number of states (Ns), determined by the HMM topology. In our case, the number of states was fixed to 17 (the same number of states we use for HMM-based recognition).

4   Experimental Results

4.1   Baseline System and Database

We have used a database consisting of 72 speakers and 11 utterances per speaker for the 10 Spanish digits. This database was recorded at 8 kHz in clean conditions. Since this database is not large enough to achieve reliable speaker-independent results, we have used a 9-fold cross validation to artificially extend it. Specifically, we have split each database into 9 balanced groups; 8 of them for training and the remaining one for testing, averaging the results afterwards. In summary, we use a total of 7,920 words for testing our systems.

The baseline HMM-based ASR system is an isolated-word, speaker independent system developed using the HTK package [16]. Left-to-right HMMs with continuous observation densities were used. Each of the whole-digit models contains a different number of states (which depends on the number of allophones in the phonetic transcription of each digit) and three Gaussian mixtures per state.

For the baseline experiment with the HMM classifier, a Hamming window with a width of 30 ms was used and the feature vectors (consisting of 12 MFCC, the log-energy, 12 delta-MFCC and the delta-log energy) were extracted once every 10 ms.

We have tested our systems in clean conditions and in presence of additive noise. For that purpose, we have corrupted our database with two kinds of noises, namely: white noise and the noise produced by a F16 plane. Both noises have been extracted from the NOISEX database [17] and added to the speech signal to achieve a signal-to-noise ratio of 12 dB. As we have used clean speech for estimating the acoustic models (in both, HMM and SVM-based recognizers), the noises are only added for testing the recognition performance.

4.2   Selecting Parameters for the SVM-Based Recognizer

In order to get the best performance possible for the system proposed, we have to select properly the values of the different parameters involved. Specifically, we must answer the following questions:

1. What is the best kernel and which are the best parameters for it?
2. How many states must we use in the HMM-segmentation procedure?
3. When must we extract the features, in the transitions between states, or between two transitions, when the voice is stationary?
4. What's the best window size? Does it depend on the length of the utterance?
5. Which parameterization should we use and what is the best normalization procedure?

It is hard to make a guess *a priori* for these questions. However, the answer will have a great impact in the recognition rates reached for our system. For this reason, we ran a set of experiments with different values for all these configuration parameters.

We have used the RBF kernel (eq. (8)) in all the experiments, finding values for γ and for the regularization parameter C of the SVM by using *grid search*. However, we didn't find a significant difference among the different values tried (about 1% in the recognition rate).

Regarding the number of the states of the HMMs used for the segmentation, we have the problem that the different words in the dictionary have different lengths. This implies that for a given number of states, some words can be oversampled while, for others, we won't have taken a number of parameters large enough. For both SVM classifiers (1-vs-all and 1-vs-1), we have finally used a 15 state HMM to produce the sampling instants in which the speech signal is analysed. Thus, in this case we use 15 feature vectors per utterance as the SVM input. The number of 15 was chosen because with less, the recognition rate was poor, and with more, the computational cost was very high, while the improvement in recognition was not so noticeable.

Once selected the most adequate number of states for the HMMs, is necessary to determine the best moment to extract parameters for the SVM. This moment could correspond with the transition between two states, which is associated with a change in the spectrum, or with the time when the voice is stationary, between two transitions. We have tried the two schemes and, even, a combination of both, but all the results were very similar. We finally decided to extract parameters in the transitions between states.

As long as we are using a fixed number of states for all words, and these words can have very different durations, one could think that it would be a good idea to adjust the size of the window used, making it wider depending on the length between samples (i.e. transitions between states). In Figure 2 is illustrated this approach. In the final implementation, however, as the results with a variable window didn't differ from those obtained with the fixed one, we used the standard fixed window of 30 ms.

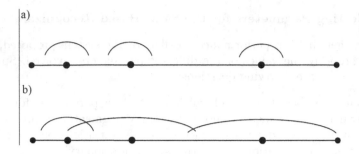

Fig. 2. Parameter extraction using a fixed window size a), and a size depending on the length between transitions b)

Table 1. Word Accuracy Rate (%) obtained with four different normalizations of the speech features: only subtracting the mean value of each parameter, dividing the result between the standard deviation of each parameter, dividing between the standard deviation of all parameters together and dividing between the maximum value of each parameter

	HMM		SVM 1-vs-All		SVM 1-vs-1	
	Clean	12dB WN	Clean	12dB WN	Clean	12dB WN
Baseline	**99.89**	36.92	**99.72**	39.44	**99.72**	37.07
$x_i = x_i - \bar{x}_i$	99.42	41.67	99.51	40.68	99.5	39.87
$x_i = \frac{x_i - \bar{x}_i}{\sigma_{x_i}}$	99.08	31.19	98.8	34.17	98.72	33.29
$x_i = \frac{x_i - \bar{x}_i}{\sigma_x}$	98.32	33.24	98.48	32.23	98.6	31.2
$x_i = \frac{x_i - \bar{x}_i}{\max(x_i)}$	99.34	**49.4**	99.35	**50.73**	99.35	**50.6**

Table 2. Recognition results obtained with the two proposed hybrid HMM-SVM-based classifiers for two types of noises (white and F16). Results obtained with the conventional HMM-based ASR system are presented as well.

	Clean	White (SNR=12 dB)	F16 (SNR=12 dB)
HMM-based ASR	99.34%	49.4%	59.31%
Hybrid HMM-SVM ASR system(1-vs-all)	99.35%	50.73%	59.47%
Hybrid HMM-SVM ASR system(1-vs-1)	99.35%	50.6%	59.42%

Concerning parameterization, we used the same as in the baseline experiment (12 MFCC + logE + Δ + Δ-logE). However, an important issue in HMMs, and even more in SVMs, is the normalization used. We tried four different types of normalization, as we can see in Table 1. Although with clean speech the rates without normalization are slightly better, with noisy speech the results obtained subtracting the mean to each parameter and dividing them between their maximum values are the best ones.

4.3 Experiments and Results

On Table 2 are shown the best word recognition rates obtained with both alternatives of the multiclass SVM-based system and in comparison to those achieved by the HMM-based system. As it can be observed, the SVM classifiers performed only slightly better than the baseline system. The explanation for this is that, when HMM fails, the segmentation obtained is far from the optimal and so, the SVMs don't have very much to do. However, even so, SVMs outperforms the HMM-based system in all experiments, getting an improvement of more than 1% in the presence of white noise.

If we compare the two approaches proposed for recognition with SVMs, we can see that results with 1-vs-all performs better than 1-vs-1. However, rates obtained with both methods are again very close.

5   Conclusions and Further Work

In this paper, we have proposed two different approaches to a multiclass SVM classifier (1-vs-all and 1-vs-1) with application to a specific ASR task. Experimental results have shown that recognition rates obtained with SVM-based systems are very close to that achieved by a conventional HMM-based ASR system in clean conditions. However, in noisy environments, differences are enlarged, getting the 1-vs-all SVM-based classifier the best results.

Although the improvement obtained for the system proposed is not very large, from our point of view, the results are very encouraging since HMM-based systems have been accurately tuned during the last three decades for automatic speech recognition, while speech recognition based on SVMs is a new field of study with a big margin for improvement.

With respect to the further work, we consider several lines: first of all, it would be desirable to find an alternative method of getting a fixed-dimension input, avoiding by this way the problem of a bad segmentation when the HMM fails. Some method based on the behaviour of the derivative of the spectral features could be considered.

Also, since the parameterization used is specially designed for a back-end based on HMMs, it would be interesting to explore alternative parameterizations. Currently, we are completing the first experiments with LSP parameters and the results outperform those showed here.

Finally, we expect to extend the SVM framework for ASR by using string kernels, which has been used with success in tasks as protein [14] and text [17] classification. These kernels, based mainly on the Fisher score and other score spaces, work in conjunction with a generative model, such as an HMM, and can deal with sequences of different length.

References

1. H. Sakoe, R. Isotani, K. Yoshida, K. Iso, and T. Watanabe. Speaker-independent word recognition using dynamic programming neural networks. *Proc. ICASSP-89*, pages 29–32, 1989.
2. K. Iso and T. Watanabe. Speaker-independent word recognition using a neural prediction model. *Proc. ICASSP-90*, pages 441–444, 1990.
3. J. Tebelskis, A. Waibel, B. Petek, and O. Schmidbauer. Continuous speech recognition using predictive neural networks. *Proc. ICASSP-91*, pages 61–64, 1991.
4. Y. Bengio. Neural networks for speech and sequence recognition. *London International Thomson Computer Press*, 1995.
5. H.A. Bourlard and N.Morgan. Connectionist speech recognition: a hybrid approach. *Kluwer Academic Publishers*, 1994.
6. B. Schölkopf and A. Smola. Learning with kernels. *M.I.T. Press*, 2001.
7. V. Vapnik. Statistical learning theory. *Wiley*, 1998.
8. P. Clarkson and P.J. Moreno. On the use of support vector machines for phonetic classification. *IEEE International Conference on Acoustics, Speech and Signal Processing*, 2:585–588, 1999.
9. A. Ganapathiraju. Support vector machines for speech recognition. *PhD Thesis, Mississipi State Universisty*, 2002.
10. N.D. Smith and M.J.F. Gales. Using SVMs and discriminative models for speech recognition. *IEEE International Conference on Acoustics, Speech and Signal Processing*, 2002.
11. J.M. García-Cabellos, C. Peláez-Moreno, A. Gallardo-Antolín, F. Pérez-Cruz, and F. Díaz de María. SVM classifiers for ASR: A discussion about parameterization. *Proceedings of EUSIPCO 2004*, pages 2067–2070, 2004.
12. E.L. Allwein, R.E. Schapire, and Y. Singer. Reducing multiclass to binary: a unifying approach for margin classifiers. *Journal of Machine Learning Research*, 1:113–141, 2000.
13. C.-W. Hsu and C.-J. Lin. A comparison of methods for multi-class support vector machines. *IEEE Transactions on Neural Networks*, 13, 2002.
14. T.K. Huang, R.C. Weng, and C.J. Lin. A generalized bradley-terry model: From group competition to individual skill. *[on-line] http://www.csie.ntu.edu.tw/~cjlin/libsvmtools/*, 2004.
15. Ch. Chih-Chung and L. Chih-Jen. LIBSVM: a library for support vector machines. *[on-line]http://www.csie.ntu.edu.tw/ cjlin/libsvm/*, 2004.
16. S. Young et al. HTK-Hidden Markov Model toolkit (ver 2.1). *Cambridge University*, 1995.
17. A.P. Varga, J.M. Steenneken, M. Tomlinson, and D. Jones. The NOISEX-92 study on the effect of additive noise on automatic speech recognition. *Tech. Rep. DRA Speech Research Unit*, 1992.

Segment Boundaries in Low Latency
Phonetic Recognition*

Giampiero Salvi

KTH, School of Computer Science and Communication,
Dept. for Speech, Music and Hearing, Stockholm, Sweden
giampi@kth.se

Abstract. The segment boundaries produced by the Synface low latency phoneme recogniser are analysed. The precision in placing the boundaries is an important factor in the Synface system as the aim is to drive the lip movements of a synthetic face for lip-reading support. The recogniser is based on a hybrid of recurrent neural networks and hidden Markov models. In this paper we analyse the look-ahead length in the Viterbi-like decoder affects the precision of boundary placement. The properties of the entropy of the posterior probabilities estimated by the neural network are also investigated in relation to the distance of the frame from a phonetic transition.

1 Introduction

The Synface system [1] uses automatic speech recognition (ASR) to derive the lip movements of an avatar [2] from the speech signal in order to improve the communication over the telephone for hearing impaired people.

The recogniser, based on a hybrid of artificial neural networks (ANN) and hidden Markov models (HMM), has been optimised for low latency processing (look-ahead lengths in the order of tens of milliseconds). The effect of limiting the look-ahead length has been investigated in [3, 4] by means of standard evaluation criteria, such as recognition accuracy, number of correct symbols and percent correct frame rate.

However, in applications such as this, where the alignment of the segment boundaries is essential, standard evaluation criteria hide important information.

In this study the recognition results from the Synface recogniser are analysed in more detail showing how the boundary placement in some cases is dependent on the look-ahead length.

The use of neural networks allows the estimation of the posterior probabilities of a class given an observation. The entropy of those probabilities is shown to assume local maxima close to phonetic transitions, which makes it a good candidate for a predictor of phonetic boundaries.

* This research was funded by the Synface European project IST-2001-33327 and carried out at the Centre for Speech Technology supported by Vinnova (The Swedish Agency for Innovation Systems), KTH and participating Swedish companies and organisations.

M. Faundez-Zanuy et al. (Eds.): NOLISP 2005, LNAI 3817, pp. 267–276, 2005.

The rest of the paper is organised as follows: Section 2 describes the recogniser and the data used in the experiments. Section 3 displays examples extracted from the sentence material in the test set. Section 4 explains the method and the measures analysed in the experiments. Finally Section 5 presents the results of the analysis and Section 6 concludes the paper.

2   The Framework

2.1   The Recogniser

The Synface recogniser is a hybrid of recurrent neural networks (RNNs), and hidden Markov models (HMMs). The input layer of the RNN contains thirteen units that represent the Mel frequency cepstral coefficients $C_0, ..., C_{12}$.

The single hidden layer contains 400 units and is fully connected with the input and output layers with direct and time delayed connections. Moreover, the hidden layer is connected to itself with time delayed connections.

The activities of the output layer represent the posterior probability $P(x_i|O)$ of each of the $N = N_p + N_n$ acoustic classes x_i, given the observation O. The acoustic classes include N_p phonetic classes and N_n noise and silence classes. The total number N of output units depends on the language, see Section 2.2.

The posterior probabilities are fed into a Viterbi-like decoder where the look-ahead length can be varied. The recognition network specified by a Markov chain defines a loop of phonemes, where every phoneme is represented by a three state left-to-right HMM.

2.2   Data

The recognizer was trained on the SpeechDat database [5] independently on three languages (Swedish, English and Flemish). The experiments in this paper refer to Swedish. The Swedish database has been divided into a training, a validation and a test set, with 33062, 500 and 4150 utterances, respectively. The Mel frequency cepstral coefficients were computed at every 10 msec.

Phonetic transcriptions have been obtained with forced alignment. For part of the test utterances, the transcriptions have been manually checked in order to obtain a more reliable reference of the test data.

The phonetic transcriptions used in the following use SAMPA symbols [6] with a few exceptions [7] presented in Table 1. The total number of acoustic classes is 50 for Swedish, with 46 phonemes and 4 kinds of noise/silence.

Table 1. Modification of the SAMPA phonetic symbols

original	}	2	{	9	@
modified	uh	ox	ae	oe	eh

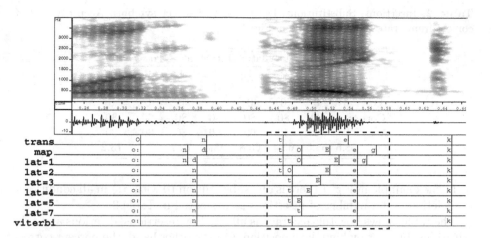

Fig. 1. Example of recognition results with varying look-ahead length. The sentence contains the phrase "Någon teknisk..." ("Some technical..."). See detailed information in the text.

3   Observations

3.1   Boundaries and Latency

Fig. 1 shows an example from the phrase "Någon teknisk..." ("Some technical..."). The spectrogram and waveform are shown together with a number of transcription panes. From the top: the reference transcription (trans), the maximum a posteriori solution (map) obtained selecting for each frame the phonetic class corresponding to the neural network output node with the highest activity, the approximated Viterbi solution (lat=1,7) with look-ahead 1 to 7, and finally the standard Viterbi solution.

It can be seen from the figure that the boundary placement is strongly dependent on the look-ahead length whenever there is ambiguity in the posterior probability estimates. For example, the transitions between 0 and n, between e and k and between k and the next symbol, do not present any difficulties. The transition between t and e ($t = 0.47$ sec), on the other hand, is more problematic as the segment e is partly confused with E and 0 (see map solution). This has an effect on the final solution that strongly depends on the look-ahead length [1].

Some standard evaluation criteria for speech recognition (e.g. accuracy and correct symbols) compute the similarity between the recognition and the reference string of symbols by aligning the two sequences and counting the number of insertions I, deletions D, and substitutions S obtained in the process. Other measures (e.g. percent correct frame rate) work with equally spaced frames in time.

---

[1] Note, however, that from the application point of view, this particular case should not be considered as an error as E and e are mapped to the same visemic class (they share the same visual properties).

Table 2. Insertions, Substitutions, Deletions, % Correct symbols, Accuracy and % correct frame rate for the example in Fig. 1

	map	lat=1	lat=2	lat=3	lat=4	lat=5	lat=7	viterbi
I	4	4	2	1	1	1	0	0
S	1	1	1	1	1	1	1	1
D	0	0	0	0	0	0	0	0
%Corr	80	80	80	80	80	80	80	80
Acc	0	0	40	60	60	60	80	80
%cfr	58.5	58.5	63.4	65.8	68.3	70.7	70.7	73.2

In the example in Fig. 1 there is one substitution (O with o:) in all conditions, no deletion and a number of insertions as indicated in Table 2.

The table shows how the accuracy is affected by insertions, deletions and substitutions, but not by boundary position. On the other hand, the percent correct frame rate, also shown in the table, measures the overlap in time of correctly classified segments, but does not take the boundaries explicitly into account. This motivates a more detailed study on the errors in boundary alignment.

3.2 Boundaries and Entropy

Fig. 2 shows an example from the phrase "...lägga fram fakta..." ("...present facts..."). In this case the entropy of the posterior probabilities estimated by the output nodes of the neural network is displayed for each frame together with the reference transcription.

It is clear that at each transition from a phonetic segment to the next, the entropy assumes a local maximum. Note in particular that in the transition between f and a the maximum is shifted backward, compared to the reference transcription. In this case the position of the reference boundary is set at the onset of voicing ($t = 1.715$ sec) whereas the "entropy" boundary is more related

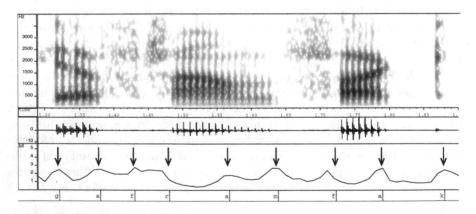

Fig. 2. Example of time evolution of the entropy. The sentence contains the phrase "...lägga fram fakta..." ("...present facts...").

to the onset of the articulation. The choice of one boundary or the other is questionable. From the application point of view the boundary indicated by the maximum in the entropy is more informative, as it is related to a change in articulation that is more visually relevant.

The figure indicates that the entropy may be a good predictor of phonetic transitions.

4   Method

The experiments in this study aim at evaluating the precision of boundary placement at different latency conditions, and at investigating whether the frame-by-frame entropy of the posterior probabilities of the phonetic classes can be used as a predictor of the phonetic boundaries.

Two kinds of measures are therefore considered: the first relates the recognition boundaries to the reference boundaries. The second relates the entropy (or a quantity extracted from the entropy) measured over a frame to the position of the frame with respect to the reference boundaries.

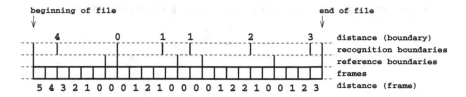

Fig. 3. Example computation of the distance between couple of boundaries and between a frame and a boundary

The way the distance is measured in the two cases is exemplified in Fig. 3. The reference boundaries are in the middle line. The line above shows the recognition boundaries with the distance from the nearest reference boundary. Note that beginning and end of the speech file are not considered as boundaries. The line below the reference shows the frames and their distance from the reference boundaries: frames adjacent to a transition have distance 0.

We use the term displacement when the distance is considered with a sign (negative displacement indicates that the recognition boundary is earlier than the reference). The displacement is interesting when evaluating whether there is a bias in the position of the recognition transitions.

4.1   Entropy

Given the posterior probabilities $P(x_i|O(n))$, $i \in [1, N]$ of the acoustic class x_i given the observation $O(n)$ at frame n, the entropy of the frame is defined as

$$e(n) = -\sum_{i=1}^{N} P(x_i|O(n)) \log_2 P(x_i|O(n))$$

The entropy varies as a function of the uncertainty in the classification of each frame. As shown by the observations in the previous section, we can expect uncertainty to be higher at phonetic boundaries, but there are many sources of uncertainty that need be considered.

For example, some phonemes are intrinsically more confusable than others, some speakers are harder to recognise, or there might be noise in the recordings that increases the entropy.

In order to reduce the effect of the first of these factors, the mean entropy was computed for each phoneme and subtracted from the frame by frame entropy in some evaluations. Fig. 4 shows the boxplot of the entropy for each of the phonemes, as a reference.

As the focus is on the evolution of the entropy in time, the first and second derivative defined as $e'(n) = e(n) - e(n-1)$ and $e''(n) = e'(n+1) - e'(n)$ have been also considered. The definition of the second derivative assures the maxima and minima of $e''(n)$ correspond to the maxima and minima of $e(n)$. Note that given the rate of variation of the entropy in function of time (frames) the first derivative $e'(n)$ should not be expected to be close to zero in correspondence of a maximum of $e(n)$. On the other hand a negative value of the second derivative $e''(n)$ is a strong indicator of a maximum in $e(n)$ for the same reason.

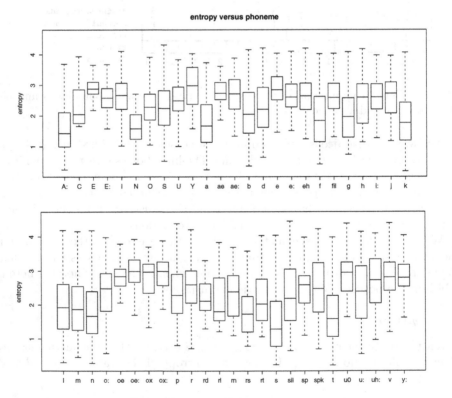

Fig. 4. Box plot of the frame entropy for each phonemic class. The maximum entropy is $\log_2 N = 5.64$ bits.

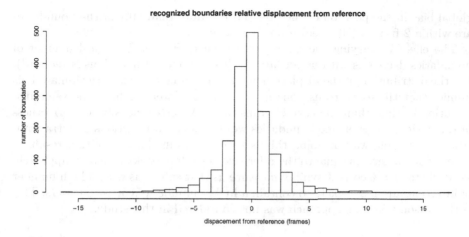

Fig. 5. Distribution of the relative displacement of the recognition boundaries to the reference

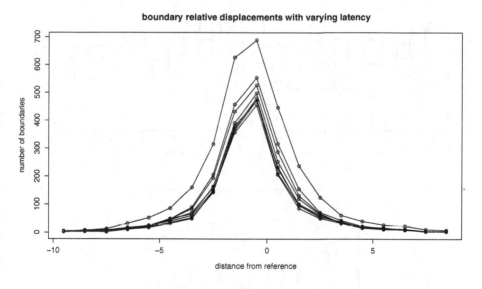

Fig. 6. Distribution of the relative displacement of the recognition boundaries to the reference for varying latency

5   Analysis

5.1   Boundaries and Latency

Fig. 5 shows the distribution of the recognition boundaries displacement (in frames) with respect to the reference. The latency in this case is three frames (30 msec), as used in the Synface prototype. The figure shows that there is no global bias in the position of the boundaries. More than 80% of the boundaries are within 2 frames (20 msec) from the reference.

The effect of varying the latency is shown in Fig. 6. The total number of boundaries decreases with larger latency (the number of insertions is reduced), but the distribution of the displacements from the reference is very similar. This implies that the extra transitions inserted at low latencies have the same distribution in time than the correct transitions. A better measure of correctness in the position of phoneme boundaries would disregard insertions of extra transitions. A simple way of doing this is to select from the recognition results a number of transitions equal to the reference. The drawback of such an approach, similarly to the % correct symbol measure, is that solutions with a high number of insertions are more likely to contain a higher number of correct answers. This is the reason why this approach was not considered in this study.

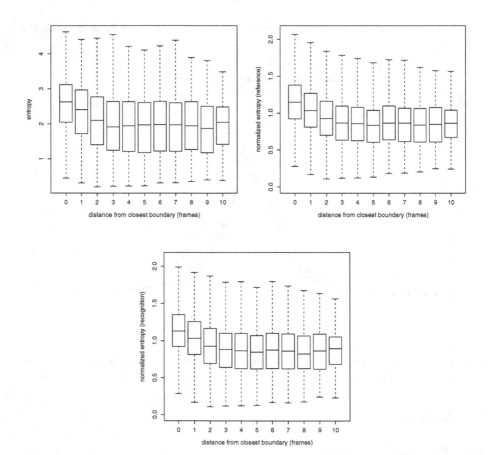

Fig. 7. Box plot of the entropy at varying distances from the nearest reference boundary. In the second and third plots the entropy is normalised to the mean for each phoneme in the reference and recognition transcription, respectively.

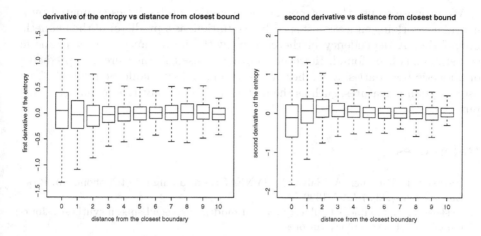

Fig. 8. Box plot of the first and second derivative in time of the entropy at varying distance from the nearest reference boundary

5.2   Boundaries and Entropy

The distribution of the entropy for different distances from the a phonetic transition are displayed in Fig. 7 with box plots. The three figures show the unnormalised entropy, the entropy normalised with the average for every phoneme in the reference transcription and the same normalisation but referring to the recognition transcription (with three frames latency). All plots show that the entropy increases in the neighbourhood of a phonetic transition (distance from the boundary equals 0). In the normalised cases the variation around the median is reduced.

Even in the normalised cases, the large variation around the median shows that the frame-by-frame entropy needs to be improved to be a good enough predictor of phonetic boundaries.

Fig. 8 shows the distributions of the first and second derivative depending on the distance from a phonetic boundary. As explained in Section 4 the first derivative assumes values far from zero at boundaries, which indicates that the entropy has a strong variation in time. The second derivative often assumes negative values, suggesting there might be a maximum of the entropy or at least a convex shape of the entropy curve.

6   Conclusions

This article analyses the phonetic transitions obtained with a low latency phoneme recogniser. It shows that the distribution of displacements of the recognition boundaries, with respect to the reference transcription, do not vary significantly with the latency, in spite of the increased number of insertions at low latency conditions.

We propose to use the entropy of the posterior probabilities estimated by a neural network in connectionist speech recognition, as a predictor of the phonetic boundaries. A dependency of the entropy with the distance from a phonetic transition has been found. However, in order to use this measure as a predictor of phonetic boundaries, a number of interfering factors should be removed. The use of dynamic features, such as the first and second derivative might serve this purpose.

References

1. Karlsson, I., Faulkner, A., Salvi, G.: SYNFACE - a talking face telephone. In: Proc. Eurospeech. (2003) 1297–1300
2. Beskow, J.: Trainable articulatory control models for visual speech synthesis. Journal of Speech Technology (in press)
3. Salvi, G.: Truncation error and dynamics in very low latency phonetic recognition. In: ISCA Tutorial and Research Workshop on Non-linear Speech Processing (NOLISP), Le Croisic, France. (2003)
4. Salvi, G.: Dynamic behaviour of connectionist speech recognition with strong latency constraints. Speech Communication (in press)
5. Elenius, K.: Experience from collecting two swedish telephone speech databases. International Journal of Speech Technology 3 (2000) 119–127
6. Gibbon, D., Moore, R., Winski, R., eds.: SAMPA computer readable phonetic alphabet, Part IV, section B. In: Handbook of Standards and Resources for Spoken Language Systems. Mouton de Gruyter, Berlin and New York (1997)
7. Lindberg, B., Johansen, F.T., Warakagoda, N., Lehtinen, G., Kačič, Z., Žgank, A., Elenius, K., Salvi, G.: A noise robust multilingual reference recogniser based on SpeechDat(II). In: 6th Intern. Conf. on Spoken Language Processing. Volume III. (2000) 370–373

Third-Order Moments of Filtered Speech Signals
for Robust Speech Recognition

Kevin M. Indrebo, Richard J. Povinelli, and Michael T. Johnson

Dept. of Electrical and Computer Engineering, Marquette University,
Milwaukee, Wisconsin, USA
{kevin.indrebo, richard.povinelli,
mike.johnson}@Marquette.edu

Abstract. Novel speech features calculated from third-order statistics of subband-filtered speech signals are introduced and studied for robust speech recognition. These features have the potential to capture nonlinear information not represented by cepstral coefficients. Also, because the features presented in this paper are based on the third-order moments, they may be more immune to Gaussian noise than cepstrals, as Gaussian distributions have zero third-order moments. Experiments on the AURORA2 database studying these features in combination with Mel-frequency cepstral coefficients (MFCC's) are presented, and some improvement over the MFCC-only baseline is shown when clean speech is used for training, though the same improvement is not seen when multi-condition training data is used.

1 Introduction

Spectral-based acoustic features have been the standard in speech recognition for many years, even though they are based on limiting assumptions of the linearity of the speech production mechanism [1]. Specifically, mel-frequency cepstral coefficients (MFCC), which are calculated using a discrete cosine transform on the smoothed power spectrum, and perceptual linear prediction (PLP) cepstral coefficients, similar to MFCCs, but based on human auditory models, are used in almost all state-of-the-art speech recognition systems [1]. While these feature sets do an excellent job of capturing linear information of speech signals, they do not encapsulate information about nonlinear or higher-order statistical characteristics of the signals, which have been shown to exist, and are not insignificant [2-4].

As successful as MFCCs have been in the field of speech recognition, performance of state-of-the-art systems remains unacceptable for many real applications. One of the largest failings of popular spectral features is their poor robustness in the face of ambient noise. Many environments in which automatic speech recognition applications would be ideal have large amounts of background additive noise that makes voice-activated systems infeasible. In this paper, we introduce acoustic features based on higher-order statistics of speech signals. It is shown that these features, when combined with MFCC's, can produce higher recognition accuracies in some noise conditions.

The rest of the paper is as follows. Section 2 gives some background on robust speech recognition and nonlinear speech recognition. In section 3, computation of the

M. Faundez-Zanuy et al. (Eds.): NOLISP 2005, LNAI 3817, pp. 277–283, 2005.

proposed features is detailed. Experiments comparing the feature sets including the third-order moment features and MFCC's are presented in section 4, and are followed by the conclusion in section 5.

2 Background

2.1 Robust Speech Recognition

Robust speech recognition research has focused on subjects such as perceptually motivated features, signal enhancement, feature compensation in noise, and model adaptation. Perceptual-based features include PLP cepstral coefficients [5] and perceptual harmonic cepstral coefficients (PHCC) [6], which have been shown to be more robust than MFCCs in the presence of additive noise. Signal enhancement and feature compensation include techniques like spectral subtraction [7] and iterative wiener filtering [8], as well as more advanced algorithms such as SPLICE (stereo-based piecewise linear compensation in environments) [9]. While these techniques focus on adapting the extracted features, model adaptation methods such as MLLR and MAP [10] attempt to adjust the model parameters to better fit the noisy signals.

Though some progress has been made, the performance of speech recognition systems in noisy environments is still far from acceptable. Word error rates for a standard large vocabulary continuous speech recognition (LVCSR) task like recognition of the 5,000 word Wall Street Journal corpus can drop from under 5% to over 20% when Gaussian white noise is added at a signal-to-noise-ratio (SNR) of +5dB, even with compensation techniques [9]. Even continuous digit recognition word error rates often exceed 10% when faced with high noise levels [11].

2.2 Nonlinear Features for Speech Recognition

Recently, work has been done to investigate the efficacy of various feature sets based on nonlinear analysis. Dynamical invariants based on chaos theory [12], such as Lyapunov exponents and fractal dimension have been used to augment the standard linear feature sets [13], as well as nonlinear polynomial prediction coefficients [14]. In [15], an AM-FM model of speech production is exploited for extraction of nonlinear features. Also, Phase space reconstruction has been used for statistical modeling and classification of speech waveforms [16].

In [17], reconstructed phase spaces built from speech signals that have been subband filtered were used for isolated phoneme classification, showing improved recognition accuracies over fullband signal phase space reconstruction features. However, this approach is infeasible for continuous speech recognition because of its high computational complexity. In this paper, nonlinear features from subbanded speech signals that are much simpler to compute are introduced.

3 Third-Order Moment Feature Computation

An approach based on time-domain filtering of speech signals is taken for computation of the nonlinear features. An utterance is parameterized by first filtering the signal into P subbands that have cutoffs and bandwidths derived from the Mel-scale. Each of these signals is then broken into frames with lengths of 25.6 ms, updated every 10 ms. The third-order moment of the signal amplitudes of each of these channels is calculated for each frame, and the set of these coefficients form a feature vector. Log energy of the unfiltered signal frame is appended to these features, which are then orthogonalized using a principle component analysis (PCA). In the experiments presented in this paper, 20 filter channels are used, and the PCA reduces the dimension third-order moment feature space to 13.

Because much of the information that distinguishes speech sounds is contained in the power spectrum, it is not expected that these features by themselves would carry enough information to compete with MFCC features. Therefore, the proposed features are appended to the baseline MFCC feature vector for modeling and recognition of speech.

There are two advantages to this approach. First, nonlinear information that may be useful for recognition that is not captured by traditional features is added to the recognizer. Also, because some types of noise have approximately Gaussian statistical distributions, and Gaussian distributions have zero third-order moments, the proposed features my be less affected by additive noise than MFCC's. This conjecture is tested by comparing a combined feature set of MFCC's and the proposed features to a baseline feature set of only MFCC's for use in noisy speech recognition.

4 Experiments

The preliminary recognition experiments are run using the AURORA2 database [18]. This corpus contains utterances of connected digits corrupted with different types and levels of noise. There are eleven words: the digits zero through nine and "oh". Two sets of experiments were run. In the first set the models were trained using clean speech signals, and tested on test set A, which contains four different types of noise

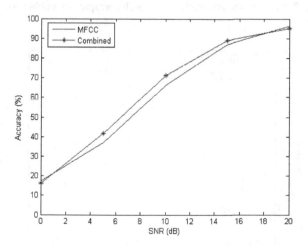

Fig. 1. Recognition accuracies for speech corrupted by subway noise

at varying SNR levels. The second set of experiments used models trained on the multi-condition training set in AURORA2, and the tests were performed on test set A and test set B, which has four different types of noise. The multi-condition training set has the same noise types as test set A, providing a matched noisy training-test scenario. The noise types in test set B are not included in any training signals. HTK [19] is the software used for experimentation. Each word is modeled using a 16-state left-to-right diagonal covariance Hidden Markov Model (HMM). Additionally, a 3-state silence model and single-state short pause model are implemented. The frame rate is 10 ms, with frame lengths of 25.6 ms.

Two types of feature sets are used. The baseline feature vector is a 39-element vector of 12 MFCC's, log energy, and the first and second time derivatives. The second feature

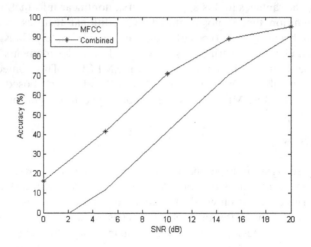

Fig. 2. Recognition accuracies for speech corrupted by babble noise

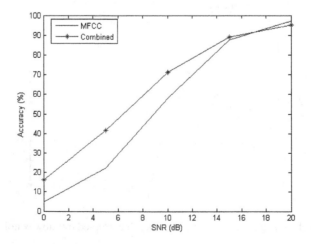

Fig. 3. Recognition accuracies for speech corrupted by car noise

set is a 45-element vector composed of the 39-coefficient MFCC vector concatenated with 6 coefficients from the PCA of the third-order moment space.

Figures 1-4 show the recognition accuracies for the two feature types on the four different noise types of AURORA's test set A, using models trained on clean speech signals. These noises are subway, babble, car noise, and exhibition hall, respectively. The accuracies are plotted against the SNR levels, ranging from 0 to 20 dB. It can be seen that, except for the babble noise case, the MFCC-only features give better recognition accuracies at 20 dB SNR. The MFCC and third-order moment concatenation feature vector, however, outperforms the MFCC-only set in most of the lower SNR cases.

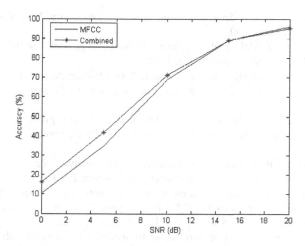

Fig. 4. Recognition accuracies for speech corrupted by exhibition hall noise

Table 1. Average recognition accuracies for models trained on corrupted speech

Feature type	Test set A	Test set B
MFCC's	88.22%	84.10%
Combined features	80.98%	61.37%

Table 1 shows the accuracies of models trained on the multi-condition training set and tested on both test sets A and B for the MFCC feature set and the combined feature set, averaged over all the noise types and SNR levels from 0 to 20 dB. This table shows that when the models are trained on speech corrupted with different types and levels of noise, the addition of the third-order moment features does not improve upon the MFCC baseline, even degrading the performance significantly.

5 Conclusion

A new type of acoustic feature extraction method was presented based on higher-order statistics of subband filtered speech signals, and tested on noisy speech signals. The results show that the combination of traditional MFCC features and these new features can improve the robustness of speech recognition systems when the speech models are trained on clean speech data. The largest improvement is seen when the speech signals used for recognition are corrupted by babble noise. However, when the speech models are trained on clean speech, the performance of the recognition degrades compared to MFCC only features. For these features to be useful in real systems, some adaptive combination may be necessary, so that information from third-order moment features is only used when it will improve the recognition estimates of the recognition system.

References

[1] B. Gold and N. Morgan, *Speech and Audio Signal Processing*. New York, New York: John Wiley and Sons, 2000.

[2] M. Banbrook and S. McLaughlin, "Is Speech Chaotic?," presented at IEE Colloquium on Exploiting Chaos in Signal Processing, 1994.

[3] M. Banbrook, S. McLaughlin, and I. Mann, "Speech characterization and synthesis by nonlinear methods," *IEEE Transactions on Speech and Audio Processing*, vol. 7, pp. 1 -17, 1999.

[4] H. M. Teager and S. M. Teager, "Evidence for nonlinear sound production mechanisms in the vocal tract," presented at NATO ASI on Speech Production and Speech Modelling, 1990.

[5] H. Hermansky, "Perceptual linear predictive (PLP) analysis for speech recognition," presented at Journal of the Acoustical Society of America, 1990.

[6] L. Gu and K. Rose, "Perceptual harmonic cepstral coefficients for speech recognition in noisy environments," presented at IEEE International Conference on Acoustics, Speech, and Signal Processing (ICASSP '01), Salt Lake City, UT, 2001.

[7] S. F. Boll, "Suppression of acoustic noise in speech using spectral subtraction," *IEEE Transactions on Acoustics, Speech, and Signal Processing*, vol. 27, pp. 113-120, 1979.

[8] K. Yu, B. Xu, M. Dai, and C. Yu, "Suppressing cocktail party noise for speech recognition," presented at 5th International conference on signal processing (WCCC-ICSP 2000), Beijing, China, 2000.

[9] L. Deng, A. Acero, M. Plumpe, and X. Huang, "Large-Vocabulary Speech Recognition Under Adverse Acoustic Environments," presented at Internation Conference on Spoken Language Processing (ICSLP), Beijing, China, 2000.

[10] S. Young, J. Odell, D. Ollason, V. Valtchev, and P. Woodland, "The HTK Book," 1997.

[11] C. Meyer and G. Rose, "Improved Noise Robustness By Corrective and Rival Training," presented at International Conference on Acoustics, Speech, and Signal Processing (ICASSP-03), 2003.

[12] E. Ott, *Chaos in dynamical systems*. Cambridge, England: Cambridge University Press, 1993.

[13] V. Pitsikalis and P. Maragos, "Speech analysis and feature extraction using chaotic models," presented at International Conference on Acoustics, Speech, and Signal Processing (ICASSP), 2002.

[14] X. Liu, R. J. Povinelli, and M. T. Johnson, "Vowel Classification by Global Dynamic Modeling," presented at ISCA Tutorial and Research Workshop on Non-linear Speech Processing (NOLISP), Le Croisic, France, 2003.

[15] D. Dimitriadis, P. Maragos, and A. Potamianos, "Modulation features for speech recognition," presented at International Conference on Acoustics, Speech, and Signal Processing (ICASSP), 2002.

[16] M. T. Johnson, R. J. Povinelli, A. C. Lindgren, J. Ye, X. Liu, and K. M. Indrebo, "Time-Domain Isolated Phoneme Classification using Reconstructed Phase Spaces," *IEEE Transactions on Speech and Audio Processing*, in press.

[17] K. M. Indrebo, R. J. Povinelli, and M. T. Johnson, "Sub-banded Reconstructed Phase Spaces for Speech Recognition," *Speech Communication*, in press.

[18] D. Pearce and H. Hirsch, "The AURORA experimental framework for the performance evaluation of speech recognition systems under noisy conditions," Beijing, China, 2000.

[19] "HTK Version 2.1," Entropic Cambridge Research Laboratory Ltd., 1997.

New Sub-band Processing Framework Using Non-linear Predictive Models for Speech Feature Extraction

Mohamed Chetouani[1], Amir Hussain[2], Bruno Gas[1], and Jean-Luc Zarader[1]

[1] Laboratoire des Instruments et Systèmes d'Ile-De-France,
Université Paris VI, Paris, France
mohamed.chetouani@upmc.fr
[2] Dept. of Computing Science and Mathematics,
University of Stirling, Scotland, UK

Abstract. Speech feature extraction methods are commonly based on time and frequency processing approaches. In this paper, we propose a new framework based on sub-band processing and non-linear prediction. The key idea is to pre-process the speech signal by a filter bank. From the resulting signals, non-linear predictors are computed. The feature extraction method involves the association of different Neural Predictive Coding (NPC) models. We apply this new framework to phoneme classification and experiments carried out with the NTIMIT database show an improvement of the classification rates in comparison with the full-band approach. The new method is also shown to give better performance than the traditional Linear Predictive Coding (LPC), Mel Frequency Cepstral Coding (MFCC) and Perceptual Linear Prediction (PLP) methods.

1 Introduction

Speech feature extraction stages are commonly based on time and frequency processing methods. Indeed, the most widely used method is the Mel Frequency Cepstral Coding (MFCC) but the Linear Predictive Coding (LPC) is also popular. The success of the MFCC is partly due to the sub-band processing based on a Mel-scale filter bank.

The filter banks are usually designed according to an auditory system [1]. Several models have been proposed for speech feature extraction [15], [8], [10], [6]. Among the many advantages of sub-band processing approaches, the enhanced robustness is a key issue. For instance in speech coding [7], [17] and in speech enhancement [12], the filter banks have been shown to significantly improve the performance in noisy environments.

From a classification point of view, the division of the whole frequency domain into sub-bands and the application of diverse strategies in different bands is a possible way of achieving error rates reduction. In speaker recognition applications [2], it is known that some sub-bands retain more speaker dependent

M. Faundez-Zanuy et al. (Eds.): NOLISP 2005, LNAI 3817, pp. 284–290, 2005.

features and similar ideas have been followed in speech recognition. Work by Allen et al. [1] has shown that the linguistic message is decoded in different sub-bands and the final decision involves merging the information from these sub-bands [11].

In this paper, we are interested in the combining sub-band processing and non-linear predictive methods. The key idea is to divide the whole frequency domain into sub-bands and then employ non-linear predictors for feature extraction. The combination of filter banks and predictors is widely used for speech coding. The purpose of this paper is to compare full-band non-linear predictors for feature extraction with the proposed sub-band based approaches.

The paper is organized as follows. Section 2 is dedicated to the description of the new sub-band based feature extractor. Then, section 3 presents the non-linear predictor used, namely: The Neural Predictive Coding (NPC). We then present the experimental conditions and the comparative performance of the system for different phoneme groups. Finally, we present some concluding remarks and future work suggestions.

2   Non-linear Predictive Sub-band Feature Extractor

Traditionally, linear and non-linear predictors are computed directly from speech signal samples. In speech coding and in speech enhancement, it has been shown that such an approach is not adapted for noisy environments. Several methods have been proposed [7] and among them the combination of filter banks and predictors. In feature extraction, similar ideas have been investigated. The

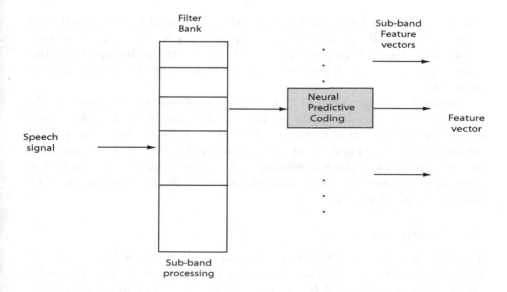

Fig. 1. Non-Linear Predictive Sub-band Based Feature Extraction

Perceptual Linear Prediction (PLP) [9] is one example which involves the modification of the spectrum (through auditory knowledge) and then using an all-pole
modelling method like in the LPC.

The principle of the proposed feature extraction method is described in
figure 1. The first stage aims to pre-process the speech signal by a filter bank.
The following stage involves extraction of the features. Instead of using energy
from the sub-bands, the Neural Predictive Coding (NPC) model is used. The
features are extracted from signals resulting from the pre-processing stage. In
this work, the sub-bands are achieved by modifying the spectra of input signals.
The FFT based analysis filter bank employed here can be efficiently implemented
using a bank of band-pass filters.

This approach is different from traditional feature extraction methods in that
we propose to pre-process the signal by a filter bank. The first stage can be
designed to reflect auditory knowledge - and feature extractors based on auditory
models usually compute the energy of the sub-bands. In our model, the second
stage is based on a non-linear feature extractor: The NPC model. The next
section is dedicated to the description of this model.

3   Neural Predictive Coding

The Neural Predictive Coding (NPC) model [5], [3] is a non-linear extension of
the well-known LPC encoder. Like in the LPC framework with the AR model, the
vector code is estimated by prediction-error minimization. The main difference
lies in the fact that the model is non-linear and it is a connectionnist model:

$$\hat{y}_k = F(\mathbf{y_k}) = \sum_j \mathbf{a_j}\sigma(\mathbf{w^T y_k}) \tag{1}$$

Where F is the prediction function realized by the neural model. $\hat{y}_k$ is the pre-
dicted sample. $\mathbf{y_k}$ the prediction context: $\mathbf{y_k} = [\mathbf{y_{k-1}, y_{k-2}, ..., y_{k-\lambda}}]^\mathbf{T}$ and λ is
the length of the prediction window. $\mathbf{w}$ and $\mathbf{a}$ represent the first and the output
layer weights. σ is a non-linear activation function, namely the sigmoid function
in our case.

The key idea is to use the NPC model as a non-linear auto-regressive model.
As in the LPC framework for the predictor coefficients, the NPC weights represent the vector code. It is well-known that the weights can be considered as a
representation of the input vector. A drawback of this method is that non-linear
models have no clear physical meanings [14]. The solution weights can be very
different for a same minimum of the prediction error. In our approach, we impose
constraints on the weights.

3.1   Description

The NPC model is a Multi-Layer Perceptron (MLP) with one hidden layer (cf.
figure 2). In our case, only the output layer weights are used as the coding
vector (instead of using all the neural weights). For this purpose we consider

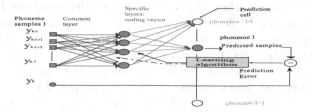

Fig. 2. Neural Predictive Coding (NPC) model

that the function F realized by the model, under convergence assumptions, can be decomposed into two functions: $G_{\mathbf{w}}$ ($\mathbf{w}$ first layer weights) and $H_{\mathbf{a}}$ ($\mathbf{a}$ output layer weights):

$$F_{\mathbf{w},\mathbf{a}}(\mathbf{y_k}) = \mathbf{H_a} \circ \mathbf{G_w}(\mathbf{y_k}) \tag{2}$$

With $\hat{y}_k = H_{\mathbf{a}}(\mathbf{z_k})$ and $\mathbf{z_k} = \mathbf{G_w}(\mathbf{y_k})$.

As one can note, the NPC structure allows the use of a different prediction window's length independent of the size of the coding vector, which is contrary to the case of the LPC structure.

For the layers' specialization, the learning phase is realized in two stages. First, the *parameterization phase* involves the learning of all the weights using the prediction error minimization criterion:

$$Q = \sum_{k=1}^{K}(y_k - \hat{y}_k)^2 = \sum_{k=1}^{K}(y_k - F(\mathbf{y_k}))^2 \tag{3}$$

With y represents the speech signal, $\hat{y}$ the predicted speech signal, k the samples index and K the number of samples respectively.

In this phase, only the first layer weights $\mathbf{w}$ (which constitute the NPC encoder parameters) are retained. Since the NPC encoder is set up by the parameters defined in the previous phase, the second phase, called the *coding phase*, involves the computation of the output layer weights $\mathbf{a}$ (vector code). This is done also by prediction error minimization but only the output layer weights are updated. One can note that the output function is linear (cf. equation 1), so once can readily employ the Levinson algorithm for the LPC model. In our case, in order to ensure consistency with the previous *parameterization phase*, the update is done using the backpropagation algorithm.

Finally, one can note that the first layer weights $\mathbf{w}$ are common to all the speech signal frames while the second layer weights $\mathbf{a}$ are specific to each frame. For each frame, a feature vector $\mathbf{a}$ is computed by prediction error minimization.

4 Evaluation and Discussion

4.1 Experimental Conditions

The NTIMIT database [13] is used in this experiment and in particular, the two first regions (DR1, DR2). By using this database, we carry out speech recognition at telephone quality. One can note that the telephone bandwidth is approximately limited to 300-3400Hz. We focus on the processing of front vowels (/ih/, /ey/, /eh/, /ae/), voiced plosives (/b/, /d/, /g/) and unvoiced plosives (/p/, /t/, /k/). This choice can be justified by the fact that the classification of these phonemes is known to be difficult and they are also often used. For training and test databases, we use the division proposed by the database.

The classification is carried out by GMM (with 16 centers and a diagonal assumption) and it operates as a frame by frame classification (with a frame size of 32ms with 16ms of overlapping). The dimension of the features is set to 12.

The proposed feature extractor is based on sub-band processing but determining the optimal number of sub-bands is still an open issue. In Mel-scale filter banks, this number is about 20. We evaluate the comparative performance of our model with 2 sub-bands (300-1140Hz, 1046-3400Hz) and 4 sub-bands (300-765Hz, 700-1640Hz, 1515-2700Hz, 2100-3400Hz) [16]. For the first case, we extract 6 NPC coefficients (in each sub-band) and for the second one, we extract 3 NPC coefficients in order to keep a dimension 12 for the feature vector.

We make comparisons with three traditional methods, namely the LPC, MFCC and PLP. And in order to evaluate the comparative performance of the sub-band NPC approach, we also implement an equivalent full-band NPC model.

4.2 Results

The classification rates for the NPC model in the different sub-bands are grouped in 1 which show that the performance of both the sub-band models is better than the full-band model. These preliminary results shows that the sub-band approach, by dividing the whole frequency domain, can be effective for phoneme classification. Indeed the phoneme dependent features are also known to be distributed non-linearly among different sub-bands [1], [11] and a diverse sub-band processing approach can therefore prove more effective for extracting such features. However, for the 4 sub-bands model, the performance decreases compared to the 2-sub band model. This needs to be further investigated and could be partly due to the lack of optimised data for each band, and/or the uniform (linear) spacing of all the bands.

Table 1. Classification rates for different sub-bands (vowels)

Phoneme	Full-band	2 sub-bands	4 sub-bands
/ih/, /ey/, /eh, /ae/	49.03	52.4	50.89

Table 2. Classification rates: improvements by non-linear and sub-band methods

Phoneme	LPC	MFCC	PLP	NPC	Sub-Band NPC
/ih/, /ey/, /eh, /ae/	35.22	48.12	45.12	49.03	**52.4**
/b/, /d/, /g/	54.13	59.23	57.21	62.24	**63.87**
/p/, /t/, /k/	44.10	51.45	46.98	49.36	**52.56**

The classification rates for the different methods are presented in table 2. For voiced phonemes, the performances can be seen to be improved by the use of non-linear methods even in the case of the plosives (NPC: 62.24%, Sub-band NPC:63.87 %).

In the case of unvoiced plosives, the performance of the sub-band method is found to be slightly better than the more complex and state-of-art MFCC method (MFCC:51.45%, Sub-band NPC: 52.56%). On the other hand, the sub-band method is seen to further outperform the full-band approach (Full-band: 49.36%, Sub-band: 52.56%).

5   Conclusion

In this paper, we propose a new framework for speech feature extraction comprising a combination of filter banks and non-linear predictors. The filter banks act to pre-process the signal in individual frequency bands. In this way, we can apply different (optimised) strategies in each band for feature extraction. The features are extracted by the help of a non-linear feature extractor, namely the Neural Predictive Coding (NPC) model. The obtained features are decorrelated since they are extracted using different models. This characteristic, decorrelation can also be very useful for classification purposes [4].

The preliminary results obtained in this paper using the NTIMIT database show that the proposed sub-band based NPC method can offer superior performance compared to both the full-band NPC approach and the three traditional methods: MFCC, LPC and PLP.

In our approach, determination of the optimal number and spacing of the sub-bands is still an open issue, which is currently being further investigated. For future work, we will incorporate an explicit discriminant criterion between the NPC models as part of the proposed sub-band framework. We also aim to look at other interesting applications such as speaker recognition and speech/music discrimination.

References

1. J.B. Allen, "How Do Humans Process and Recognize Speech?," *IEEE Trans. on Speech and Audio Processing*, Vol. 2:4, pp. 567-577 (1994).
2. L. Besacier and J.F. Bonastre, "Subband approach for automatic speaker recognition: Optimal division of the frequency," *Lecture Notes in Computer Science, Audio and Video-based Biometric Person Authentification*, Springer, pp. 195-202 (1997).
3. M. Chetouani, "Codage neuro-prédictif pour l'extraction de caractéristiques de signaux de signaux de parole," *Université Paris VI* (2004).
4. Richard O. Duda and Peter E. Hart and David G. Stork, "Pattern Classification," *Wiley-Interscience Publication*, (2001).
5. B. Gas and J.L. Zarader and C. Chavy and M. Chetouani. "Discriminant neural predictive coding applied to phoneme recognition," *Neurocomputing*, 56, pp. 141-166 (2004).

6. O. Ghitza, "Auditory Models and Human Performance in Tasks Related to Speech Coding and Speech Recognition," *IEEE Trans. on Speech and Audio Processing*, Vol. 2:1, pp. 115-132 (1994).
7. B. Gold and N. Nelson, "Speech and Audio Signal Processing : Processing and Perception of Speech and Music," *John Wiley and Sons, INC* (2000).
8. S. Greenberg, "Representation of speech in the auditory periphery," *Journal of Phonetics, Special Issue*, Vol. 16:1, January (1994).
9. H. Hermansky, "Perceptual linear predictive (PLP) analysis of speech," *The Journal of the Acoustical Society of America*, pp. 1738-1752 (1990).
10. H. Hermansky, "Auditory Modeling in Automatic Recognition of Speech," *Proc. Keele Workshop* (1996).
11. H. Hermansky and S. Tibrewala and M. Pavel, "Towards ASR on Partially Corrupted Speech," *Proc. ICSLP* (1996).
12. A. Hussain and D.R. Campbell, "Binaural Sub-Band Adaptive Speech Enhancement Using Artificial Neural Networks," *Speech Communication*, pp. 177-186 (1998).
13. C. Jankowski and A. Kalyanswamy and S. Basson and J. Spitz, "NTIMIT: A Phonetically Balanced, Continous Speech, Telephone Bandwidth Speech Database," *Proc. IEEE Int. Conf. Acoustics, Speech, Signal Processing (ICASSP)*, 1, 109–112 (1990).
14. W. B. Kleijn. Signal Processing Representations of Speech. IEICE Trans. Inf. and Syst. **E86-D**, 3, March, (2003), 359–376
15. K.K. Paliwal, "Spectral Subband Centroid Features for Speech Recognition," *Proc. IEEE Int. Conf. Acoustics, Speech, Signal Processing (ICASSP)*, **2**, pp. 617-620 (1988).
16. S. Tibrewala and H. Hermansky, "Sub-band Based Recognition of Noisy Speech," *Proc. IEEE Int. Conf. Acoustics, Speech, Signal Processing (ICASSP)*, **2**, pp. 1255-1258 (1997).
17. Rongshan Yu and C. C. Ko, "A Warped Linear-Prediction-Based Subband Audio Coding Algorithm," *IEEE Trans. on Speech and Audio Processing*, Vol. 10:2, pp. 1-8 (2002).

Noise Robust Automatic Speech Recognition with Adaptive Quantile Based Noise Estimation and Speech Band Emphasizing Filter Bank

Casper Stork Bonde, Carina Graversen, Andreas Gregers Gregersen,
Kim Hoang Ngo, Kim Nørmark, Mikkel Purup,
Thomas Thorsen, and Børge Lindberg

Department of Communication Technology, Aalborg University,
Niels Jernes Vej 12, DK-9220 Aalborg Ø
lindberg@kom.aau.dk

Abstract. An important topic in Automatic Speech Recognition (ASR) is to reduce the effect of noise, in particular when mismatch exists between the training and application conditions.

Many noise robutness schemes within the feature processing domain use as a prerequisite a noise estimate prior to the appearance of the speech signal which require noise robust voice activity detection and assumptions of stationary noise. However, both of these requirements are often not met and it is therefore of particular interest to investigate methods like the Quantile Based Noise Estimation (QBNE) mehtod which estimates the noise during speech and non-speech sections without the use of a voice activity detector. While the standard QBNE-method uses a fixed pre-defined quantile accross all frequency bands, this paper suggests adaptive QBNE (AQBNE) which adapts the quantile individually to each frequency band.

Furthermore the paper investigates an alternative to the standard mel frequency cepstral coefficient filter bank (MFCC), an empirically chosen Speech Band Emphasizing filter bank (SBE), which improves the resolution in the speech band.

The combinations of AQBNE and SBE are tested on the Danish SpeechDat-Car database and compared to the performance achieved by the standards presented by the Aurora consortium (Aurora Baseline and Aurora Advanced Fronted). For the High Mismatch (HM) condition, the AQBNE achieves significantly better performance compared to the Aurora Baseline, both when combined with SBE and standard MFCC. AQBNE also outperforms the Aurora Baseline for the Medium Mismatch (MM) and Well Matched (WM) conditions. Though for all three conditions, the Aurora Advanced Frontend achieves superior performance, the AQBNE is still a relevant method to consider for small foot print applications.

1 Introduction

Car equipment control and the rapidly growing use of mobile phones in car environments have developed a strong need for noise robust automatic speech

M. Faundez-Zanuy et al. (Eds.): NOLISP 2005, LNAI 3817, pp. 291–302, 2005.

recognition systems. However, ASR in car environments is still far from achieving sufficient performance in the presence of noisy conditions.

While significant efforts are needed both in the acoustic modelling domain and in the feature processing domain of speech research, the present paper focusses on the latter in the context of standard Hidden Markov modelling of acoustic units.

One feature processing method proposed by Stahl et al. is Quantile Based Noise Estimation (QBNE) which estimates the noise during speech and non-speech sections without the explicit use of a voice activity detector [1].

In QBNE the noise estimate is based on a predefined quantile (q-value) being constant for all frequencies and independent of the characteristics of the noise. In an attempt to adapt to the noise characteristics of the data, this paper suggests adaptive QBNE (AQBNE) in which the q-value is determined independently for each frequency according to the characteristics of the data for that particular frequency. This gives a non-linear noise estimate compared to QBNE.

While the standard MFCC is motivated mainly from perceptual considerations, we suggest in this paper to consider a Speech Band Emphasizing (SBE) filter bank which has the purpose to better focus on the speech information available in the signal.

The paper is organised as follows. In section 2 QBNE, spectral subtraction, AQBNE and the SBE filter bank are described in further details. In section 3, the experimental framework, based on the Danish SpeechDat-Car database, is described. In section 4 the experimental results are presented and compared to the performance achieved by the Aurora consortium. In section 5 the conclusions are drawn.

2   Methods

2.1   Quantile Based Noise Estimation and Spectral Subtraction

QBNE, as proposed in [1], is a method based on the assumption that each frequency band contains only noise at least the q'th fraction of time, even during speech sections. This assumption is used to estimate a noise spectrum $N(\omega, i)$ from the spectrum of the observed signal $X(\omega, i)$ by taking the maximum value of the q-quantile in every frequency band. For every frequency band ω the frames of the entire utterance $X(\omega, i)$, $i = 0, \cdots, I$, are sorted so that $|X(\omega, i_0)| \leq |X(\omega, i_1)| \leq \cdots \leq |X(\omega, i_I)|$, where i denotes the frame number. This means that for each frequency band the data is sorted by amplitude in ascending order. From this the q-quantile noise estimate is defined as:

$$\hat{N}(\omega) = |X(\omega, i_{\lfloor qI \rfloor})| \tag{1}$$

Fig. 1 shows a plot of the sorted data for a sample utterance. The curves in the figure corresponds to different frequencies. The chosen q-value can be seen as a vertical dashed line in the figure. The intersection between the vertical

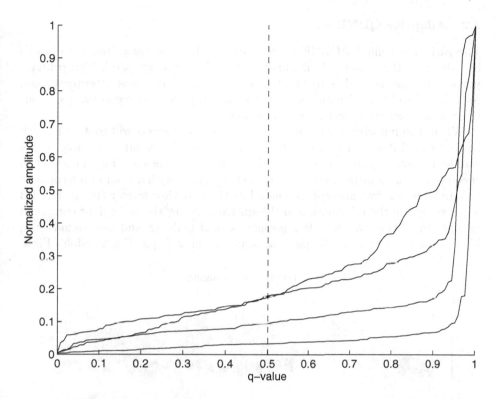

Fig. 1. Data sorted by amplitude for different frequency bands ($q = 0.5$). Only a few frequencies are plotted.

line and each frequency curve is the noise estimate for the corresponding frequency.

Ideally, the noise can be eliminated from the observed signal by applying amplitude spectral subtraction in which the observed signal $X(\omega)$ is modeled as a speech signal $S(\omega)$ to which uncorrelated noise $N(\omega)$ is added.

The amplitude spectrum of the speech is estimated by subtracting a weigthed amplitude spectrum of the estimated noise from the amplitude spectrum of the observed signal:

$$|\hat{S}(\omega)| = |X(\omega)| - \eta|\hat{N}(\omega)| \tag{2}$$

where η, the weighting constant, in the present experiments has been choosen to $\eta = 2.5$. This value was found to be an optimal value in [1]. To avoid negative amplitude, the implemented spectral subtraction in its final form is as follows:

$$|\hat{S}(\omega)| = \max\left(|X(\omega)| - \eta|\hat{N}(\omega)| \,,\; \gamma|\hat{N}(\omega)|\right) \tag{3}$$

The value of γ, the fraction of the estimated noise, is set to 0.04, which has been found to be a reasonable value in [1].

2.2   Adaptive QBNE

The rationale behind AQBNE is to adapt to the utterances and noise levels by adjusting the quantile individually for each frequency band. The primary purpose of the method is to be able to train with low noise utterances (lab. recordings) and test with high noise utterances (application recordings) without the typical associated performance degradation.

When training with low noise data the recognizer models will contain a much more detailed description of the utterance compared to what is possible to obtain by eliminating the noise from a high noise test utterance. Fig. 2 shows two spectograms of an utterance synchroneously recorded with a headset microphone (A) and a hands free microphone placed at the rear view mirror (B). In Fig. 2B a larger part of the information in the speech is below the noise floor compared to Fig. 2A. These two signals represent typical training and test signals in a high mismatch scenario. The purpose of introducing AQBNE is two-fold. First,

A. Headset microphone

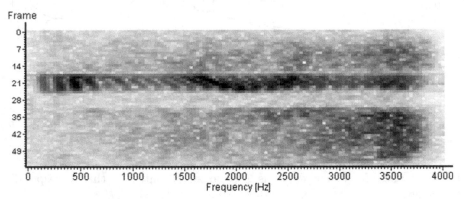

B. Hands free microphone

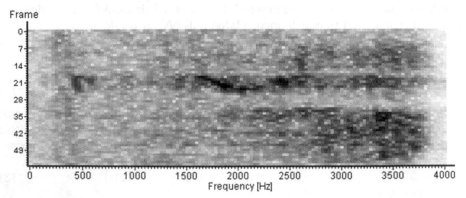

Fig. 2. Two spectograms of the same utterance recorded with two different microphones. A. with a headset microphone. B. with a hands free microphone at the rear view mirror.

the purpose is to provide a better noise-estimate in the high-noise signals. Second, the purpose is to let the signal (A) be more equal to the signal (B) during training, resulting in a model that does not describe the utterance at a level of detail that is unobtainable under test. For the second purpose, however, experiments have then to reveal to what extent the increased performance under mismatched conditions is at the expense of decreased performance under well matched conditions.

AQBNE is developed by examining quantile plots as shown in Fig. 3. First observe that frequency bands with high noise contains more energy in the majority of the utterance than low noise frequency bands. This leads to the assumption that a smaller q-value is desired for the high noise frequency bands and a higher q-value is desired for the low noise frequency bands. Statistically a low fixed q-value corresponds to a low noise estimate for all frequency bands and a higher fixed q-value to a higher noise estimate. By adapting the q-value to the noise level of the frequency band the low and high noise utterances will converge to similar representations when the noise is eliminated, which subsequently should lead to better ASR performance.

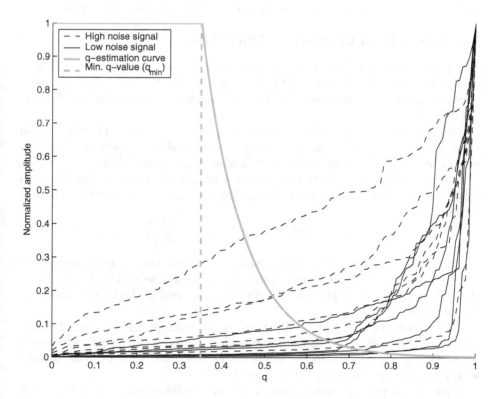

Fig. 3. Data sorted by amplitude for different frequency bands for a low noise (solid black) and high noise (dashed black) speech signal. The solid gray curve is the q-estimation curve, and the dashed gray line is the minimum allowed q-value.

In contrast to the fixed q-value method in QBNE (represented by a vertical line in Fig. 1 in subsection 2.1), the q-value in AQBNE is determined by a q-estimation curve as shown in Fig. 3. The intersection between this curve and each frequency curve is the noise estimate for the corresponding frequency. The vertical dashed line in Fig. 3 refers to the minimum desired q-value.

The q-estimation curve is defined as follows:

$$f(q) = e^{(q_{\min}-q)\tau} \tag{4}$$

where $q_{\min}$ is the minimum allowed q-value and τ is the slope of the curve.

Define $\tilde{q}$ as the q-value associated with the intersection of the q-estimation curve and the frequency curve, then $\tilde{q}$ is the solution of:

$$f(\tilde{q}) = |X(\omega, i_{\lfloor \tilde{q}I \rfloor})| \tag{5}$$

The noise estimate is then defined by

$$\hat{N}(\omega) = |X(\omega, i_{\lfloor \tilde{q}I \rfloor})| \tag{6}$$

and subsequently used as the noise estimate during spectral subtraction.

2.3   Speech Band Emphasizing Filter Bank

The predominant parametric representation of features in speech is Mel Frequency Cepstrum Coefficients (MFCC), introduced in 1980 by Davis and Mermelstein as an improvement to Linear Frequency Cepstrum Coefficients (LFCC) [2].

The method compresses the spectral information by applying the psycho acoustic theory of critical bands combined with a mel scale warping of the frequency axis, in order to more closely resemble human perception.

The method is implemented by applying a mel filter bank to the speech signal consisting of half overlapping triangular filters linearly distributed on the mel scale:

$$\text{Hertz2Mel}\{f\} = 2595 \log \left(1 + \frac{f}{700} \right) \tag{7}$$

The output from each filter is integrated to produce the reduced and warped spectrum which is then transformed into cepstrum coefficients. Fig. 4 (top) shows the mel filter bank with 23 triangular filters as specified in the ETSI ES 201 108 standard [3].

While both the mel-scale and critical band assumptions are well founded theoretically as well as experimentally they do not necessarily translate to robust ASR. As Fig. 4 (top) reveals the filters in the mel filter bank are concentrated with maximum resolution at low frequencies.

With the purpose to better focus on the speech information avaliable in the signal, we will investigate a speech band emphasizing (SBE) filter bank, where the filter concentration is empirically chosen to be highest at 1500 Hz and to decrease with higher and lower frequencies. To define the distribution using the

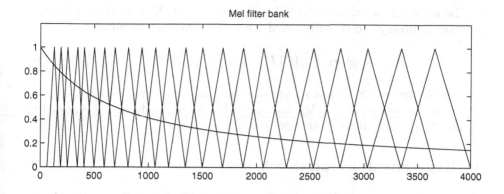

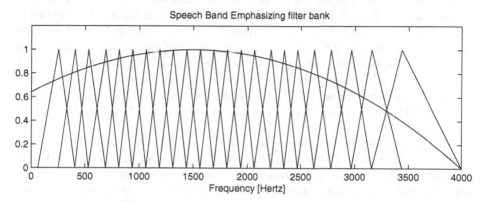

Fig. 4. The mel filter bank compared to the Speech Band Emphasizing filter bank (SBE). The continuous lines are the normalized derivatives of the mapping functions of Eq. 7 and 8indicating the concentration of triangular filters (importance functions).

existing framework the mapping function between Hertz and mel is replaced by a new mapping function between Hertz and the imaginary unit SBE:

$$\text{Hertz2SBE}\{f\} = 12\,000\,000 f + 4500 f^2 - f^3 \qquad (8)$$

Instead of distributing the triangular filters linearly on the mel scale the filters are now distributed linearly on the SBE scale, which is defined by Eq. 8. The resulting SBE filter bank is illustrated in Fig. 4 (bottom).

The mapping function in Eq. 8 is obtained by first defining an *importance function*. This function attains values between 0 and 1, where 1 indicates "most important" and 0 indicates "least important". The importance function of the SBE scale is defined by the polynomial (depicted as a continuos line in Fig. 4 (bottom)):

$$i(f) = \frac{-(f - 1500\text{Hz})^2}{(4000\text{Hz} - 1500\text{Hz})^2} + 1 \qquad (9)$$

The mapping function is the indefinite integral of the importance function with the arbitrary constant discarded (or set to a convenient value):

$$Hertz2SBE\{f\} = I(f) = \int i(f)df \tag{10}$$

Because the unit SBE has no physical meaning or interpretation, any scale factor can be applied to simplify the expression. In particular Eq. 8 is scaled by 18 750 000 to make the highest order coefficient unity.

3   Experimental Framework

To evaluate the methodology an experimental framework based on the Danish SpeechDat-Car Digits Database is used, which is part of the Aurora-3 database [4]. The corpus comprises 2457 utterances recorded in a car under different noise conditions, 265 utterances for quiet (motor idling, car stopped), 1513 utterances for low noise (town traffic or low speed on rough road) and 679 utterances for high noise (high speed on rough road). Each utterance is recorded synchroneously by two microphones Close Talking (CT) and Hands Free (HF) using a sample rate of 16 kHz and a 16 bit quantization. This results in a total of 4914 speech recordings.

Training and test definitions proposed by the Aurora consortium is used which divide the test in three parts: Well Match (WM), Medium Mismatch (MM) and High Mismatch (HM). In WM training and test are performed under all conditions for both microphones. Because of the matched conditions under training and test the WM scenario does not reveal the frontends ability to adapt to unexpected noise during test. In MM training is performed at low noise level and tested at high noise level, both with the HF microphone. MM is primarily a test of the frontends ability to suppress additive noise. In HM training is performed at all noise levels with the CT microphone and tested at high noise level with the HF microphone. HM is primarily a test of independence of microphone type, placement and distance relative to the speaker.

The speech recognizer used in Aurora is based on whole word HMMs. The structure of each HMM is a simple left-to-right model 16-state 3-mixture per state for each digit. The silence model is a 3-state 6-mixture and the short pause model is a 1-state 6-mixture. The HMM parameters are estimated using Viterbi training and Baum-Welch re-estimation procedure. To build and manipulate the HMM's the Hidden Markov Model Toolkit (HTK) is used [5].

4   Experimental Results

The baseline reference is a 32 bit floating point precision C implementation of the Aurora standard frontend (WI007) which is supplied with SpeechDat-Car and fully specified in the ETSI ES 201 108 standard [3].

To enable easy implementation of the various methods an extensible clone of this frontend has been written in Java using 64 bit floating point precision.

Table 1. Results with mel filter bank

Frontend	WM	MM	HM	Average	AI
Aurora Baseline	80.13	51.56	33.61	58.50	-
Aurora Adv. Frontend	93.37	81.49	79.59	85.77	-
QBNE					
q=0.40	**84.60**	66.80	21.72	62.65	3.73
q=0.45	84.19	**68.10**	35.95	**66.50**	**14.99**
q=0.50	83.77	65.23	38.05	65.85	14.40
q=0.55	81.58	63.80	38.63	64.62	12.77
q=0.60	80.46	63.15	**40.68**	64.46	13.29
Adaptive QBNE					
q_{min}=0.30, τ=10	83.77	64.06	**53.68**	**69.35**	**25.23**
q_{min}=0.30, τ=15	**85.12**	67.45	36.32	66.74	15.29
q_{min}=0.35, τ=10	82.60	65.36	48.46	68.03	21.65
q_{min}=0.35, τ=15	83.10	66.67	41.22	66.88	17.40
q_{min}=0.40, τ=10	81.58	64.97	51.95	68.36	23.47
q_{min}=0.40, τ=15	82.24	65.49	48.54	67.95	21.61
q_{min}=0.45, τ=10	80.33	65.23	18.84	59.67	-1.61
q_{min}=0.45, τ=15	81.05	66.02	49.81	67.98	22.32

Table 2. Results with SBE filter bank

Frontend	WM	MM	HM	Average	AI
SBE baseline	82.04	54.43	31.59	59.76	1.40
SBE + Adv. Frontend	N/A	N/A	N/A	-	-
QBNE					
q=0.40	**84.94**	67.97	12.88	60.99	-1.88
q=0.45	83.49	**71.74**	32.21	66.56	14.33
q=0.50	83.75	70.83	**46.15**	**69.83**	**24.22**
q=0.55	81.22	68.49	23.20	62.26	4.29
q=0.60	80.78	65.76	36.82	64.53	12.35
Adaptive QBNE					
q_{min}=0.30, τ=10	81.82	69.14	45.33	68.26	21.49
q_{min}=0.30, τ=15	**83.92**	**70.57**	31.47	66.14	13.20
q_{min}=0.35, τ=10	82.00	66.93	48.79	68.42	22.66
q_{min}=0.35, τ=15	81.84	69.66	36.65	66.28	15.40
q_{min}=0.40, τ=10	81.22	62.37	**60.63**	69.48	**27.98**
q_{min}=0.40, τ=15	82.69	68.36	49.81	69.45	24.73
q_{min}=0.45, τ=10	79.27	66.02	42.86	65.53	16.27
q_{min}=0.45, τ=15	81.84	67.45	52.90	**69.57**	25.99

Average is weighted 0.4WM+0.35MM+0.25HM. **AI** is the weighted average of the improvements for each of the three test conditions compared to *baseline*. Best accuracy in each category is in boldface.

This frontend is the baseline for all relative improvement calculations, and will be denoted *baseline*. This frontend has also been implemented with the SBE filter bank in place of the mel filter bank. It is denoted *SBE baseline*. All tested methods are implemented as extensions to these two frontends.

The results are reported in table 1 for methods implemented on the *baseline* frontend, and in table 2 for methods implemented on the *SBE baseline* frontend. Table 1 also shows the equivalent results obtained using the Aurora Advanced Frontend as reported in [6].

The first three columns are the accuracies obtained for each of the three test conditions. The averages are calculated as a weighting of these three results with the equation 0.4WM+0.35MM+0.25HM. The rightmost column is a weighted average of the improvement in each of the three results compared to *baseline*, using the same weighting. For each method the best accuracies are typeset in boldface.

The SBE filter bank performs slightly better than the mel filter bank. QBNE shows even greater improvements, especially with the SBE filter bank. Adaptive QBNE which sought to improve high mismatch performance succeeds ind this regard, yielding even better results than QBNE, again especially in combination with the SBE filter bank.

Fig. 5 shows a graphical approach of method comparison. For each method and test condition, the best score is selected, and the improvement to *baseline* is calculated. The white columns are methods implemented on *baseline* (mel filter bank) and grey columns are methods implemented on *SBE baseline* (SBE filter bank).

AQBNE is a clear advantage over QBNE in the high mismatch scenario with an 80.39% improvement compared to a 21.04% improvement for QBNE with mel

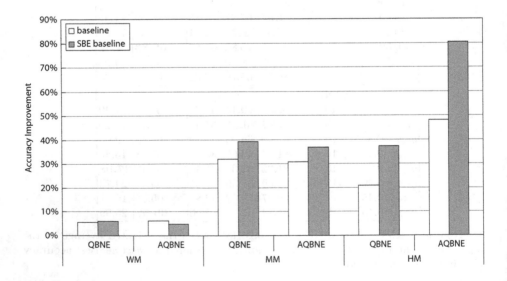

Fig. 5. Comparison of methods by improvement to *baseline*. For each method, the best result for each of the three test conditions is selected.

filter bank. It achieves this without significantly compromising the performance in WM and MM, reflected in an average improvement of 27.98% compared to 14.99% for QBNE with mel filter bank. Given the very small computational impact of the method, compared to QBNE, it is a relevant improvement in all ASR applications.

Furthermore it is evident that the SBE filter bank in general improves recognition performance, and especially combines well with both QBNE and AQBNE.

Finally, it is observed that for MFCC the Aurora Advanced Frontend achieves superior performance compared to AQBNE. So far the SBE frontend has not been integrated with the Aurora Advanced Frontend in order to compare the two feature representations.

5 Conclusion

In this paper the task of eliminating unwanted noise from a speech signal recorded in a noisy car environment has been investigated. Two methods have been considered seperately and in combination.

The first method is Quantile Based Noise Estimation (QBNE), which estimates the noise during both speech and non-speech sections. Adaptive QBNE (AQBNE) is suggested and concluded to improve the standard QBNE. In addition an empirically chosen Speech Band Emphasizing filter bank (SBE) is suggested as an alternative to the standard mel filter bank.

The experiments conducted on the Danish SpeechDat-Car database show that SBE generally show an improvement over the standard mel filter bank (Aurora baseline), achieving significant increases in recognition performance when combined with both QBNE and AQBNE.

Specifically AQBNE with SBE achieved remarkable results under highly mismatched training and test conditions with an 80.39% improvement compared to a 21.04% improvement for QBNE with mel filter bank. The average improvement for SBE with AQBNE was as high as 27.98% compared to 14.99% for QBNE with mel filter bank.

Though for all test conditions, the Aurora Advanced Frontend achieves superior performance compared to AQBNE, the AQBNE is still a relevant method to consider for small foot print applications. In addition, it is relevant to investigate the integration of the AQBNE and SBE methods into the Aurora Advanced Frontend.

References

1. Volker Stahl, Alexander Fischer and Rolf Bippus, "Quantile Based Noise Estimation for Spectral Subtraction and Wiener Filtering", pp. 1-4, ICSLP 2000.
2. Steven B. Davis and Paul Mermelstein, "Comparison of parametric representations for monosyllabic word recognition in continuously spoken sentences", pp. 357-366, IEEE Trans. Acoustics, Speech, Signal Processing, vol. 28(4).
3. European Telecommunications Standards Institute, "ES 201 108 v.1.1.2", http://www.etsi.org/, 2000.

4. Asuncin Moreno, Boerge Lindberg, Christoph Draxler, Gal Richard, Khalid Choukri, Stephan Euler and Jeffrey Allen, "SpeechDat-Car. A Large Speech Database for Automotive Environments", pp. 1-6, LREC 2000.
5. Steve Young, Gunnar Evermann, Thomas Hain, Dan Kershaw, Gareth Moore, Julian Odell, Dave Ollason, Dan Povey, Valtcho Valtchev and Phil Woodland, "The HTK Book (for HTK Version 3.2.1)", http://htk.eng.cam.ac.uk, 2002.
6. Dusan Macho, Laurent Mauurary, Bernhard No, Yan Ming Cheng, Doug Ealey, Denis Jouver, Holly Kelleher, David Pearce and Fabien Saadoun, "Evaluation of a Noise-Robust DSR Front-end on Aurora Databases", pp. 17-21, Proc. ICSLP 2002, Denver, Colorado

Spotting Multilingual Consonant-Vowel Units of Speech Using Neural Network Models

Suryakanth V. Gangashetty, C. Chandra Sekhar,
and B. Yegnanarayana

Speech and Vision Laboratory,
Department of Computer Science and Engineering,
Indian Institute of Technology Madras, Chennai - 600 036, India
{svg, chandra, yegna}@cs.iitm.ernet.in

Abstract. Multilingual speech recognition system is required for tasks that use several languages in one speech recognition application. In this paper, we propose an approach for multilingual speech recognition by spotting consonant-vowel (CV) units. The important features of spotting approach are that there is no need for automatic segmentation of speech and it is not necessary to use models for higher level units to recognise the CV units. The main issues in spotting multilingual CV units are the location of anchor points and labeling the regions around these anchor points using suitable classifiers. The vowel onset points (VOPs) have been used as anchor points. The distribution capturing ability of autoassociative neural network (AANN) models is explored for detection of VOPs in continuous speech. We explore classification models such as support vector machines (SVMs) which are capable of discriminating confusable classes of CV units and generalisation from limited amount of training data. The data for similar CV units across languages are shared to train the classifiers for recognition of CV units of speech in multiple languages. We study the spotting approach for recognition of a large number of CV units in the broadcast news corpus of three Indian languages.

1 Introduction

The main objective of continuous speech recognition system is to provide an efficient and accurate mechanism to transcribe human speech into text. Typically, continuous speech recognition is performed in the following two steps: (1) speech signal to symbol (phonetic) transformation, and (2) symbol to text conversion. Two approaches are commonly used for subword unit based continuous speech recognition. The first approach is based on segmentation and labelling [1]. In this approach, the continuous speech signal is segmented into subword unit regions and a label is assigned to each segment using a subword unit classifier. The main limitation of this approach is the difficulty in automatic segmentation of continuous speech into subword unit regions of varying durations. Because of imprecise articulation and coarticulation effects, the segment boundaries are manifested poorly. The second approach to speech recognition is based on building word

M. Faundez-Zanuy et al. (Eds.): NOLISP 2005, LNAI 3817, pp. 303–317, 2005.

models as compositions of subword unit models, and recognising sentences by performing word-level matching and sentence level matching using word models and language models respectively [1]. The focus of this approach is on recognising higher level units of speech such as words and sentences rather than on recognising subword units.

In this paper, we propose an approach for multilingual speech recognition by spotting subword units. Specifically, we consider a method for spotting subword units using vowel onset points (VOPs) as anchor points and labelling the regions around these VOPs using suitable classifiers. The important features of spotting approach are that there is no need for automatic segmentation of speech and it is not necessary to use models for higher level units to recognise the subword units. The symbols that capture the phonetic variations of sounds are suitable units for signal to symbol transformation. Pronunciation variation is more systematic at the level of syllables compared to the phoneme level. Syllable-like units such as consonant-vowel (CV) units are important information-bearing sound units from production and perception point of view [2]. Therefore, we consider CV units of speech as the basic subword units for speech recognition. In Indian languages, the CV units occur with high frequency.

The distribution capturing ability of autoassociative neural network (AANN) models is explored for detection of VOPs in continuous speech [3]. An important issue for the development of a suitable classification system for the recognition of CV units in Indian languages is the large number of these units. Combination of more than 30 consonants and 10 vowels of a language result in a set of about 300 CV units. Further, there are many regional languages across the country. Difficulties in the development of multilingual speech recognition systems are due to the presence of several new classes, degree of overlapping of classes and frequency of occurrence of a given class in different languages. The difficulties in designing a multilingual system are also due to variability among the data set, amount of training data and large number of CV classes. Also, many of the CV units have similar acoustic features. Additionally, the number of examples available in a corpus is not the same for all the units. There may be many units for which only a small number of examples are available. We consider a data sharing approach for development of classification system by combining same type of CV classes across the Indian languages [4]. We consider support vector machine (SVM) based classifiers due to their ability of generalization from limited training data and also due to their inherent discriminative learning [5]. The variability among the data set and more number of classes in multiple languages has less effect on the recognition performance when SVMs are used for classification [4]. However, the application of SVMs to speech recognition problems has been limited to smaller vocabulary tasks due to computational complexity. To reduce the computational complexity, we propose nonlinear compression of large dimensional input pattern vectors using the dimension reduction capability of autoassociative neural network models [6] [7]. We demonstrate the CV spotting based approach to continuous speech recognition for sentences in multiple Indian languages.

The paper is organised as follows: In Section 2, we discuss the issues in spotting CV units. The description of speech corpus used in the studies is given in Section 3. Studies on detection of VOPs in continuous speech utterances are described in Section 4. In Section 5, we present the studies on recognition of multilingual CV units. The system for spotting multilingual CV units in continuous speech is described in Section 6. In this section the spotting approach is illustrated with an example. Studies on recognition of CV units by processing the segments around the hypothesised VOPs in continuous speech utterances are also presented in this section.

2   Issues in Spotting Multilingual CV Units

Strategies for spotting subword units in continuous speech have been based on training the classifiers to recognise only the segments of the continuous speech signal belonging to subword units and reject all other segments. The models thus trained to classify or reject are then used to scan the speech signal continuously and hypothesise the presence or absence of the corresponding subword units. This strategy is similar to the keyword spotting approaches [8]. The main limitation of this strategy based on scanning is that a large number of spurious hypotheses are given by the spotting system [9]. For spotting CV units in continuous speech, we consider an approach based on detection of VOPs and labelling the segments around the VOPs using SVM based CV classifier [4] [10]. The main issues in spotting CV units in the proposed approach are development of a method for detection of VOPs with good accuracy and development of an SVM based classifier capable of discriminating large number of CV classes.

2.1   Detection of Anchor Points

Figure 1 shows the significant events in the production of a typical CV unit. Utterances of CV units consists of all or a subset of the following significant speech production events: Closure, burst, aspiration, transition and vowel [11]. The vowel onset point (VOP) is the instant at which the consonant part ends and the vowel part begins in a CV utterance [12]. It is obvious that all the CV units have a distinct VOP in their production [11] [13]. Because every CV utterance has a VOP, the VOPs can be used as anchor points for CV spotting. This approach requires detection of VOPs in continuous speech with a good accuracy. The VOPs of all CV segments in a continuous speech utterance should be detected with minimum deviation. Since labelling will be done only for the segments around the VOPs detected, the effect of any VOP not being detected is that the CV segment around that VOP will not be recognised. Therefore it is important to minimise the number of missing errors by the VOP detection method. The effect of spurious VOPs being detected is that segments around them will also be given to the CV classifier for labelling.

In the method proposed in [13], a multilayer feedforward neural network (MLFFNN) model is trained to detect the VOPs by using the trends in the speech signal parameters at the VOPs. The input layer of the network contains

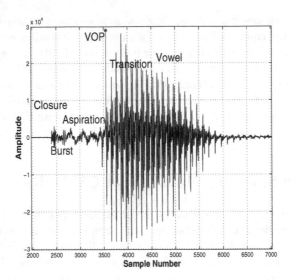

Fig. 1. Significant events in the production of the CV unit /kha/ with VOP at sample number 3549

9 nodes and the output layer has 3 nodes. One of the output nodes is labeled as VOP node to indicate the presence of the VOPs, and the other two nodes are labeled as pre-VOP and post-VOP to indicate the absence of VOPs. The signal energy, residual energy and spectral flatness parameters extracted from two frames around the VOP and the ratio of the parameters in the two frames are used to form an input vector. Two other such vectors are also extracted from each CV utterance. One vector is derived from two frames in the region before the VOP for representing the pre-VOP region. Another vector is derived from two frames in the region after the VOP for representing the post-VOP region. An MLFFNN classifier is trained using the vectors extracted from the three different regions of each utterance. For detection of VOP in a CV utterance using the network trained as above, a parameter vector extracted at every 10 msec is given as input to the network. The parameter vector is extracted from two frames, with one frame starting at the point under consideration and another frame starting 20 msec after this point. Thus the speech signal of a CV utterance is scanned by the network to detect the VOP. The point at which the output for the VOP node of the network is maximum is hypothesised as the VOP of the CV utterance. This method requires a large number of training examples to capture the trends in speech signal parameters at the VOP.

In another method for detection of VOPs, we consider AANN models [3]. A five layer AANN model, shown in Fig. 2, with compression layer in the middle has important properties suitable for distribution capturing, data compression, and extraction of higher order correlation tasks [14] [7]. We explore the distribution capturing of feature vectors by the AANN models to hypothesise the consonant and vowel regions and then detect VOPs in continuous speech. In Section 4, we

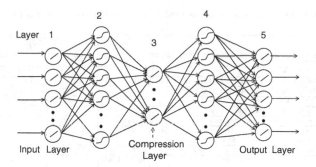

Fig. 2. Five layer AANN model

describe the method used for VOP detection in continuous speech using AANN models.

2.2 Classifier for Recognition of Multilingual CV Segments

Hidden Markov models (HMM) are used in most speech recognition systems. These models use maximum likelihood (ML) approach for training. The incremental model optimization approach in ML framework simplifies the training process, but loses discriminative information in the process [15]. This is due to the fact that training data corresponding to other models are not considered during the optimization of parameters for a given model. Training by optimization over the entire pattern space gives better discriminative power to the models since the models now learn patterns that need to be discriminated. Multilayer feed-forward neural network (MLFFNN) models and support vector machine (SVM) models are good at this type of learning since the training involves optimization over entire pattern space [5]. The MLFFNN models have been shown to be suitable for pattern recognition tasks because of their ability to form complex decision surfaces. In order to obtain a better classification performance it is necessary to tune the design parameters such as structure of network, number of epochs, learning rate parameter and momentum. For better generalization, it is necessary to have large amount of training data. But arriving at optimal parameters for complex recognition problem using MLFFNN models is a difficult proportion. SVM models have attained prominence due to their inherent discriminative learning and generalization capabilities from the limited training data. These models learn the boundary regions between patterns belonging to two classes by mapping the input patterns into a higher dimensional space, and seeking a separating hyperplane so as to maximize its distance from the closest training examples. In the next section, we describe the speech corpus used in the studies.

3   Speech Data and Representation

Speech corpus consisting of recording of television broadcast news bulletins for three Indian languages namely, Tamil, Telugu and Hindi is used in our studies.

Table 1. Description of broadcast news speech corpus used in studies

Description	Language			
	Tamil	Telugu	Hindi	Multilingual
Number of bulletins	33	20	19	72
News readers (Male:Female)	(10:23)	(11:9)	(6:13)	(27:45)
Number of bulletins used for training (Male:Female)	27 (8:19)	16 (9:7)	16 (5:11)	59 (22:37)
Number of bulletins used for testing (Male:Female)	6 (2:4)	4 (2:2)	3 (1:2)	13 (5:8)
Number of CVs (Cs:Vs)	324 (27:12)	432 (36:12)	360 (36:10)	432 (36:12)
Number of CV units used for the study	123	138	103	196
Number of CV segments in the training data	44,612	43,491	22,109	1,10,212
Number of CV segments used for training which are covered by number of CV units used for the study	43,541	41,725	20,236	1,05,502
Percentage of number of CV segments used	97.53%	95.93%	91.52%	95.72%
Range of frequency of occurrence for the units in the training data	39 to 1,633	40 to 2,037	40 to 1,264	40 to 2,826
Number of CV segments in the test data covered by number of CV units used for the study	10,293	11,347	4,137	25,777
Speech sentences considered for testing	1,416	1,348	630	3,094

Each bulletin (session) contains 10 to 15 minutes of speech from a single (male or female) speaker. The CV utterances in the corpus are excised and labeled manually. A brief description of the speech corpus used in our studies is given in Table 1. Interpretation of the contents of the table for Tamil language data is as follows: On the whole, 33 bulletins, read by 10 male and 23 female speakers are collected. The data in 27 bulletins read by 8 male speakers and 19 female speakers is used for training. The data in the remaining 6 bulletins by 2 male speakers and 4 female speakers is used for testing. There are 27 consonants and 12 vowels leading to a total of 324 CV units. The CV units have different frequencies of occurrence in the speech corpus. The CV units that occur at least 50 times in the corpus are considered in our studies. This results in a set of 123 CV classes for Tamil language. Out of a total of 44,612 CV segments in the training data, 43,541 segments (*i.e.*, 97.53%) belong to these 123 CV classes. The frequency of occurrence for these classes in the training data varies from 39 to 1,633. The test data includes about 10,293 CV segments belonging to the 123 CV classes. There are 1,416 continuous speech sentences available for testing. A similar description for the speech corpora of Telugu, Hindi and multiple languages is also given in Table 1.

Short-time analysis of the speech signal of the CV utterances is performed using a frame size of 20 msec duration with a shift of 5 msec. Each frame is represented by a parametric vector consisting of 12 mel-frequency cepstral coefficients (MFCC), energy, their first order derivatives and their second order derivatives [16] [17]. The dimension of the parametric vector for each frame is 39.

Models based on SVMs are suitable for classification of fixed dimensional patterns. However, durations of CV utterances vary not only for different classes, but also for a particular CV class. It is necessary to develop a method for representing the CV utterances by fixed dimensional patterns. It is useful to identify the region before the VOP as corresponding to the manner of articulation (MOA), the transition region after VOP to the place of articulation (POA), and the remaining portion to the steady vowel (V). Generally it is difficult to isolate these regions precisely. Moreover, the acoustic characteristics of each region will influence the other regions. Thus, all the three regions need to be represented together as a single pattern vector [11] [18]. Since the vowel region is prominent in the signal due to its large amplitude characteristics, and also due to its periodic excitation property, it is easy to locate this event compared to other speech production events [13]. The information necessary for classification of CV utterances can be captured by processing a portion of the CV segment containing parts of the closure and vowel region, and all of the burst, aspiration, and transition regions. The closure, burst and aspiration regions are present before the VOP. The transition and vowel regions are present after the VOP. To capture the acoustic characteristics of the CV units, it is necessary to represent each of these units as a sequence of frames, and extract the spectral information corresponding to each frame. A segment of typically 50 to 100 msec duration around the VOP contains most of the information necessary for classification of the CV utterances. This segment can be processed to derive a fixed dimensional pattern, automatically from a varying duration segment of a CV unit [13]. Portions of a CV utterance in the beginning and the end are not included in the fixed duration segment, since they may be affected by the coarticulation effects. From the analysis of broadcast news data it is observed that, the average minimum duration of segments for a CV class is 80 msec. Therefore a 65 msec segment around the VOP is used to represent each CV segment. Once the VOP is detected, five overlapping frames are considered to the left of VOP and five to the right. Thus each CV segment is represented by a 390-dimensional pattern vector. In the next section, we describe the method used for VOP detection in continuous speech using AANN models.

4   System for Detection of VOPs in Continuous Speech Utterances

A five layer AANN model to capture distribution of feature vectors is shown in Fig. 2. In this model the input and output layers have the same number of units, and all these units are linear. For each CV class, two AANN models (one corresponding to the consonant region and the other to the vowel region) are developed. For training the AANN model corresponding to the consonant region, the fifth frame to the left of the manually marked VOP frame is selected from each of the training examples. For training the AANN model corresponding to the vowel region we consider the VOP frame and the fourth frame to the right of VOP frame. The model corresponding to a region of a CV class captures the distribution of feature vectors. The distribution is expected to be different for the consonant and vowel regions of a class. The distribution of feature vectors of a region is captured using a network structure *39L 60N 4N 60N 39L*, where *L* refers to linear units and *N* refers to nonlinear units. The integer value indicates the number of units in that particular layer. The activation function for the nonlinear units is a hyperbolic tangent function. The network is trained using error backpropagation algorithm in pattern mode for 1000 epochs.

For detection of VOPs in continuous speech, each frame is given as input to the pairs of AANN models of all the CV classes. From the evidence available

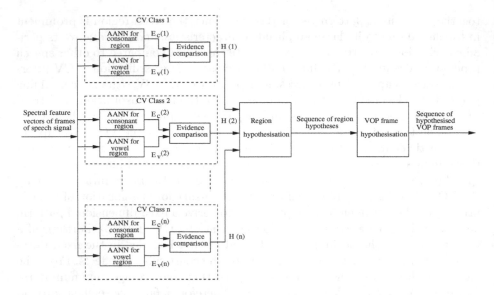

Fig. 3. Block diagram of the system for detection of VOPs in continuous speech. $E_c(k)$ and $E_v(k)$ are the evidence obtained from consonant and vowel region models of kth class respectively. $H(k)$ is hypothesised region of the current frame by the models of class k.

in the outputs of the models of a class, the hypothesised region of the frame is obtained as the region of the model with higher evidence. The hypotheses from the models of different CV classes are used to assign the frame to the consonant or vowel region. In this way we obtain a sequence of region labels for the sequence of frames of the continuous speech utterance. VOP frames are identified as those frames, at which there is a change of labels from consonant to vowel. The block diagram of the system for detection of VOPs in continuous speech utterances is shown in Fig. 3.

We study the performance of the VOP detection method using AANN models. For comparison we consider the method based on MLFFNN model described in Section 2.1 [13]. The performance is measured in terms of the number of matching, missing and spurious hypotheses of VOPs. The VOPs detected with a deviation upto 25 msec are considered as the matching hypotheses. When the deviation of hypothesised VOP is more than 25 msec or there is no hypothesised VOP around an actual VOP, the VOPs of such segments are considered as the missing hypotheses. When there are multiple hypotheses within 25 msec around an actual VOP or the hypothesised VOP does not fall in this range, such hypotheses are considered as spurious ones. For testing we consider the utterances of 120, 120 and 60 sentences selected at random from 1416, 1348, and 630 sentences for Tamil, Telugu and Hindi languages, respectively. These 300 sentences consist of a total number of 3924 syllable-like units corresponding to 1580, 1648 and 696 actual VOPs from sentences of Tamil, Telugu and Hindi languages, respectively. These VOPs have been marked manually. For each utterance the hypothesised VOPs are determined by the MLFFNN and AANN based

Table 2. Comparison of the average performance of different methods for detection of VOPs in continuous speech. The performance is given as a percentage of total number of VOPs in the continuous speech utterances, for the matching, missing and spurious hypotheses.

Method	VOP detection performance (in %)		
	Matching hypotheses	Missing hypotheses	Spurious hypotheses
MLFFNN	68.80	31.19	33.10
AANN	68.62	31.37	6.21

methods. The average performance of different VOP detection methods for the data of three languages is given in Table 2. It is seen from Table 2 that the performance of both the methods is nearly the same for matching case. However, the VOP detection method based on AANN gives significantly less number of spurious VOPs. Many of the missing VOPs in case of AANN based method are observed to be for CV units whose consonants are semivowels, fricatives and nasals.

5 Classification System for Recognition of Multilingual CV Units

In this section, we describe a multilingual system in which data sharing approach is considered for recognition of frequently occurring CV units of three Indian languages. This approach is motivated by the commonality among CV classes across Indian languages. The similar CV classes from different languages are derived from Indian language TRANSliteration (ITRANS) code [19]. The ITRANS code was chosen, as it uses the same symbol across the Indian languages to represent a given sound. A summary of the description of the database used for the development of multilingual CV recognition system is given in the last column of Table 1. The number of CV classes with at least 50 examples in the data set is 123, 138, and 103 for Tamil, Telugu and Hindi respectively, leading to a total of 364 classes. Out of these 364 classes, 27, 25, and 28 classes are unique to Tamil, Telugu and Hindi, respectively. The number of CV classes common to any two languages is 64. There are 52 CV classes common to all the three languages. The union of the set of CV classes in three languages gives a set of 196 CV classes for multilingual data. The number of segments available for training the models of these classes is 1,05,502, and the number of segments in the test data set is 25,777. Thus sharing of data across languages leads to availability of large training data sets, but variability in the data of a class is also increased.

As explained in Section 3, each CV utterance is represented by a pattern vector of dimension 390. To reduce computational complexity, we propose nonlinear compression of the large dimensional input pattern vectors using AANN models [6][7]. The block diagram of the system for recognition of multilingual CV units is shown in Fig. 4. It consists of three stages. In the first stage, the 390-dimensional input pattern vector **x** is compressed to a 60-dimensional vector, using an AANN

312   S.V. Gangashetty, C. Chandra Sekhar, and B. Yegnanarayana

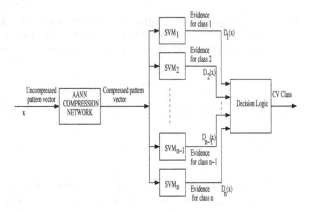

Fig. 4. Block diagram of the multilingual CV recognition system for labelling region around the VOP

with structure *390L 585N 60N 585N 390L*. These compressed pattern vectors are used to train the SVM classifier. One-against-the-rest approach is used for decomposition of the learning problem in n-class pattern recognition into several two-class learning problems [20]. SVM models are generated by assigning one model to each class, and training this model by considering data from all the three languages. An SVM is constructed for each class by discriminating that class against the remaining $(n-1)$ classes. The recognition system based on this approach consists of n number of SVMs. The set of training examples $\{\{(\mathbf{x}_i, k)\}_{i=1}^{N_k}\}_{k=1}^{n}$ consists of N_k number of examples belonging to k^{th} class, where the class label $k \in \{1, 2, \ldots, n\}$. All the training examples are used to construct an SVM for a class. The SVM for the class k is constructed using a set of training examples and their desired outputs, $\{\{(\mathbf{x}_i, y_i)\}_{i=1}^{N_k}\}_{k=1}^{n}$. The examples with $y_i = +1$ are called positive examples, and those with $y_i = -1$ are called negative examples. An optimal hyperplane is constructed to separate positive examples from negative examples. The separating hyperplane (margin) is chosen in such a way as to maximize its distance from the closest training examples of different classes [5]. The support vectors are those data points that lie closest to the decision surface, and therefore are the most difficult to classify. For a given pattern $\mathbf{x}$ around a VOP, the evidence $D_k(\mathbf{x})$ is obtained from each of the SVMs. In the decision logic, the class label k associated with the SVM that gives maximum evidence is hypothesised as the class of the pattern $\mathbf{x}$ representing the CV segment around VOP.

The recognition system is developed using the SVM models trained with compressed pattern vectors. The recognition system is also developed using the SVM models trained with 390-dimensional uncompressed vectors. The recognition performance of the SVM models trained with 390-dimensional uncompressed vectors and the models trained with 60-dimensional compressed vectors is given in Table 3. In comparison with uncompressed case, the classification performance is nearly the same for reduced dimension. Thus it is possible to compress the

Table 3. Classification performance of CV recognition systems using compressed and uncompressed pattern vectors in multiple languages

Language	Classification performance (in %)	
	Compressed	Uncompressed
Multilingual	45.31	45.10

Table 4. Comparison of the k-best classification performance for multilingual CV recognition systems

System	k−best classification performance (in %)				
	$k=1$	$k=2$	$k=3$	$k=4$	$k=5$
HMM	41.32	47.46	50.80	52.91	54.57
SVM	45.31	57.62	64.00	68.08	71.03

390-dimensional pattern vectors to 60-dimensional vectors without affecting the classification performance.

The studies in this section show that it is possible to compress large dimensional pattern vectors to reduced dimensional vectors without affecting the classification performance of the SVM based classifiers. The compression also leads to a significant reduction in the computational complexity of the kernel operations in the SVM models.

The k-best recognition performance of the multilingual system for 196 CV classes is given in Table 4. For comparison, the performance of the hidden Markov model (HMM) based systems is also obtained. A CV segment is analyzed frame by frame, with a frame size of 20 msec and a frame shift of 5 msec. Each frame is represented by a parametric vector consisting of 12 MFCC coefficients, energy, their first order derivatives, and their second order derivatives. In this case, the dimension of each frame is 39. A 5-state, left-to-right, continuous density HMM using multiple mixtures with diagonal covariance matrix is trained for each CV class. The number of mixtures is 2 for the CV classes with a frequency of occurrence less than 100 in the training data. The number of mixtures is 4 for those CV classes whose frequency of occurrence lies between 100 and 500. For other classes, the number of mixtures is 8. All the frames of a CV segment are used in training and testing the HMM based system. The recognition performance of CV segments for the HMM based system is also given in Table 4. It is seen from Table 4 that the SVM based multilingual system performs significantly better than that based on HMMs. The SVMs use discriminative information in the process of learning, whereas HMM models are trained using the maximum likelihood (ML) methods. The ML framework does not use discriminative information. Due to this fact, there is a significant difference (71.03% vs 54.57%) in the 5-best classification performance of SVM and HMM based systems. In the next section, we describe CV recognition system using SVM models for classifying the CV segments around hypothesised VOPs.

6   Spotting CV Units in Continuous Speech

The block diagram of the integrated system for spotting multilingual CV units in continuous speech utterances is given in Fig. 5. The speech signal is given as input to the VOP detection module to locate VOPs in it. The short-time analysis is performed on 65 msec segment around each of the hypothesised VOPs to extract 390-dimensional MFCC based pattern vectors. This pattern vector is compressed using an AANN model. The compressed pattern vector is given to the multilingual CV recognition system to hypothesise the CV class of the current segment. Thus a sequence of hypothesised CV units is obtained for a given speech utterance.

For illustration, we consider a Tamil language continuous speech utterance /kArgil pahudiyilirundu UDuruvalkArarhaL/ consisting of 16 syllables (kAr, gil, pa, hu, di, yi, li, run, du, U, Du, ru, val, kA, rar, haL) whose waveform is shown in Fig. 6(a). The hypothesised region labels obtained using the VOP detection system are shown in Fig. 6(b). The label C corresponds to the consonant region and V to the vowel region. Using the procedure described in Section 4, the VOPs are detected. The hypothesised locations in terms of sample numbers (320, 720, 2440, 3760, 4800, 5560, 6200, 7480, 9480, 11120, 12080, 13240, 14560, 16960) are shown in Fig. 6(c). For comparison we consider manually marked VOP locations (280, 2360, 3800, 4920, 5480, 6320, 7400, **8200**, 9440, 11160, 12080, **12520**, 13200, 14520, **15840**, 16960) shown in Fig. 6(d).

It is seen that there are three VOPs (their sample numbers are indicated in boldface) that have been missed around the locations 8200, 12520, and 15840 corresponding to the syllables /ru/, /ru/, and /ra/, respectively. The VOP at location 720 is hypothesised as spurious VOP. For the segments around the hypothesised VOPs, the five CV class alternatives given by the multilingual CV recognition system (developed in Section 5) are given in Table 5. It is seen that for most of the segments the actual CV class of the segment is present among the alternatives. The correctly identified classes in the CV lattice are written in boldfaces. The segment around the hypothesised location 11120 has been hypothesised as /mu/, where as the actual syllable is /U/. This belongs to the case in which the vowel is in the initial portion of a word. Recognition of only

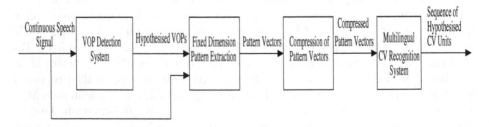

Fig. 5. Block diagram of the multilingual continuous speech recognition system based on spotting CV units

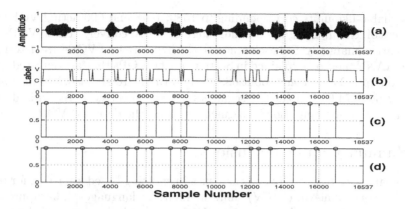

Fig. 6. Plots of the (a) Waveform of the speech signal, (b) Hypothesised region labels for each frame, (c) Hypothesised VOPs, and (d) Manually marked (actual) VOPs for the Tamil language sentence /kArgil pahudiyilirundu UDuruvalkArarhaL/

Table 5. The classes hypothesised by the multilingual CV classifier for a continuous speech utterance /kArgil pahudiyilirundu UDuruvalkArarhaL/. The alternative classes for the segment around a hypothesised VOP are given in a row of the lattice. The entries in the last column represent position of actual CV in hypothesised alternatives.

VOP locations (in sample numbers)		Lattice of hypothesised CVs					Actual	Position
Actual	Hypothesised	1	2	3	4	5	syllable	
280	320	pA	**kA**	vA	ha	shu	kAr	2
—	720	kA	pA	hA	na	pa	—	—
2360	2440	**gi**	yE	hi	ya	yai	gil	1
3800	3760	hA	pA	**pa**	sA	sa	pa	3
4920	4800	**hu**	gu	mu	vu	pu	hu	1
5480	5560	bI	vi	Ti	Ni	**dI**	di	5
6320	6200	**yi**	lA	li	zi	tI	yi	1
7400	7480	**li**	ni	ru	ja	lai	li	1
8200	—	VOP Missed					run	—
9440	9480	**du**	Ru	ja	dE	rA	du	1
11160	11120	**mu**	kU	va	pO	vA	U	1
12080	12080	**Du**	da	dA	nA	tu	Du	1
12520	—	VOP Missed					ru	—
13200	13240	**va**	da	kai	hi	vA	val	1
14520	14560	**kA**	ka	ga	cha	zA	kA	1
15840	—	VOP Missed					rar	—
16960	16960	**ha**	kA	ka	ga	sa	haL	1

vowels is not addressed in the current studies. All the classes hypothesised by the recognition system are of type CV.

We study the performance of the spotting approach for recognition of CV units for a large number of sentences in three Indian languages. For testing we consider 300 sentences from different languages consisting of a total number of

3924 syllable-like units corresponding to 1580, 1648 and 696 actual VOPs from sentences of Tamil, Telugu and Hindi languages, respectively. These VOPs have been marked manually. For each sentence the hypothesised VOPs are determined by the AANN method explained in Section 4. The VOPs that are detected with a deviation upto 25 msec are about 68.62% and there are about 6.21% of spurious VOPs. About 74.63% of the CV segments have been correctly recognised in five alternatives by spotting the CV segments around the detected VOPs.

7   Summary and Conclusions

In this paper, we have addressed the issues in spotting based approach for recognition of consonant-vowel (CV) units in multiple languages. The approach is based on using the vowel onset points (VOPs) as anchor points and then classifying the segments around VOPs using a classifier. Autoassociative neural network (AANN) models are used for detecting VOPs in continuous speech. The methods for minimising the number of missing VOPs have to be explored. We use support vector machine (SVM) based classifier for recognition of CV segments around the hypothesised VOPs. To reduce the computational complexity of kernel operations in the SVM models, we perform nonlinear compression using AANN models for compression of pattern vectors. The results show that it is possible to compress the 390-dimensional pattern vectors to 60-dimensional vectors without affecting the classification performance. We proposed a data sharing approach for the development of multilingual CV recognition system. Though the variability among the data of a class is more and the number of CV classes is larger for the multilingual system, it has less effect on the recognition performance when SVMs are used for classification. However, classification performance of the hidden Markov model (HMM) based system is affected more by the large number of classes. The hypothesised CV sequence can be processed to perform word-level matching and sentence-level matching to recognise complete sentences.

References

1. L. R. Rabiner and B. -H. Juang: Fundamentals of Speech Recognition. PTR Prentice Hall, Englewood Cliffs, New Jersey (1993)
2. P. Eswar, S. K. Gupta, C. Chandra Sekhar, B. Yegnanarayana, and K. Nagamma Reddy: An acoustic-phonetic expert for analysis and processing of continuous speech in Hindi. In: Proc. European Conf. Speech Technology, Edinburgh. (1987) 369–372
3. S. V. Gangashetty, C. Chandra Sekhar, and B. Yegnanarayana: Detection of vowel onset points in continuous speech using autoassociative neural network models. In: Proc. Eighth Int. Conf. Spoken Language Processing (INTERSPEECH 2004 - ICSLP). (2004) 1081–1084
4. S. V. Gangashetty, C. Chandra Sekhar, and B. Yegnanarayana: Acoustic model combination for recognition of speech in multiple languages using support vector machines. In: Proc. IEEE Int. Joint Conf. Neural Networks (Budapest, Hungary). Volume 4(4). (2004) 3065–3069

5. S. Haykin: Neural Networks: A Comprehensive Foundation. Prentice-Hall International, New Jersey (1999)
6. S. V. Gangashetty, C. Chandra Sekhar, and B. Yegnanarayana: Dimension reduction using autoassociative neural network models for recognition of consonant-vowel units of speech. In: Proc. Fifth Int. Conf. Advances in Pattern Recognition (ISI Calcutta, India). (2003) 156–159
7. K. I. Diamantaras and S. Y. Kung: Principal Component Neural Networks, Theory and Applications. John Wiley and Sons, Inc., New York (1996)
8. S. Roukos, R. Rohlicek, W. Russel, and H. Gish: Continuous hidden Markov modelling for speaker-independent word spotting. In: Proc. IEEE Int. Conf. Acoust., Speech and Signal Processing. (1989) 627–630
9. C. Chandra Sekhar and B. Yegnanarayana: Neural network models for spotting stop consonant-vowel (SCV) segments in continuous speech. In: Proc. Int. Conf. Neural Networks. (1996) 2003–2008
10. S. V. Gangashetty, C. Chandra Sekhar, and B. Yegnanarayana: Spotting consonant-vowel units in continuous speech using autoassociative neural networks and support vector machines. In: Proc. IEEE Int. Workshop on Machine Learning for Signal Processing (Sao Luis, Brazil). (2004) 401–410
11. C. Chandra Sekhar: Neural Network Models for Recognition of Stop Consonant-Vowel (SCV) Segments in Continuous Speech. PhD thesis, Department of Computer Science and Engineering, Indian Institute of Technology Madras (1996)
12. S. V. Gangashetty and S. R. Mahadeva Prasanna: Significance of vowel onset point for speech recognition using neural network models. In: Proc. Fifth Int. Conf. Cognitive and Neural Systems (Boston, USA). (2001) 24
13. J. Y. Siva Rama Krishna Rao, C. Chandra Sekhar, and B. Yegnanarayana: Neural networks based approach for detection of vowel onset points. In: Proc. Int. Conf. Advances in Pattern Recognition and Digital Techniques, Calcutta. (1999) 316–320
14. B. Yegnanarayana and S. P. Kishore: AANN-An alternative to GMM for pattern recognition. Neural Networks 15 (2002) 459–469
15. H. Bourlard and N. Morgan: Connectionist Speech Recognition: A Hybrid Approach. Kluwer Academic Publishers, Boston (1994)
16. S. B. Davis and P. Mermelstein: Comparison of parametric representations for monosyllabic word recognition in continuously spoken sentences. IEEE Trans. Acoust., Speech, and Signal Processing 28 (1980) 357–366
17. S. Furui: On the role of spectral transition for speech perception. J. Acoust. Soc. Am. 80(4) (1986) 1016–1025
18. C. Chandra Sekhar and B. Yegnanarayana: A constraint satisfaction model for recognition of stop consonant-vowel (SCV) utterances. IEEE Trans. Speech and Audio Processing 10 (2002) 472–480
19. A. Chopde: ITRANS Indian Language Transliteration Package Version 5.2. (Source, http://www.aczone.com/itrans/)
20. C. Chandra Sekhar, K. Takeda, and F. Itakura: Recognition of consonant-vowel (CV) units of speech in a broadcast news corpus using support vector machines. In: Proc. Int. Workshop on Pattern Recognition using Support Vector Machines (Niagara Falls, Canada). (2002) 171–185

Novel Sub-band Adaptive Systems Incorporating Wiener Filtering for Binaural Speech Enhancement

Amir Hussain[1], Stefano Squartini[2], and Francesco Piazza[2]

[1] Department of Computing Science & Mathematics, University of Stirling,
Stirling FK9 4LA, Scotland, UK
ahu@cs.stir.ac.uk
http://www.cs.stir.ac.uk/~ahu/
[2] Dipartimento di Elettronica, Intelligenza Artificiale e Telecomunicazioni,
Università Politecnica delle Marche, Via Brecce Bianche 31, 60131 Ancona, Italy
http://www.deit.univpm.it/

Abstract. In this paper, new Wiener filtering based binaural sub-band schemes are proposed for adaptive speech-enhancement. The proposed architectures combine a Multi-Microphone Sub-band Adaptive (MMSBA) system with Wiener filtering in order to further reduce the in-coherent noise components resulting from application of conventional MMSBA noise cancellers. A human cochlear model resulting in a non-linear distribution of the sub-band filters is also employed in the developed schemes. Preliminary comparative results achieved in simulation experiments using anechoic speech corrupted with real automobile noise show that the proposed structures are capable of significantly outperforming the conventional MMSBA scheme without Wiener filtering.

1 Introduction

The goal of speech enhancement systems is either to improve the perceived quality of the speech, or to increase its intelligibility. Classical methods based on full-band multi-microphone noise cancellation implementations which attempt to model acoustic path transfer functions can produce excellent results in anechoic environments with localized sound radiators [1], however performance deteriorates in reverberant environments. Adaptive sub-band processing has been found to overcome these limitations [2] in general time-varying noise fields. However the type of processing for each sub-band must take effective account of the characteristics of the coherence between noise signals from multiple sensors. Several experiments have shown that noise coherence can vary with frequency, in addition to the environment under test and the relative locations of microphones. The above evidence implies that processing appropriate in one sub-band, may not be so in another, hence supporting the idea of involving the use of diverse processing in frequency bands, with the required sub-band processing being identified from features of the sub-band signals from the multiple sensors. Dabis et al. [1] used closely spaced microphones in a full-band adaptive noise cancellation scheme involving the identification of a differential acoustic path transfer function during a noise only period in intermittent speech. A Multi-Microphone Sub-Band Adaptive (MMSBA) speech enhancement system has been described which extends this method by applying it within a set of linearly spaced sub-bands provided by a filter-bank [2]-[4]. Nevertheless, it must be noted that the MMSBA scheme assumes

M. Faundez-Zanuy et al. (Eds.): NOLISP 2005, LNAI 3817, pp. 318–327, 2005.

noisy speech input to both (or all) system sensors, in contrast to the practically restrictive 'classical' full-band speech enhancement schemes, where speech signal occurs only at the primary input sensor [1]. This makes the MMSBA solution more practically realizable. However, a proper method for detecting noise-only periods is assumed available within the MMSBA scheme. In this paper, the novel use of Wiener filtering (WF) within a binaural MMSBA scheme is investigated, in order to more effectively deal with residual incoherent noise components that may result from the application of conventional MMSBA schemes. This work originally extends that recently reported in [5] where a sub-band adaptive noise cancellation scheme utilizing WF was developed for the monaural case. Performance of the proposed binaural WF based approach is compared with the conventional MMSBA scheme (without WF) quantitatively and qualitatively using informal subjective listening tests, for the case of a real speech signal corrupted with simulated noise.

2   MMSBA Schemes Employing WF

Two or more relatively closely spaced microphones may be used in an adaptive noise cancellation scheme [1], [3] to identify a differential acoustic path transfer function during a noise only period in intermittent speech. The extension of this work, termed the Multi-Microphone sub-band Adaptive (MMSBA) speech enhancement system, applies the method within a set of sub-bands provided by a filter bank as shown in Figure 1a). The filter bank can be implemented using various orthogonal transforms or by a parallel filter bank approach. In this work, the sub-bands are distributed non-linearly according to a cochlear distribution, as in humans, following the Greenwood [6] model, in which the spacing of the sub-band filters is given by:

$$F(x) = A(10^{ax} - k). \tag{1}$$

where x is the proportional distance from 0 to 1 along the cochlear membrane and $F(x)$ are the upper and lower cut-off frequencies for each filter obtained by the limiting value of x. For the human cochlea, values of A=165.4, a=2.1 and k=0.88 are recommended and chosen here. The conventional MMSBA approach considerably improves the mean squared error (MSE) convergence rate of an adaptive multi-band LMS filter compared to both the conventional wideband time-domain and frequency domain LMS filters, as shown in [3][4]. It is assumed in this work that the speaker is close enough to the microphones so that room acoustic effects on the speech are insignificant, that the noise signal at the microphones may be modelled as a point source modified by two different acoustic path transfer functions, and that an effective voice activity detector (VAD) is available.

In the proposed MMSBA architecture, Wiener filtering (WF) operation has been applied in two different ways: at the output of each sub-band adaptive noise canceller as shown in Fig.1a, and at the global output of the original MMSBA scheme as shown in Fig. 1b. In the rest of this paper, the new MMSBA scheme employing WF in the sub-bands is termed MMSBA-WF, whereas the one employing wide-band (WB) WF is termed MMSBA-WBWF. In both the proposed architectures, the role of WF is to further mitigate the residual noise effects on the original signal to be recovered, following application of MMSBA noise-cancellation processing.

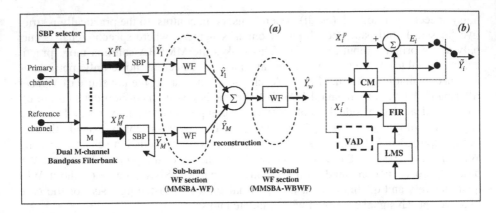

Fig. 1. (a) WF based MMSBA systems. (b) Subband Processing unit(SBP).

2.1 Diverse SBP Options

A significant advantage of using sub-band processing (SBP) for speech enhancement within the MMSBA scheme is that it allows for diverse processing in each sub-band in order to simultaneously effectively cancel both the coherent and incoherent noise components present in real reverberant environments. The SBP can be accomplished in a number of ways (Fig.1b), as follows:

– *No Processing.* Examine the noise power in a sub-band and if below (or the SNR above) some arbitrary threshold, then the signal in that band need not be modified.
– *Intermittent coherent noise canceller.* If the noise power is significant and the noise between the two channels is significantly correlated in a sub-band, then perform adaptive intermittent noise cancellation, wherein an adaptive filter may be determined which models the differential acoustic-path transfer function between the microphones during the noise alone period. This can then be used in a noise cancellation format during the speech plus noise period (assuming short term constancy) to process the noisy speech signal.
– *Incoherent noise canceller.* If the noise power is significant but not highly correlated between the two channels in a sub-band, then the incoherent noise cancellation approach of Ferrara-Widrow [7] be applied during the noisy speech period.

In this paper, we employ the above three SBP options and implement the processing using the Least Mean Squares (LMS) algorithm to perform the adaptation. For the derivation of the WF theory in the next section, we define $\tilde{X}_j, \tilde{S}_j, \tilde{N}_j$ as the global output, the reconstructed signal and the residual noise component at the j-th SBP output (or, equivalently, the adaptive noise canceller output of j band) respectively. The following relationship can be assumed to hold due to un-correlation between the noise and the desired signal at each band:

$$\tilde{X}_j = \tilde{S}_j + \tilde{N}_j. \tag{2}$$

In the original MMSBA, all $\tilde{x}_j$ sub-band noise canceller outputs are summed (at the reconstruction section) to yield the global MMSBA output $\tilde{y}$ (in the following capitalized letters will denote the corresponding variables in the frequency domain):

$$\tilde{Y} = \sum_j \tilde{S}_j + \sum_j \tilde{N}_j = \tilde{S} + \tilde{N}. \tag{3}$$

2.2 Wiener Filtering (WF)

The coefficient of a Wiener filter [8] are calculated to minimize the average squared distance between the filter output and a desired signal, assuming stationarity of the involved signals. This can be easily achieved in the frequency domain yielding:

$$W(f) = \left(P_{DY}(f) / P_{YY}(f) \right). \tag{4}$$

where $D(f)$ is the desired signal, $\hat{S}(f) = W(f)Y(f)$ is the Wiener filter output, $Y(f)$ the Wiener filter input and $P_{YY}(f)$, $P_{DY}(f)$ are the power spectrum of $Y(f)$ and the cross power spectrum of $Y(f)$, $D(f)$ respectively. If we apply such a solution to the case where the global signal is given by addition of noise and signal (to be recovered), and moving from the assumption that noise and signal are uncorrelated (as $\tilde{S}_j, \tilde{N}_j$ are) we can derive the following from (4):

$$W_j(f) = \left(P_{\tilde{S}_j \tilde{S}_j}(f) / P_{\tilde{S}_j \tilde{S}_j}(f) + P_{\tilde{N}_j \tilde{N}_j}(f) \right). \tag{5}$$

where $P_{\tilde{S}_j \tilde{S}_j}(f)$, $P_{\tilde{N}_j \tilde{N}_j}(f)$ are the signal and noise power spectra. Note that, in this task, the desired signal is $\tilde{S}_j$. It must be observed that such a formulation can be easily extended to the case when involved signals are not stationary, by simply periodically recalculating the filter coefficients for every block 1 of N_s signal samples. In this way the filter adapts itself to the average characteristics of the signals within the blocks and becomes block-adaptive. Moreover, the presence of VAD is a pre-requisite to making the Wiener filtering operation effective: in noise alone period, a precise estimation of noise power spectrum can be performed and then used in (5), assuming that its properties are still the same when the signal power spectrum is calculated during the noisy speech period. The former approximation is carried out iteratively by using the power spectrum of Wiener filter global output $\hat{S}_j(f)$.

Note that the above derivations are readily applicable to MMSBA-WF architecture as follows. Similar to (2) and (3), the following holds at j-th band Wiener filter output:

$$\hat{X}_j = \hat{S}_j + \hat{N}_j, \qquad \hat{Y} = \sum_j \hat{S}_j + \sum_j \hat{N}_j = \hat{S} + \hat{N} \tag{6}$$

where $\hat{y}$ is the new global output yielded from the reconstruction section. However, same considerations can be made when MMSBA-WBWF is dealt with, simply adapting the equations to the new situation where WF occurs after the reconstruction section. Specifically, taking (3) into account, implies:

$$\hat{Y}_w = W_w \tilde{Y} = \hat{S}_w + \hat{N}_w. \tag{7}$$

where w stands for wide-band processing, since WF operation is applied directly to MMSBA output $\tilde{y}$ to form the new Wiener filtered output $\hat{y}_w$.

2.3 Recursive Magnitude Squared Coherence (MSC) Metric for Selecting SBP

The Magnitude Squared Coherence (MSC) has been applied by Bouquin and Faucon [9] to noisy speech signals for noise reduction and also successfully employed as a VAD for the case of spatially uncorrelated noises. In this work, following [4] we use a modified MSC as a part of a system for selecting an appropriate SBP option within the MMSBA system. Assuming that the speech and noise signals are independent, the observations received by the two microphones are:

$$x_p = s_p + n_p \quad primary; \qquad x_r = s_r + n_r \quad reference. \tag{8}$$

where $s_{p,r}$, $n_{p,r}$ represent the clean speech signal and the additive noise, respectively. For each block l and frequency bin f_k ;the coherence function is given by:

$$\rho(f_k, l) = P_{X_p X_r}(f_k, l) / \sqrt{P_{X_p X_p}(f_k, l) P_{X_r X_r}(f_k, l)} \tag{9}$$

where $P_{X_p X_r}(f_k, l)$ is the cross-power spectral density, $P_{X_p X_p}(f_k, l)$ and $P_{X_r X_r}(f_k, l)$ are the auto-power spectral

$$P_{X_p X_r}(f_k, l) = \beta P_{X_p X_r}(f_k, l-1) + (1-\beta) X_p(f_k, l) X_r^*(f_k, l). \tag{10}$$

where β is a forgetting factor. During the noise alone period, for each overlapped and Hanning windowed block l we compute the Magnitude Squared Coherence (MSC) averaged over all the overlapped blocks (at each frequency bin) as

$$\overline{MSC}(f_k) = \frac{1}{l} \sum_{i=1}^{l} \left[\rho(f_k, i) \right]^2. \tag{11}$$

The above recursively averaged MSC criterion can thus be used as a means for determining the level of correlation between the disturbing noise sources within the various frequency bands (by averaging the above MSC over each respective linearly or non-linearly spaced sub-band), during the noise alone period in intermittent speech, and consequently selecting the right SBP option, as discussed in section 2.1.

On initial trials, a threshold value around 0.55 for the adaptive MSC has been chosen for distinguishing between highly correlated and weakly correlated sub-band noise signals. For 50% block overlap, a forgetting factor of $\beta = 0.8$ has been found to be adequate. The above MSC metric was successfully tested for a range of realistic SNR values (from -3dB to 25dB) using both simulated and real reverberant data.

3 Simulation Results

In this section the two new MMSBA-based WF approaches are compared to the original MMSBA approach (without WF) in order to investigate their relative effectiveness. For experimental purposes, a real anechoic speech signal $s(k)$ is used as the desired signal, whilst the noise signals are generated according to the two following schemes.

1. reference noise signal $n(k)$ is chosen to be a random signal, from which two different noise sources (one for primary and one for the reference channel) are derived and summed with $s(k)$ to form x_1, x_2 as in (8).

2. $n_1(k)$, $n_2(k)$ are chosen to be real stereo car noise sequences recorded in a Ferrari Mondial T (1991 Model), using an Audio Technica AT9450 stereo microphone mounted on a SONY DCR-PC3-NTSC video camera and a sampling frequency of 44.1 kHz; the noise sequences were manually added to the anechoic speech sentence to manufacture different SNR cases.

The value of the initial SNR, namely SNR_i, is used as a reference for the three SNR improvements calculated at the output of each of the speech enhancement structures under study, namely: the original MMSBA (without WF), MMSBA-WF and MMSBA-WBWF. Taking into account the un-correlation between noise and signal on the same channel, we can define the SNR at the output level as:

$$SNR_o(f) = \left[P_{\tilde{Y}\tilde{Y}}(f) - P_{\tilde{N}\tilde{N}}(f)\right]/P_{\tilde{N}\tilde{N}}(f). \tag{12}$$

where the involved power spectra are related to signals described by (3). Similar formulas can be derived considering power spectra in (6) and (7), for MMSBA-WF and MMSBA-WBWF respectively. Moreover it has to be said that $P_{\tilde{N}\tilde{N}}(f)$ is calculated over a sub-range of the noise alone period where noise cancellers are assumed to have converged, since this is the noise power spectrum expected to occur when the desired signal is present. On this basis, $P_{\hat{N}\hat{N}}(f)$ and $P_{\hat{N}_f\hat{N}_f}(f)$ are obtained from Wiener filtered versions for the two different schemes addressed. Choices for various experimental parameter values were selected on a trial and error basis: speech signal number of samples corresponding to a 2s long speech sentence; noise signal number of samples (in the manually defined noise alone period) corresponding to 0.2s of noise (for both situations addressed); number of iterations of WF operation: 5; number of sub-bands: 4; number of taps or order of FIR adaptive sub-band filters: 32.

Let us consider the results relative to the synthetic noise case study.

- **Coherent Noise:** The intermittent coherent noise-canceller approach is only employed as the SBP option in each band. Table 1 summarizes the results obtained using the three MMSBA approaches: from which it can be seen that MMSBA-WF and MMSBA-WBWF both deliver an improved SNR performance over the original MMSBA approach.

- **Incoherent Noise:** In this case the value of the recursive MSC metric is used to employ both intermittent and FW SBP options, with the former option used in the first sub-band (with a high MSC) and the latter in the other three bands (with a low MSC). This is justified by the coherence characteristics of available stereo noise signal. It can be seen from Table 2 that the choice of sub-band WF (within the MMSBA-WF scheme) gives the best results in this case, due to its operation in the sub-bands with diverse SBP, resulting in more effective noise cancellation in the frequency domain, compared to the wide-band WF processing (within the MMSBA-WBWF scheme).

Now we can focus our attention to the in-car recorded noise case study.

- **Coherent Noise:** In the first experimental case study, simulated coherent noise over all four bands is used (with a MSC>0.55 in each band), for which the intermittent coherent noise-canceller approach is thus employed as the SBP option in each band. Table 3 summarizes the results obtained using the three MMSBA approaches: from which it can be seen that MMSBA-WF and MMSBA-WBWF both deliver an improved SNR performance over the original MMSBA approach. It is also evident that the choice of sub-band WF (within the MMSBA-WF scheme) gives the best results, as expected, due to its operation in the sub-bands resulting in more effective noise cancellation in the frequency domain, compared to the wide-band WF processing (within the MMSBA-WBWF scheme).

- **Incoherent Noise:** In this more realistic case, simulated incoherent noise over two of the four bands is used. Accordingly in this test case, both intermittent and Ferrara-Widrow SBP options are utilized, the former in the first two sub-bands with highly correlated noises (MSC>0.55 in each band), and the former for the other two bands (with MSC<0.55). Table 4 summarizes the results, from which an even stronger impact of WF operation in the sub-bands (MMSBA-WF processing) is evident.

Note that application of the classical wide-band noise cancellation approach, namely the MMSBA with number of bands set to one and a wideband FIR filter order of 256 (equivalent to product of number of sub-bands and sub-band filter order) was actually found to degrade the speech quality resulting in a negative SNR improvement value, which is hence not shown in the Table 1-4. This finding of the inability of classical wideband processing to enhance the speech in real automobile environments is consistent with the results reported in [2][3]. Finally, informal listening tests using random presentation of the processed and unprocessed signals to three young male adults of normal hearing, also confirmed the MSSBA-WF processed speech to be both enhanced in SNR and of significantly better perceived quality than that obtained by all the other conventional wide-band and sub-band methods.

4 Conclusions

Two multi-microphone sub-band adaptive speech enhancement systems employing Wiener filtering and a human cochlear model filterbank have been presented. Preliminary comparative results achieved in simulation experiments demonstrate that the proposed WF based MMSBA schemes are capable of improving the output SNR of speech signals with no additional distortion apparent, compared to the conventional MMSBA scheme (without WF). The MMSBA-WF architecture employing sub-band WF seems to be the most promising whose improved performance is due to the ability of WF to further reduce the residual in-coherent sub-band noise components resulting from MMSBA application. A detailed theoretical analysis is now proposed to define the attainable performance, in addition to employing other perceptive evaluation measures such as the perceptually weighted segmental SNR, Bank Spectral Distortion (BSD) and Perceptual Evaluation of speech Quality (PESQ) scores. What is also needed is further extensive testing (using formal subjective listening tests) with a variety of real data (i.e., acquired through recordings in various real environments), in order to further assess and quantify the relative advantages of the new speech enhancement schemes.

Table 1. Case A. Synthetic noise. Relative average SNR improvements for all architectures involved (over 10 runs). Standard deviation values are directly depicted on the bars.

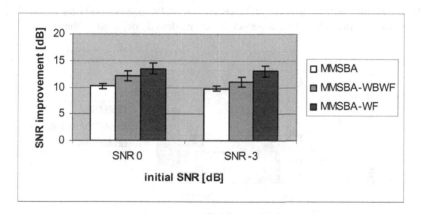

Table 2. Case B. Synthetic noise. Relative average SNR improvements for all architectures involved (over 10 runs). Standard deviation values are directly depicted on the bars.

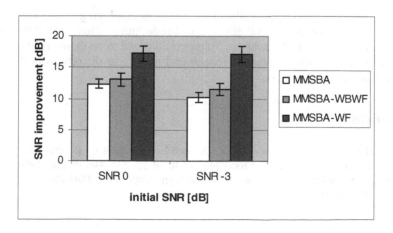

Table 3. Case A. Real in-car noise. Relative average SNR improvements for all architectures involved (over 10 runs). Standard deviation values are directly depicted on the bars.

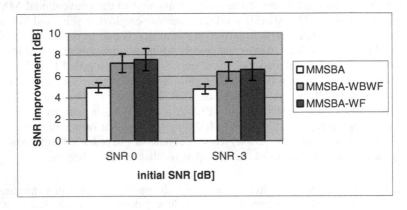

Table 4. Case B. Real in-car noise. Relative average SNR improvements for all architectures involved (over 10 runs). Standard deviation values are directly depicted on the bars.

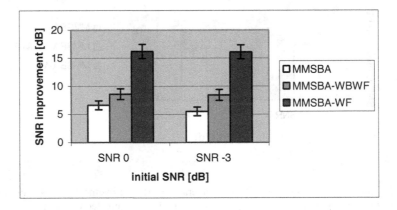

Finally, further work is currently in progress on employing non-linear sub-band adaptive filtering and cross-band effects (to mimic human lateral inhibition effects) within the binaural MMSBA scheme. These could prove to be more effective in dealing with the non-Gaussian nature of speech and non-linear distortions in the electro-acoustic transmission systems.

References

1. Dabis, H.S., Moir, T.J., Campbell, D.R.: Speech Enhancement by Recursive Estimation of Differential Transfer Functions. Proceedings of ICSP, pp. 345-348, Beijing, 1990
2. Toner, E.: Speech Enhancement using Digital Signal Processing. PhD Thesis, University of Paisley, UK, 1993

3. Darlington, D.J., Campbell, D.R.: Sub-Band Adaptive Filtering Applied to Hearing Aids, Proc.ICSLP'96, pp. 921-924, Philadelphia, USA, 1996
4. Hussain, A., Campbell, D.R.: A Multi-Microphone Sub-band Adaptive Speech Enhancement System Employing Diverse Sub-Band Processing, International Journal of Robotics & Automation, vol. 15, no. 2, pp. 78-84, 2000
5. Abutalebi, H. R., Sheikhzadeh, H., Brennan, R.L., Freeman, G. H.: A Hybrid Sub-Band System for Speech Enhancement in Diffuse Noise Fields. IEEE Sig. Process. Letters, 2003
6. Greenwood, D.D.: A Cochlear Frequency-Position Function for Several Species-29 Years Later. J. Acoustic Soc. Amer., vol. 86, no. 6, pp. 2592-2605, 1990
7. Ferrara, E.R., Widrow, B.: Multi-Channel Adaptive Filtering for Signal Enhancement, IEEE Trans. on Acoustics, Speech and Signal Proc., vol. 29, no. 3, pp. 766-770, 1981
8. Vaseghi, S.V.: Advanced Signal Processing and Digital Noise Reduction (2nd ed.). John Wiley & Sons, 2000
9. Le Bouquin, R., Faucon, G.: Study of a Voice Activity Detector and Its Influence on a Noise Reduction System. Speech Communication, vol. 16, pp. 245-254, 1995

Maximum a Posterior Probability and Cumulative Distribution Function Equalization Methods for Speech Spectral Estimation with Application in Noise Suppression Filtering

Tran Huy Dat[1], Kazuya Takeda[1], and Fumitada Itakura[2]

[1] Graduate School of Information Science, Nagoya University, Furo-cho,
Chikusa-ku, Nagoya 464-8603, Japan
{dat, takeda}@sp.is.nagoya-u.ac.jp
[2] Graduate school of Information Engineering, Meijo University, Shiogamaguchi,
Tempaku-ku, Nagoya 468-8502, Japan
itakuraf@ccmfs.meijo-u.ac.jp

Abstract. In this work, we develop and compare noise suppression filtering systems based on maximum a posterior probability (MAP) and cumulative distribution function equalization (CDFE) estimation of speech spectrum. In these systems, we use a double-gamma modeling for both the speech and noise spectral components, in which the distributions are adapted to the actual parameters in each frequency bin. The performances of the proposed systems are tested using the Aurora database they are shown to be better than conventional systems derived from the MMSE method. Whereas the MAP-based method performed best in the SNR improvement, the CDFE-based system provides a lower musical noise level and shows a higher recognition rate.

1   Introduction

Noise reduction is an important problem in speech and audio processing. Among single channel approaches, the statistical methods for speech spectrum estimation have been frequently used [1]. The MMSE and MAP estimations for the Gaussian model of the speech spectrum were proposed by Ephraim and Malah [2] and Wolfe and Godsill [3], respectively. Later, a MAP based on the supergaussian modeling of speech was derived by Lotter[4] and the MMSE based on gamma modeling was investigated by Martin [5],[6]. However, in both cases, the prior distribution parameters were chosen blindly without any adaptation. In previous work, we proposed an improved version of MAP estimation for the speech spectral magnitude by using generalized gamma modeling of the speech spectral magnitude [6]. However, that work was limited by the Gaussian assumption of the noise spectrum and therefore, was not effective under certain noise conditions. In this work, we extend gamma modeling for both speech and noise spectra and derive the MAP and cumulative distribution equalization (CDFE) estimation for the spectral components. As in our previous work [7], the prior

M. Faundez-Zanuy et al. (Eds.): NOLISP 2005, LNAI 3817, pp. 328–337, 2005.

distribution is adapted from observed signals and the estimations are derived for an arbitrary set of distribution parameters. The reason for applying MAP or CDFE instead of MMSE is that, the last generally provides non closed form solution, which is complicated even for numerical methods. Cumulative distribution equalization has frequently been used in data-driven approaches, where the empirical histogram is used. In this work, we show that, this method can also be usefully applied in the model-based manner, where the cumulative distribution function (cdf) is used. To overcome the difficulties of applying cdf, we develop an cdf estimation method via the characteristic function, which implies a multiplication for the additive model. The organization of this paper is as follows. In section 2, we describe the double-gamma modeling of the speech and noise spectral components. Section 3 contains a review of the MMSE estimation of speech spectral estimation. In sections 4 and 5, we develop the MAP and CDFE estimation of speech spectral components using the proposed modeling of speech and noise. In section 6, we reports an experimental evaluation of implemented noise suppression filtering systems, and section 7 is a summary of the present work.

2   Statistical Modeling of Speech and Noise Spectral Components

2.1   Double-Gamma Modeling of Speech and Noise Spectra

Consider the additive model of the noisy speech as below:

$$\mathbf{X}[k, m] = \mathbf{S}[k, m] + \mathbf{N}[k, m], \tag{1}$$

where $\mathbf{X}[k, m]$, $\mathbf{S}[k, m]$, and $\mathbf{N}[k, m]$ are noisy, clean speech and noise complex spectrum. The pair $[k, m]$ indicates the frequency-frame index. Each complex spectrum is presented in terms of the spectral components (real and imaginary parts) as follows:

$$\mathbf{C}[k, m] = C_R[k, m] + jC_I[k, m]. \tag{2}$$

The following assumptions are assumed for speech and noise spectral components: (1) spectral components are independent and zero-mean, (2) spectral component pdf is symmetrical, (3) The variances of spectral components are power density and determined at each frequency-frame index $[k, m]$. In this work, we investigated double-gamma modeling for both speech and noise.

$$p_{double-gamma}\left(C\left[k, m\right]\right) = \frac{ba^{b-1}}{2\sigma_C^b\left[k, m\right]\Gamma\left(a\right)}C^{b-1}\left[k, m\right]\exp\left(-b\frac{C\left[k, m\right]}{\sigma_C\left[k, m\right]}\right) \tag{3}$$

As an alternative, the conventional Gaussian model is also investigated and noted as follows:

$$p_{gauss}\left(C\left[k, m\right]\right) = \frac{1}{\sqrt{2\pi}\sigma_C\left[k, m\right]}\exp\left(-\frac{C^2\left[k, m\right]}{2\sigma_C^2\left[k, m\right]}\right), \tag{4}$$

where $C\left[k, m\right]$ denotes the spectral component (real or imaginary part) and $\sigma_C^2\left[k, m\right]$ denotes the local power density at each frequency-frame index $[k, m]$.

Note that the normalization condition $\langle C^2 \rangle = \sigma_C^2$ implies the following relationship between a and b:

$$\frac{a(a+1)}{b^2} = 1 \to b = \sqrt{a(a+1)}, \tag{5}$$

Since the spectral components are assumed to be identical independent variables, the additive model of the complex noisy speech spectral (3) can be simply denoted in terms of the spectral component as

$$X = S + N, \tag{6}$$

where each symbol in (5) corresponds to the real and imaginary parts of complex spectrum. The following three models of the speech and noise distributions are consequently investigated in this work: Gaussian/Gaussian (Model 1), gamma/Gaussian (Model 2) and gamma/gamma (Model 3).

2.2   Actual Adaptation of the Modeled Distribution Parameters

Since the prior distributions of speech and noise are scaled by their local power densities (3), which are estimated separately, the prior parameter should be adapted from each observed noisy speech. In this work, we develop a parameter estimation method, in which the prior pdf is adapted in each frequency bin. As done in our previous work [7], the high-order moments of observed noisy speech spectrum are used to derive estimation equation. In this case, it is done for both the speech and noise prior pdf. For the gamma-speech, Gaussian-noise model (Model 1), the four moments of the noisy speech spectral component are expressed as

$$\langle X^4 \rangle = \bar{\sigma}_S^4 M_4(a_S) + \bar{\sigma}_N^4 M_4(a_N) + 6\bar{\sigma}_S^2 \bar{\sigma}_N^2, \tag{7}$$

where the fourth moments of the noise and speech spectral components are given below following the Gaussian distribution,

$$M_4(a_N) = 3, \tag{8}$$

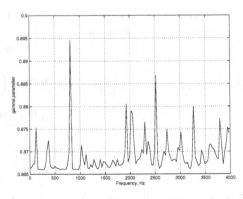

Fig. 1. Example of double-gamma estimation of speech spectral components

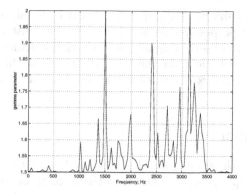

Fig. 2. Example of double-gamma estimation of noise spectral components

and through gamma distribution as

$$M_4\left(a_S\right) = \frac{(a_S + 2)(a_S + 3)}{b_S^2}. \tag{9}$$

Substituting (8) and (9) into (7) and taking (5) into account, the speech prior distribution parameter is given in a closed-form solution. Analogously, for the speech-gamma and noise-gamma model (model 3), the distribution parameter is estimated using the pair of fourth-order and sixth-order moments of the observed noisy spectrum. Figures 1 and 2 show examples of double-gamma parameter estimation for a noisy speech signal under the 5dB street noise condition.

3   MMSE Estimation

In general, the MMSE estimation is given by the conditional expectation,

$$\hat{S} = E\left[S|X\right] = \frac{\int\limits_{-\infty}^{\infty} Sp\left(X, S\right) dS}{p\left(X\right)} = \frac{\int\limits_{-\infty}^{\infty} Sp\left(X|S\right) p\left(S\right) dS}{\int\limits_{-\infty}^{\infty} p\left(X|S\right) p\left(S\right) dS}, \tag{10}$$

where the conditional pdf $p\left(X|S\right)$ is given by the noise pdf and the prior distribution $p\left(S\right)$ is the Gaussian (3) or double-gamma distribution (4). The MMSE estimation of the speech spectral components for the Gaussian modeling of noise and speech spectra yields the conventional Wiener filtering:

$$\hat{S} = \frac{\sigma_S^2}{\sigma_S^2 + \sigma_N^2} X. \tag{11}$$

The MMSE estimation using gamma prior was investigated by R.Martin [5],[6], for two special cases of double-gamma distribution, including the Laplacian distribution, in which a closed-form solutions are given. However, the MMSE in other cases of gamma modeling does not yield a closed-form solution of the estimation. For our proposed system, where the prior distribution parameters are adapted from each observed signal, the numerical calculation of integral (10) should be implemented. However the main drawback of this method is highly expensive computational cost and therefore we don't use this method for our system.

4   MAP Estimation

MAP is a general estimation method and is used in this work to estimate the speech spectral components. In contrast to the estimations presented in [3] and [7], where the spectral magnitude domain was uses, in this work, we use the spectral components domain to derive the estimation. The advantage of using this domain is that exactly matches the additive model of noisy speech and the estimation is given not only for the Gaussian model of noise spectrum. The general form of MAP estimation

$$\hat{S} = \arg\max_{S} \log\left(p\left(S|X\right)\right) = \arg\max_{S} \log\left(p\left(X|S\right)p\left(S\right)\right), \tag{12}$$

yields an equation of the derivatives

$$\frac{\partial}{\partial S}\left[\log\left(p\left(X|S\right)\right) + \log\left(p\left(S\right)\right)\right] = 0. \tag{13}$$

Since the MAP estimation for the model 1 implies the classical Wiener filter, we begin this section with model 2.

4.1   Model 2: Gamma Speech and Gaussian Noise

For this model, the conditional and prior distributions are derived as follows:

$$\frac{\partial}{\partial S}\left[\log\left(p\left(X|S\right)\right)\right] = \frac{X - S}{\sigma_N^2}, \tag{14}$$

$$\frac{\partial}{\partial S}\left[\log\left(p\left(S\right)\right)\right] = \frac{(a_S - 1)}{S} - sign\left(S\right)\frac{b_S}{\sigma_S}. \tag{15}$$

Equations (14) and (15) imply the following second-order equation for the gain function $G = \frac{\hat{S}}{X}$:

$$G^2 - G\left(1 - \frac{sign\left(X\right)b_S}{\sqrt{\gamma\xi}}\right) + \frac{(a_S - 1)}{\gamma} = 0, \tag{16}$$

where: $\gamma = \frac{X^2}{\sigma_N^2}$, and $\xi = \frac{\sigma_S^2}{\sigma_N^2}$ are posterior and prior SNRs, respectively which are estimated separately [2]. Obtaining a closed form solution for the MAP estimation is important because then the global maximum of posterior probability in (12) can be found strictly:

$$G = \max\left\{u \pm \sqrt{u^2 + v}, 0\right\}, \tag{17}$$

where

$$u = \left(0.5 - \frac{sign\,(X)\,b_S}{\sqrt{4\gamma\xi}}\right), \tag{18}$$

$$v = \frac{(a_S - 1)}{4\gamma}. \tag{19}$$

4.2   Model 3: Gamma Speech and Gamma Noise

For this model, the conditional distribution in (13) can be expressed as

$$\frac{\partial}{\partial S}\left[\log\left(p\left(X|S\right)\right)\right] = -\frac{(a_N - 1)}{X - S} - sign\,(X - S)\,\frac{b_N}{\sigma_N}. \tag{20}$$

Analogously, a second order equation for the gain function is derived.

$$G^2 - G\left(1 - \frac{(a_S + a_N - 2)}{\sqrt{\gamma}sign\,(X)\left(b_N - \frac{b_S}{\sqrt{\xi}}\right)}\right) + \frac{(a_S - 1)}{\sqrt{\gamma}sign\,(X)\left(b_N - \frac{b_S}{\sqrt{\xi}}\right)} = 0. \tag{21}$$

The solution of Eq.(21) is given in the same manner as for (17).

5   Cumulative Distribution Function Equalization

One remaining problem of the above MAP estimation is the relative sensitivity to the "poor fit" prior estimation, or other words it requires a sufficiently "good" prior. Therefore, in addition to the MAP estimation, we investigate an alternative estimation based on cumulative distribution function equalization (CDFE).

5.1   Cumulative Distribution Function Equalization

This method (CDFE) was originally called as histogram equalization and has been used in data-driven approaches. In this work, we investigate the use of cdf for the model-based approaches, in which modeled distributions are used. The principle of this method is to identify a non-linear transform from noisy to clean features, which matches the cumulative distribution function. Denoting the general equalization

$$\widehat{s} = g\left(x\right), \tag{22}$$

the criterion for our estimation here is expressed as

$$F_{g(x)}\left(g\left(x\right)\right) = F_s\left(s\right). \tag{23}$$

The key point of the method is that, the cumulative distribution function (cdf) is invariant though arbitrary nonlinear functional, that is,

$$F_{g(x)}\left(g\left(x\right)\right) = F_x\left(x\right). \tag{24}$$

From (23) and (24), the "best" nonlinear transform is obtained by equalizing cdf of noisy to clean signals.

$$g\left(x\right) = F_s^{-1}\left(F_x\left(x\right)\right) \tag{25}$$

5.2   Model 1: Gaussian Speech and Gaussian Noise

For Gaussian modeling of both noise and speech spectral components, the noisy speech spectral components are also Gaussian

$$X \sim \mathrm{N}\left(0, \sigma_S^2 + \sigma_N^2\right). \qquad (26)$$

Since both cdf $F_X(.)$, and $F_S(.)$ are Gaussian, the CDFE operation is carried out without any difficulties.

5.3   CDF Estimation Via the Characteristic Function

CDFE has the following main problem. Excepting the Gaussian model considered above, the cdf of noisy speech, presented as an addition of speech and noise, has no analytical form. To overcome this problem, we develop a cdf estimation method by using the characteristic function as follows:

$$F(x) = \begin{cases} 1 & x \geq m + 4\sigma \\ \frac{1}{2} - sign\,(x)\, \frac{2}{\pi} \int\limits_{\varepsilon}^{2\pi} \frac{f(u)}{u} \sin\,(ux)\,du & m + 4\sigma > x > m - 4\sigma, \\ 0 & x \leq m - 4\sigma \end{cases} \qquad (27)$$

where $f(u)$ denotes the characteristic function [8] of noisy speech spectral components. The main point here is that the characteristic function of the additive model (6) is multiple and therefore convenient for implementation. Note that, according to the symmetrical assumption of the pdf of speech and noise spectral components, the characteristic function of noisy speech is always a real function.

5.4   Model 2: Gamma Speech and Gaussian Noise

Denoting characteristic function of the Gaussian distribution of noise and double-gamma distribution of speech spectral components, respectively, as follows:

$$f_N(u) = e^{-\frac{u^2 \sigma_N^2}{2}}, \qquad (28)$$

$$f_S(u) = Re\left[\left(\frac{a_S}{a_S - iu}\right)^{b_S}\right] = \cos\left(b_S a \cos\left(\frac{a_S^2}{a_S^2 + u^2}\right)\right), \qquad (29)$$

the characteristic function of the noisy speech spectral component is obtained by multiplying (28) and (29). The CDFE is then derived using (27) and (25).

5.5   Model 3: Gamma Speech and Gamma Noise

For gamma modeling of both speech and noise spectral components, the cdf of noisy speech spectral components is estimated from the characteristic functions of speech and noise and is denoted by

$$f_X(u) = \cos\left(b_N a \cos\left(\frac{a_N^2}{a_N^2 + u^2}\right)\right) \cos\left(b_S a \cos\left(\frac{a_S^2}{a_S^2 + u^2}\right)\right). \qquad (30)$$

6   Experiment

The proposed noise suppression filtering systems are tested using AURORA2 to determine the ASR performance [10], where the speech enhancement is applied for both testing and training databases. The noise and signal powers are estimated using the minimum statistic [9] and decision directed [2]. The three models of speech and noise modeling described above are investigated. Each system is identified according to the estimation method (MMSE,MAP ,CDFE) and the assumed models (1, 2, 3). For reference, the Ephraim-Malah LSA version based on Gaussian modeling [1] is also implemented. The reference MMSE versions

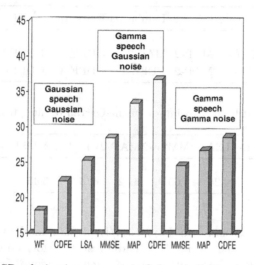

Fig. 3. ASR relative improvement of clean training: overall results

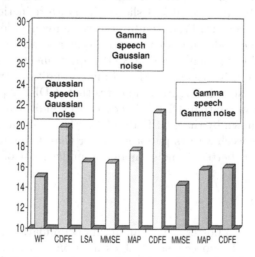

Fig. 4. ASR relative improvement of multi-conditions training: overall results

Table 1. Listening test: Q1-Which one is less distorted? Q-2 Which one is less noisy? Q-3 Which one is best?

Q	Subway	Babble	Car	Exhibition	Restaurant	Street	Airport	Station
1	CDFE-1	CDFE-2	CDFE-3	CDFE-3	CDFE-3	CDFE-3	CDFE-2	CDFE-2
2	MAP-2	MAP-2	MAP-2	MAP-2	CDFE-3	CDFE-3	MAP-2	MAP2
3	CDFE-1	CDFE-2	MAP-2	CDFE-3	CDFE-3	CDFE-1	MAP-2	CDFE-2

Table 2. Best ASR performance in each noise condition

Cond.	Subway	Babble	Car	Exhibition	Restaurant	Street	Airport	Station
CL	CDFE-1	CDFE-2	MAP-2	MAP-2	CDFE-3	CDFE-3	MAP-2	CDFE-2
MT	CDFE-1	MAP-2	MAP-2	CDFE-1	CDFE-3	CDFE-3	MAP-2	CDFE-2

Table 3. SNR improvement-Overall results [dB]

Meth	WF	CDFE-1	LSA	MMSE-2	MAP-2	CDFE-2	MMSE-3	MAP-3	CDFE-3
dB	3.25	3.52	4.25	5.65	6.32	6.07	5.45	6.11	6.12

using Laplacian/Gaussian and Laplacian/Laplacian modeling of speech and noise [5] are implemented. A simple listening test is performed with four subjects listening to 25 random chosen utterances of each noise conditions. Table 1 shows the results of the listening test, table 2 shows the best method in terms of ASR for each noise condition, and table 3 shows the noise reduction comparison in terms of SNR improvement. From the tables, we can conclude that, CDFE-3 is superior to other methods only for the restaurant and street noise conditions. Meanwhile, the MAP-2 is dominated in SNR improvements. The overall results of ASR performance using clean HMM and multi-conditions training are shown in Figure 3-4. The results in Figure 3 indicates that, CDFE-2 performs the best, as CDFE-3 is even worse than CDFE-1. For multi conditions training, the best performances are shown by the CDFE-1 and CDFE-2. This means that, double-gamma model for speech always performs better than Gaussian model but the Gaussian modeling is better for noise modeling under most of noise conditions.

7   Conclusion

We develop a maximum posterior probability and cumulative distribution equal-ization method for the speech spectral estimation using the double-gamma modeling of speech and noise spectral components. The main point of the proposed method is that a solution is given for an arbitrary set of prior distributions and therefore it is possible to combine the estimation method to a prior adaptation to improve the performances of system. Double-gamma modeling of speech and noise spectral component was shown to be adaptable to the actual distribution, without any use of a training data. The results of the experimental evaluation shows the advantage of the proposed MAP and CDFE comparing to the conventional MMSE method. Gamma modeling is superior to the Gaussian for the speech spectral modeling in all cases, but is better for noise modeling only for some particular noise conditions (restaurant and street). The CDFE shows the best ASR performance, while the MAP is better in noise reduction.

References

1. Y. Ephraim, "Statistical model based speech enhancement systems," *IEEE Proc.*, vol. 80, pp. 1526-1555, 1992.
2. Y. Ephraim, and D. Malah, "Speech enhancement using a MMSE log-spectral amplitude estimations," *IEEE Trans. ASSP*, Vol. 33, No. 2, pp.443-445, 1985.
3. P. Wolfe and S. Godsill, "Simple alternatives to the Ephraim suppression rule for speech enhancement", *IEEE Workshop on Statistical Signal Processing*, 2001.
4. T. Lotter and P. Vary "Noise Reduction by Maximum A Posteriori Spectral Amplitude Estimation with Supergaussian Speech Modeling," *in Proc. IWAENC*, Kyoto, Japan, 2003.
5. R. Martin, B. Colin, "Speech enhancement in DFT domain using Laplacian priors," *in Proc. IWAENC*, Kyoto, Japan, 2003.
6. R. Martin, "Speech enhancement using MMSE Short Time Spectral Estimation with Gamma Speech Prior," *in Proc. ICASSP 02*, Orlando Florida,USA, 2002.
7. T.H. Dat, K. Takeda, and F. Itakura, "Generalized gamma modeling of speech and its online estimation for speech enhancement," *in Proc. ICASSP*, Philadelphia, USA,2005.
8. E Parzen, "On estimation of a probability density function and mode," emp Ann. Math. Statist V.33 pp.1065-1076, 1962.
9. R. Martin, "Noise power spectral estimation based on optimal smoothing and minimum statistics" *IEEE Trans. ASSP*, Vol. 9, No.5, pp.504-512, 2001.
10. H. Hirsch, D. Pearce, "The AURORA experimental framework for the performance evaluation of speech recognition systems under noisy conditions," *in ISCA ITRW ASR*, 2000.

Modeling Fluctuations of Voiced Excitation for Speech Generation Based on Recursive Volterra Systems

Karl Schnell and Arild Lacroix

Institute of Applied Physics, Goethe-University Frankfurt,
Frankfurt am Main, Germany
{Schnell, Lacroix}@iap.uni-frankfurt.de

Abstract. For the modeling of the speech production system linear models are widely used. However, not all features of speech can be covered by linear systems. Therefore nonlinear systems are interesting in speech processing. In this contribution a time variable recursive Volterra system is used to model the fluctuations of the voiced excitation while a linear system models the resonances of the speech production system. The estimation of the Volterra system is performed by a prediction algorithm. The prediction problem is solved with the aid of an approximation by series expansion. Speech examples show that the use of a time variable Volterra system improves the naturalness of the synthetic speech.

1 Introduction

Linear systems provide adequate modeling of the resonances of the vocal tract. The parameters of the linear system can be estimated by linear prediction or inverse filtering. The speech signal is described by the linear model only partially [1] and [2], therefore an estimation by a nonlinear system is performed with respect to the residual signal, obtained after applying linear prediction to the speech signal. The estimation for the Volterra system is based on nonlinear prediction. Since in the residual signal the linear relations of the speech signal are almost eliminated, the nonlinear predictor consists of nonlinear terms only.

2 Nonlinear Prediction

The prediction $\hat{x}(n)$ of a signal value $x(n)$ is performed by a combination of products of previous signal values $x(n-i) \cdot x(n-k)$ with $i, k > 0$. For a signal x the prediction error e is defined as the difference between the estimated value $\hat{x}$ and the actual value x:

$$e(n) = x(n) - \hat{x}(n) = x(n) - \sum_{i=1}^{M} \sum_{k=1}^{i} h_2'(i,k) \cdot x(n-k)x(n-i). \tag{1}$$

In (1) the second-order kernel h_2 is assumed symmetrically resulting in $h_2'(i,k) = h_2(i,k)$ for $i=k$ and $h_2'(i,k) = 2 \cdot h_2(i,k)$ for $i \neq k$. The coefficients of the

M. Faundez-Zanuy et al. (Eds.): NOLISP 2005, LNAI 3817, pp. 338–347, 2005.

predictor are optimal if the expected value $E[e(n)^2]$ is minimized; this is approximated by $\sum e(n)^2 \to \min$, representing the least squares approach.

2.1 Prediction Based on Vectors

If the analyzed signal $x(n)$ is a finite signal of length L the prediction (1) can be described in a vector notation

$$x = \hat{x} + e = \sum_{i=1}^{M} \sum_{k=1}^{i} h_2'(i,k) \cdot x_{i,k} + e \qquad (2)$$

with the vectors:
$$x = \left(x(0), x(1), \ldots, x(L-1) \right)^{\mathrm{T}}$$
$$x_{i,k} = \left(x(-i)x(-k),\ x(-i+1)x(-k+1),\ \ldots \right)^{\mathrm{T}}.$$

The indices i and k of the vectors $x_{i,k}$ are assigned to a single index λ resulting in

$$x_\lambda' = x_{i,k} \text{ and } a(\lambda) = h_2'(i,k) \quad \text{with} \qquad (3)$$
$$\lambda = k + (i-1)M - (i-2)(i-2)/2$$

$$x = \hat{x} + e = \sum_{\lambda=1}^{\Lambda} a(\lambda) \cdot x_\lambda + e. \qquad (4)$$

The error of the approximation $|e| = |x - \hat{x}|$ is to be minimized. Equation (4) represents a vector expansion of the vector x by the vectors x_λ representing a vector space of dimension $\Lambda = M(M+1)/2$. It is assumed that the vectors x_λ are independent; if a vector depends on other vectors, the vector and the corresponding coefficient can be omitted from the procedure reducing the dimension; however, this is most unlikely for the intended application. The basis vectors x_λ are not necessarily orthogonal among each other. The determination of the optimal parameters can be performed with the aid of a transformation, converting the vectors x_λ into an orthogonal basis v_λ with $\langle v_i, v_k \rangle = 0$ for $i \neq k$. This can be achieved by using the formula

$$v_\lambda = x_\lambda' - \sum_{i=1}^{\lambda-1} \frac{\langle x_i', v_i \rangle}{|v_i|^2} \cdot v_i \quad \text{for} \quad \lambda = 1 \ldots \Lambda, \qquad (5)$$

recursively starting with $v_1 = x_1'$. The orthogonalization procedure is known under the name Gram-Schmidt. The vector expansion of equation (4) can be formulated with the orthogonal vectors v_λ

$$x = \hat{x} + e = \sum_{\lambda=1}^{\Lambda} b(\lambda) \cdot v_\lambda + e. \qquad (6)$$

Therefore the optimal coefficients b of the vector expansion can be easily obtained by the dot product $\langle , \rangle$ of the vector x and the orthogonal basis vectors v_λ

$$b(\lambda) = \frac{\langle x, v_\lambda \rangle}{|v_\lambda|^2}, \qquad (7)$$

yielding $|e| = \min$. Then the coefficient vector $b = (b(1), b(2), ... b(\Lambda))^{\mathrm{T}}$ of the basis $\{v_\lambda\}$ is converted back into the coefficient vector $a = (a(1), a(2), ... a(\Lambda))^{\mathrm{T}}$ of the basis $\{x'_\lambda\}$. This is performed with the aid of the matrix $\Phi = (\varphi_1, \varphi_2, ..., \varphi_\Lambda)^{\mathrm{T}}$ by $a = \Phi \cdot b$. The matrix can be calculated iteratively starting with the identity matrix. The iteration steps are performed by

$$\varphi_i := \varphi_i - \sum_{k=1}^{i-1} \frac{\langle x'_i, v_k \rangle}{|v_k|^2} \cdot \varphi_k \qquad \text{for} \quad i = 1 ... \Lambda . \qquad (8)$$

Finally the resulting coefficients $a(\lambda)$ of the vector a are mapped into the coefficients $h'_2(i, k)$ by inverting the assignment of equation (3).

3   Analysis of Residual of Speech

The nonlinear system is used to model the fluctuations of voiced speech. These fluctuations are caused mainly by the voiced excitation and its interaction with the vocal tract. Therefore at first the speech signal is inverse-filtered by LPC-analysis representing the conventional linear prediction. The parameters for the nonlinear system are estimated from the residual of speech; to enable a time variability of the parameters the analysis is performed blockwise in short time intervals. For that purpose the residual signal is segmented into overlapping segments. The segments consist of a whole number of pitch periods. The overlapping of the segments is about one period. The nonlinear prediction yields for every segment a diagonal matrix $h'_{2\lambda}(i, k)$ of estimated coefficients; the index λ represents the analyzed λ-th segment. Since the statistic of the segments varies, the estimated coefficients vary, too. The variations of the coefficients from segment to segment model the fluctuations of the speech. The estimation of the coefficients is affected by windowing, too. Therefore it is favorable that the segments are sufficiently wide, to limit the fluctuations caused by segmentation. Segment lengths are used between two and five periods.

4   Synthesis

4.1   Synthesis Structure of Volterra Systems

The prediction error filter (1) is suitable for the analysis of signals while the inverse system of the prediction error filter can be used for synthesis. The inverse system H_2^{-1} of (1) has a recursive structure:

$$y(n) = x(n) + \sum_{i=1}^{M}\sum_{k=1}^{i} h'_2(i,k)\cdot y(n-k)y(n-i). \tag{9}$$

The recursive Volterra system is used to model the fluctuations of the voiced excitation. To introduce variations, the parameters of the nonlinear system are time variable. The whole system of the voiced excitation is shown in figure 1. The recursive Volterra system H_2^{-1} is excited by an impulse train corresponding to the fundamental frequency. The subsequent de-emphasis system $P^{-1}(z) = 1/((1-k_1 z^{-1})(1-k_2 z^{-1}))$ is a linear filter containing two real poles. The nonlinear prediction described in section 2 and 3 yields for every segment a set $h'_{2\lambda}(i,k)$ of coefficients with λ representing the analyzed λ-th segment. During the synthesis the parameters of the system H_2^{-1} are controlled by the parameter matrices $h'_{2\lambda}(i,k)$ consecutively indicated in fig. 1. Since the coefficients are updated for every period, the pulse train u_1 varies from period to period. Fig. 2 shows the output of the system caused by the excitation of the impulse train before and after the de-emphasis. The variations of the coefficients of H_2^{-1} cause fluctuations of the processed impulses u_1 depicted in fig. 2(a). Applying de-emphasis to u_1 these fluctuations produce changing ripples in the signal u_2, as can be seen in fig. 2(b). The signal u_2 can be interpreted as glottal pulses. The occurrence of ripples in the glottis signal is a known phenomenon, caused by nonlinearities due to glottal flow and vocal tract interaction [2]. The sets of coefficients $h'_{2\lambda}(i,k)$ for fig. 2 are obtained from the analysis of the residual of the vowel /a:/, with a sampling rate of 16 kHz. The order M of the nonlinear prediction is 18 and the analyzed segments consist of 4 periods. The absolute values of the impulse response of H_2^{-1} converge rather fast towards values close to zero, so that the effective length of the impulse response is shorter than one pitch period, whereas the impulse response is actually infinite. Since the overlapping of the impulse responses can be neglected,

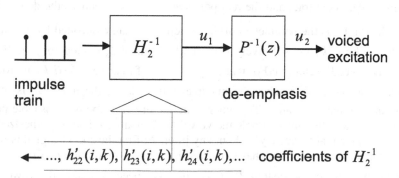

Fig. 1. Generation of voiced excitation: Excitation of the time variable recursive Volterra system H_2^{-1} by an impulse train resulting in u_1, filtering by the linear system $P^{-1}(z)$ for de-emphasis resulting in u_2

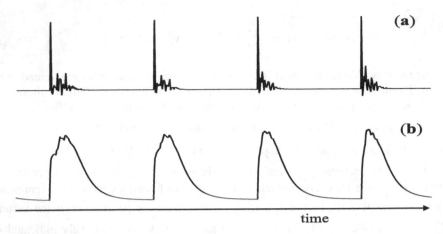

Fig. 2. Time signals of voiced excitation: (a) Impulse train u_1 processed by the time variable nonlinear system; (b) voiced excitation u_2 applying de-emphasis to u_1

no problems with instabilities occurred in our investigation. The greater the order M is chosen for analysis and synthesis the longer is the effective impulse response. Large numbers of M may cause instable systems. Since H_2^{-1} is a nonlinear system, the amplitude of the exciting impulse train affects also the shape of the signals u_1 and u_2. To ensure stable systems the amplitude of the exciting impulse train should not be too high. The degree of periodicity can be adjusted by the choice of M, the analyzed segment length, and by the amplitude of the impulse train.

4.2 Synthesis of Stationary Sounds

The recursive Volterra system H_2^{-1} is a component in the speech production system which can be seen in fig. 3. The linear system $H(z)$ in fig. 3 models the resonance structure of the vocal tract and the real pole system $P(z)^{-1}$ realizes the de-emphasis. During the synthesis the parameters of the system H_2^{-1} are controlled by the parameter matrices $h_{2\lambda}'(i,k)$ consecutively with $\lambda = [1, 2, 3, ..., \lambda_{max}]$, modeling the fluctuations of the voiced excitation. For that purpose a set of matrices $h_{2\lambda}'(i,k)$ is stored. If the increasing index λ reaches the maximal index λ_{max}, the index λ is set to a random integer smaller than 5. To demonstrate the effect of the time variable nonlinear system in the following example the vowel /a:/ is analyzed and resynthesized with constant fundamental frequency without any jitter. At first the speech signal is filtered by a repeated adaptive pre-emphasis realized by a linear prediction of first order determining the poles of the system $P^{-1}(z)$. Then the speech signal of the vowel /a:/ is filtered by a conventional linear prediction and the residual is analyzed blockwise as described before. The linear system $H(z)$ is in this case the standard

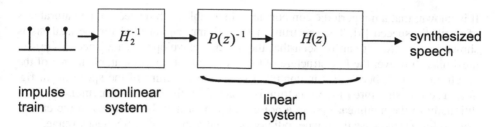

Fig. 3. Systems involved in the synthesis of voiced speech: Excitation of the time variable recursive Volterra system H_2^{-1} by an impulse train, filtering by a linear system consisting of $P(z)^{-1}$ for de-emphasis and $H(z)$ for the resonances of the vocal tract

all-pole model obtained from linear prediction. To show the impact of the nonlinear system, the spectra of the resynthesized vowel /a:/ are shown in fig. 4, synthesized with and without the recursive Volterra system. The use of the time variable Volterra system causes nonperiodicities, which can be seen in fig. 4(b). In contrast the synthesized signal shown in fig. 4(a) is exactly periodic without the nonlinear system.

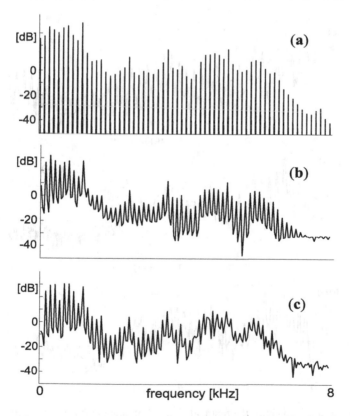

Fig. 4. Spectra of the vowel /a:/: (a) Synthesized without the nonlinear system; (b) synthesized with the nonlinear system; (c) original signal

It is known, that a nonperiodic component is favorable with respect to the naturalness of synthetic speech [3]. The spectrum of the original signal used for the analysis is shown in fig. 4(c). It can be seen that the spectral envelopes of the spectra 4(a)-(c) aresimilar, however the fine structure of the spectrum 4(b) is comparable to that of the spectrum of real speech 4(c) in contrast to the fine structure of the spectrum in fig. 4(a). The fine structure of the spectrum is affected by the degree of harmonicity. Additionally to the nonlinear system a noise component may be included in the excitation, to further increase the nonperiodicity especially in the high frequency range.

It should be mentioned that the time variable system modeling the fluctuations consists of nonlinear terms only, without additional linear terms. Synthesis results have been generated also with linear terms alone and in combination with nonlinear terms in H_2 and H_2^{-1} respectively. However, the inclusion of linear terms decreases the speech synthesis quality.

4.3 Synthesis of Words

Besides of resynthesis of stationary vowels, the excitation is used for a parametric synthesis which includes a lossy tube model as linear system $H(z)$ in fig. 3. The

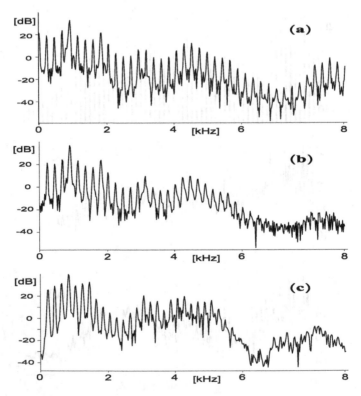

Fig. 5. Spectra of a segment of the first part of the diphthong /aI/: (a) Generated by parametric synthesis without time variable Volterra system; (b) generated by parametric synthesis with time variable Volterra system; (c) obtained from a segment of diphone [v-aI]

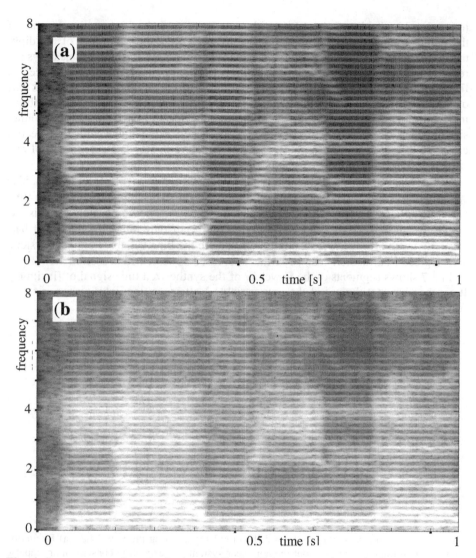

Fig. 6. Spectrograms of synthesized word [lavi:n@] by parametric synthesis with lossy tube model: (a) Excitation consists of an impulse train without the nonlinear system; (b) excitation consists of an impulse train including the nonlinear system

lossy tube model is based on the standard lossless tube model realized in lattice structure extended by the frequency dependent losses of the vocal tract and the lip termination. The parameters of the lossy tube model are reflection coefficients which can be converted into pseudo vocal tract areas. The reflection coefficients are estimated from diphones by an optimization algorithm [4]. For the analysis the diphones are segmented into overlapping frames. The estimation algorithm yields a parameter vector for each frame. For the synthesis the parameter vectors are used one after the other. To smooth the discontinuities at the diphone joints, the model parameters are linearly interpolated in the description of logarithmic areas. Fig. 5 shows short-time spectra of the parametric synthesis and of real speech representing the beginning of the diphthong /aI/ which is part of a synthesized word. The time variable nonlinear system reduces the periodicity especially in the high frequency range which can be seen in fig. 5(b). Fig. 5(c) shows a spectrum of a segment of the diphthong /aI/ which is part of the diphone analyzed for the parametric synthesis. Since the parametric synthesis makes use of transitions between the diphones, the spectral envelopes of the synthesized speech in fig. 5(a), (b) are only approximately comparable with the spectral envelope of real speech in fig. 5(c). The fine structure in 5(b) is closer to that of real speech 5(c) in contrast to the fine structure in 5(a).

In fig. 6 spectrograms of the synthesized German word [lavi:n@] are shown. The spectrograms result from the parametric diphone synthesis with and without the time variable Volterra system. The fundamental frequency is chosen constant for this example. It can be seen that especially in the high frequency range the degree of harmonicity in fig 6(b) is lower than in fig 6(a) due to the time variable nonlinear system. The degree of harmonicity can be further decreased as described in section 4.1.

Fig. 7 shows segments of the vowel /a/ of the synthesized time signal of [lavi:n@]. It can be seen that the use of the time variable Volterra system introduces variations from period to period.

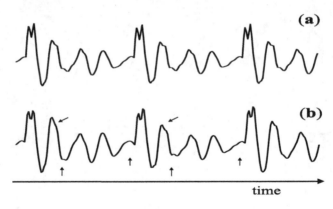

Fig. 7. Segment of the synthesized word [lavi:n@] representing the vowel /a/: (a) Excitation consists of an impulse train without the nonlinear system; (b) excitation consists of an impulse train with the nonlinear system. Arrows indicate fluctuations resulting from the time variable Volterra system.

5 Conclusions

The fluctuations of voiced speech from period to period can be modeled by a time variable recursive Volterra system which is excited by an impulse train. The output of the recursive Volterra system is a processed pulse train which varies from period to period due to the time variability of the nonlinear system. The parameters of the nonlinear system are estimated from the residual of an isolated spoken vowel. The degree of fluctuations is affected by the length of the analyzed segments, by the number of coefficients of the Volterra system, and by the amplitude of the exciting impulse train. It has been explained how the nonlinear system can be integrated into the speech production system of a parametric synthesis. Examples show that the introduced fluctuations are favorable with respect to the naturalness of the synthesized speech.

References

1. McLaughlin, S.: Nonlinear Speech Synthesis. Proc. EUSIPCO-2002, Toulouse France, (2002) 211–218.
2. Bavegard, B., Fant, G.: Notes on glottal source interaction ripple. STL-QPSR (4), Royal Institute of Technology Stockholm, (1994) 63-77.
3. Sambur, M.R., Rosenberg, A.E., Rabiner, L.R., McGonegal, C.A.: On reducing the buzz in LPC synthesis. J.Acoust.Soc.Am. 63, (1978) 918-924.
4. Schnell, K., Lacroix, A.: Speech Production Based on Lossy Tube Models: Unit Concatenation and Sound Transitions. Proc. INTERSPEECH-2004 ICSLP, Vol. I, Jeju-Island Korea, (2004) 505-508.

A Simple, Quasi-linear, Discrete Model of Vocal Fold Dynamics

Max Little[1,*], Patrick McSharry[2], Irene Moroz[1], and Stephen Roberts[3]

[1] Applied Dynamical Systems Research Group,
Oxford Centre for Industrial and Applied Mathematics, Oxford University, UK
littlem@maths.ox.ac.uk
http://www.maths.ox.ac.uk/ads
[2] Oxford Centre for Industrial and Applied Mathematics, Oxford University, UK
[3] Pattern Analysis Research Group, Engineering Science, Oxford University, UK

Abstract. In current speech technology, linear prediction dominates. The linear vocal tract model is well justified biomechanically, and linear prediction is a simple and well understood signal processing task. However, it has been established that, in voiced sounds, the vocal folds exhibit a high degree of nonlinearity. Hence there exists the need for an approach to modelling the behaviour of the vocal folds. This paper presents a simple, nonlinear, biophysical vocal fold model. A complementary discrete model is derived that reflects accurately the energy dynamics in the continuous model. This model can be implemented easily on standard digital signal processing hardware, and it is formulated in such a way that a simple form of nonlinear prediction can be carried out on vocal fold signals. This model could be of utility in many speech technological applications where low computational complexity synthesis and analysis of vocal fold dynamics is required.

1 Introduction

The linear signal processing of speech is a well developed science, having a long history of association with the science of *linear acoustics*. Referring to Fig. 1, the use of linear acoustics is well justified biophysically, since a realistic representation of the vocal organs is obtained by assuming that the influence of the vocal tract is that of an *acoustic tube* that acts as a *linear resonator*, amplifying or attenuating the harmonic components of the vocal folds during voiced sounds. This resonator can be represented in discrete-time as a *digital filter* [1].

Access to biophysical speech parameters enables certain technology such as communications (e.g. wireless mobile telephone systems), clinical, therapeutic and creative manipulation for multimedia. For example, *linear prediction* [2] can be used to find vocal tract parameters: thus much effort has been directed towards the application of this particular analysis tool to speech signals. The results of such work are an efficient set of techniques for linear prediction of speech

---

* Supported by doctoral training grant provided by the Engineering and Physical Research Council, UK.

M. Faundez-Zanuy et al. (Eds.): NOLISP 2005, LNAI 3817, pp. 348–356, 2005.

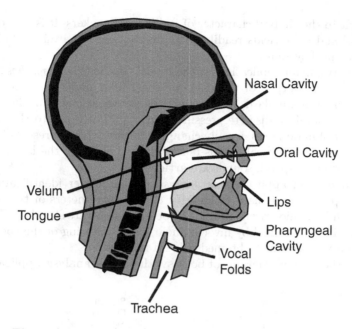

Fig. 1. Arrangement of vocal organs inside the head and neck

that now form a fundamental part of current technology [3]. However, empirical and numerical investigation and modelling of the vocal folds has revealed a high degree of *nonlinearity*, therefore the use of the same linear tools is inappropriate. Hence there exists a need for a similar approach to modelling the behaviour of the vocal fold oscillation.

For typical technological applications, where computational power is at a premium *lumped* models are to be preferred over full, continuum mechanical models because they capture only the important, relevant dynamical effects. Progress on such lumped models has been steady [4], and there exist a range of models of varying complexity (see, for example, the two mass models of [5] and [6]).

This paper presents a simple, practical, continuous model of vocal fold behaviour, and an appropriate discrete counterpart, as described in [7]. The model has only five parameters and can be integrated using only simple computational operations of existing digital signal processing hardware. The discrete counterpart is derived using a specialised integration technique that replicates the long-term energy properties of the continuous model, thus alleviating problems of numerical discretisation error. Furthermore, the model is *quasi-linear* and thus forms a natural extension to linear prediction, for which efficient, parametric identification techniques have already been developed. It exhibits nonlinear oscillation due to the existence of a *stable limit cycle* that captures the energy balancing inherent to typical stable intonations in continuous speech. The model also captures the observed asymmetry of flow rate output behaviour that is important to the timbral character of individual speakers. It is also only two-dimensional and thus yields readily to straightforward analysis techniques for planar dynamical systems.

There are two main applications of the model. The first is to *synthesise* speech signals. Here, the output of the discrete vocal fold model u_n is fed directly into the input of a standard, linear model of the vocal tract and lip radiation impedance, to obtain a discrete pressure signal p_n. The input to the vocal fold model is a set of parameters, usually changing in time, that represent the change in configuration of the vocal organs (such as the muscles of the larynx and the lungs) as particular speech sounds are articulated.

The second main application is *analysis* by parameter identification. Here quasi-linear prediction is used to identify the five parameters of the model directly from discrete measurements of the vocal fold flow rate signal u_n. Typically, this signal will be obtained by inverse linear digital filtering of the speech pressure signal from recordings obtained using a microphone.

Figure 2 shows flow diagrams of both synthesis and analysis applications.

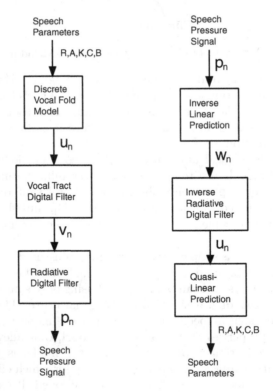

Fig. 2. (Left) Typical arrangement of forward signal processing when the model is used for speech synthesis applications, (Right) Similarly for parameter identification (analysis) applications

2   Deriving the Continuous Model

Figure 3 shows the geometric setup and configuration of the model of the vocal folds. We assume inviscid, laminar air flow in the vocal tract. The areas of the two points A and B in the vocal folds are:

$$a_A = 2lx_A, \quad a_B = 2lx_B \tag{1}$$

where x_A, x_B are the positions of points A and B. Point A is assumed to be stationary, therefore x_A is a parameter of the model.

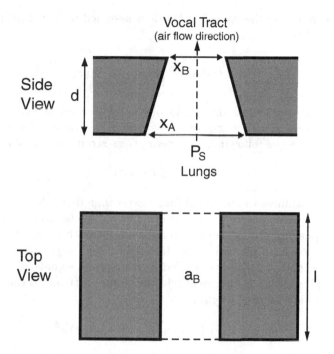

Fig. 3. Geometry and configuration of the nonlinear vocal fold model

The Bernoulli pressure at both points is:

$$\frac{1}{2}\rho_0\frac{U^2}{a_A^2} + P_A = P_S, \quad \frac{1}{2}\rho_0\frac{U^2}{a_B^2} + P_B = P_S \tag{2}$$

where U is the air flow rate through the vocal folds, ρ_0 is the equilibrium density of air, and P_S is the (static) lung pressure.

At the top of the vocal folds, a jet is assumed to form such that there is no air pressure. Therefore $P_B = 0$ and so:

$$U = 2l\sqrt{\frac{2P_S}{\rho_0}}x_B\Phi(x_B) \tag{3}$$

where the Heaviside step function $\Phi(x)$ is used to indicate that there is no air flow when point B is negative (the vocal folds are completely closed at the top). Therefore the air flow rate is proportional to the position of x_B when positive:

$$U \propto x_B, \quad x_B > 0 \tag{4}$$

The pressure at point A is:

$$P_A = P_S - \frac{1}{2}\rho_0 \frac{U^2}{a_A^2} = P_S \left[1 - \Phi(x_B) \frac{x_B^2}{x_A^2}\right] \tag{5}$$

and the force acting on the vocal fold tissue is assumed to be the average of that at points A and B:

$$F = \frac{1}{2}(P_A + P_B)\,ld = \frac{1}{2}ldP_S \left[1 - \Phi(x_B) \frac{x_B^2}{x_A^2}\right] \tag{6}$$

where l and d are the length and height of the folds respectively.

From now on we write $x = x_B$ for convenience. For the vocal folds, the tissue is assumed to have the following, nonlinear stress-strain relationship [8]:

$$s(x, \dot{x}) = kx + ax\dot{x} \tag{7}$$

where k is the stiffness of the vocal fold tissue that depends highly upon the tightness of the vocal muscles in the larynx. The parameter a controls the extent of velocity-dependent stiffness of the vocal folds. It is this velocity-dependence [8] of the relationship that causes the important *time asymmetry* of the vocal fold flow rate signal U which is observed in real speech signals [9].

With damping effects of the vocal fold tissue proportional to the velocity the equation of motion for the system is:

$$m\ddot{x} + r\dot{x} + s(x, \dot{x}) = F = b - c\Phi(x)x^2 \tag{8}$$

where $b = P_S ld/2$, $c = P_S ld/(2x_A^2)$ and r is the frictional damping constant that depends upon the biomechanical properties of vocal fold tissue.

3   Deriving the Discrete Model

Making use of the *discrete variational calculus* [10] we can derive the discrete equations of motion as:

$$m\left(\frac{x_{n+1} - 2x_n + x_{n-1}}{\Delta t^2}\right) + r\left(\frac{x_n - x_{n-1}}{\Delta t}\right) +$$

$$ax_n\left(\frac{x_n - x_{n-1}}{\Delta t}\right) - b + kx_n + c\Phi(x_n)x_n^2 = 0 \tag{9}$$

where n is the time index and Δt is the time difference between samples of a speech signal. Such a discretisation has a *discrete energy expression* that represents the mechanical energy in the vocal folds:

$$E_n = \frac{1}{2}\left(x_{n+1} - x_n\right)^2 + \frac{1}{2}Kx_n^2 \tag{10}$$

and the corresponding rate of change of discrete energy is:

$$dE_n = -\left(x_{n+1} - x_n\right)\left[R\left(x_{n+1} - x_n\right) + Ax_n\left(x_{n+1} - x_n\right) - B + Cx_n^2\right] \tag{11}$$

where:

$$R = \frac{r\Delta t}{m}, \quad A = \frac{a\Delta t}{m}, \quad B = \frac{b\Delta t^2}{m}, \quad K = \frac{k\Delta t^2}{m}, \quad C = \frac{c\Delta t^2}{m} \tag{12}$$

The discrete equations of motion (9) can be used as an explicit integrator for the model:

$$x_{n+1} = 2x_n - x_{n-1} - R\left(x_n - x_{n-1}\right) - \\ Ax_n\left(x_n - x_{n-1}\right) + B - Kx_n - C\Phi\left(x_n\right)x_n^2 \tag{13}$$

This is a quasi-linear discrete system for which the method of *quasi-linear prediction*, described in the next section, can be used to obtain the parameters from

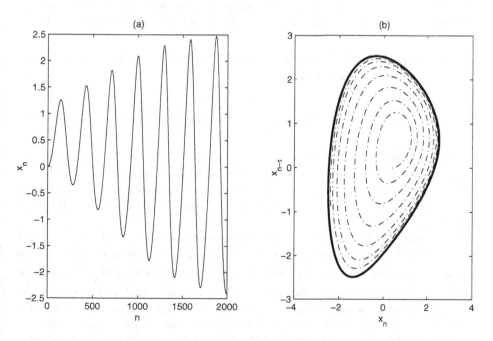

Fig. 4. Typical vocal fold model behaviour with $R = 0.001$, $A = -0.007$, $B = 0.00025$, $K = 0.00026$ and $C = 0.00024$. (a) Time series x_n, (b) Two-dimensional embedding of x_n with embedding delay $\tau = 60$. Thick black line is the limit cycle.

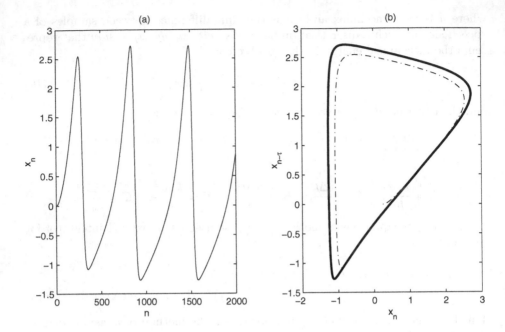

Fig. 5. Typical vocal fold model behaviour x_n with $R = 0.03125$, $A = -0.0375$, $B = 0.000234$, $K = 3.906e - 6$ and $C = 0.0002343$. (a) Time series x_n, (b) Two-dimensional embedding of x_n with embedding delay $\tau = 60$. Thick black line is the limit cycle.

a recording of the vocal fold behaviour using a straightforward matrix inversion. The discrete output flow rate is:

$$u_n = 2l\sqrt{\frac{2P_S}{\rho_0}} x_n \Phi(x_n) \tag{14}$$

Figures 4 and 5 show typical behaviour of the vocal fold model for certain ranges of parameters. Figure 4 shows oscillation in a limit cycle, and Fig. 5 shows asymmetric oscillation, with the rising slope being slower than the falling slope. This is typical of vocal fold behaviour as identified from real, voiced speech signals [9].

In general, when we have obtained a recorded pressure signal p_n we cannot know the real scale. In other words, the recording equipment leads to unknown amplification. This means that, subsequently, the scaling factor $2l\sqrt{\frac{2P_S}{\rho_0}}$ in equation (14) cannot be known. Similarly, there is an additional ambiguity for the mass parameter m which cannot be resolved. This implies that the parameters are scaled by some unknown factor. Therefore we can only compare values obtained from different recordings, and the parameters have no absolute physical interpretation.

4   Quasi-linear Prediction for Parametric Identification

Since the position of point A in the vocal folds, by equation (4), is proportional to the discrete flow rate signal u_n, we assume, initially that $x_n = u_n$. Then we exclude all values of u_n that are negative. Then, by defining the *residual error* as:

$$e_n = x_{n+1} - 2x_n + x_{n-1} + R(x_n - x_{n-1}) +$$
$$Ax_n(x_n - x_{n-1}) - B + Kx_n + C\Phi(x_n)x_n^2 \quad (15)$$

we can extend the linear prediction process [2], assuming that e_n has a zero-mean, independent Gaussian distribution. This leads to the least-squares solution to find the best fit parameters of the model.

For a non-negative speech signal of length N the system that is the solution to the least-squares problem is:

$$\sum_{n=2}^{N-1} \mathbf{M}\mathbf{a} = -\sum_{n=2}^{N-1} \mathbf{d} \quad (16)$$

where:

$$\mathbf{a} = \begin{bmatrix} R & A & B & K & C \end{bmatrix}^T \quad (17)$$

The 5×5 system matrix is:

$$\mathbf{M} = \begin{bmatrix} x_n^2 - 2x_nx_{n-1} + x_{n-1}^2 & x_n^3 - 2x_n^2x_{n-1} + x_nx_{n-1}^2 & x_{n-1} - x_n \\ x_n^3 - 2x_n^2x_{n-1} + x_nx_{n-1}^2 & x_n^4 - 2x_n^3x_{n-1} + x_n^2x_{n-1}^2 & x_nx_{n-1} - x_n^2 \\ x_{n-1} - x_n & x_nx_{n-1} - x_n^2 & 1 \\ x_n^2 - x_nx_{n-1} & x_n^3 - x_n^2x_{n-1} & -x_n \\ \Phi(x_n)x_n^2(x_n - x_{n-1}) & \Phi(x_n)x_n^3(x_n - x_{n-1}) & -\Phi(x_n)x_n^2 \end{bmatrix} \cdots$$

$$\cdots \begin{matrix} x_n^2 - x_nx_{n-1} & \Phi(x_n)x_n^2(x_n - x_{n-1}) \\ x_n^3 - x_n^2x_{n-1} & \Phi(x_n)x_n^3(x_n - x_{n-1}) \\ -x_n & -\Phi(x_n)x_n^2 \\ x_n^2 & -\Phi(x_n)x_n^3 \\ \Phi(x_n)x_n^3 & -\Phi(x_n)x_n^4 \end{matrix} \quad (18)$$

and:

$$\mathbf{d} = \begin{bmatrix} x_{n+1}x_{n-1} - x_{n+1}x_n + 2x_n^2 - 3x_nx_{n-1} + x_{n-1}^2 \\ x_{n+1}x_nx_{n-1} - x_{n+1}x_n^2 + 2x_n^3 - 3x_n^2x_{n-1} + x_nx_{n-1}^2 \\ x_{n+1} - 2x_n + x_{n-1} \\ 2x_n^2 - x_{n+1}x_n - x_nx_{n-1} \\ -\Phi(x_n)x_n^2(x_{n+1} - 2x_n + x_{n-1}) \end{bmatrix} \quad (19)$$

The coefficients, $\mathbf{a}$ of the quasi-linear model, taken together with the residual e_n and the initial conditions x_1, x_2 as a set, form a one-one representation of the modelled data x_n, and we can exactly reconstruct x_n using only this information.

5 Discussion and Conclusions

This paper has introduced a new, simplified model of vocal fold dynamics. Making use of variational integration methods, a corresponding discrete counterpart is derived. This discrete model can then be used to synthesise vocal fold flow rate signals or used to identify model parameters from estimated vocal fold flow rate signals obtained from speech pressure recordings.

The main advantage of this model is that it captures the overall features of vocal fold oscillation (limit cycles, open/close quotient and pulse asymmetry) whilst the computational complexity is low, meaning that it can be implemented on standard digital signal processing hardware. The corresponding disadvantage is that certain interesting vocal pathologies cannot be replicated, for example *creaky voice* which is suggested to originate from period doubling bifurcations [11].

In a related study, this model has been used to identify and resynthesise estimated vocal fold flow rate signals. Discovering that the model was capable of replicating actual flow rate signals with some successes and some failures, the suggested method for parametric identification is to ensure positivity of all parameters (non-negative least squares [12]), and subsequently reversing the sign of the parameter A. This guarantees that the parameters conform to the modelling assumptions.

References

1. Markel, J.D., Gray, A.H.: Linear prediction of speech. Springer-Verlag (1976)
2. Proakis, J.G., Manolakis, D.G.: Digital signal processing: principles, algorithms, and applications. Prentice-Hall (1996)
3. Kleijn, K., Paliwal, K.: Speech coding and synthesis. Elsevier Science, Amsterdam (1995)
4. Story, B.H.: An overview of the physiology, physics and modeling of the sound source for vowels. Acoust. Sci. & Tech. **23** (2002)
5. Ishizaka, K., Flanagan, J.: Synthesis of voiced sounds from a two-mass model of the vocal cords. ATT Bell System Tech Journal **51** (1972) 1233–1268
6. Steinecke, I., Herzel, H.: Bifurcations in an asymmetric vocal-fold model. J. Acoust. Soc. Am. **97** (1995) 1874–1884
7. Little, M.A., Moroz, I.M., McSharry, P.E., Roberts, S.J.: System for generating a signal representative of vocal fold dynamics. (2004) UK patent applied for.
8. Chan, R.W.: Constitutive characterization of vocal fold viscoelasticity based on a modified arruda-boyce eight-chain model. J. Acoust. Soc. Am **114** (2003) 2458
9. Holmes, J.: Speech synthesis and recognition. Van Nostrand Reinhold (UK) (1988)
10. Marsden, J., West, M.: Discrete mechanics and variational integrators. Acta Numerica (2001) 357–514
11. Mergell, P., Herzel, H.: Bifurcations in 2-mass models of the vocal folds – the role of the vocal tract. In: Speech Production Modeling 1996. (1996) 189–192
12. Lawson, C., Hanson, R.: Solving least square problems. Prentice Hall, Englewood Cliffs NJ (1974)

Blind Channel Deconvolution of Real World Signals Using Source Separation Techniques

Jordi Solé-Casals[1] and Enric Monte-Moreno[2]

[1] Signal Processing Group, University of Vic,
Sagrada Família 7, 08500 Vic, Catalonia
jordi.sole@uvic.es
http://www.uvic.es/eps/recerca/processament/ca/inici.html
[2] Polytechnic University of Catalonia, Campus Nord,
UPC, Barcelona, Catalonia
enric@gps.tsc.upc.es
http://gps-tsc.upc.es/veu/personal/enric/enric.html

Abstract. In this paper we present a method for blind deconvolution of linear channels based on source separation techniques, for real word signals. This technique applied to blind deconvolution problems is based in exploiting not the spatial independence between signals but the temporal independence between samples of the signal. Our objective is to minimize the mutual information between samples of the output in order to retrieve the original signal. In order to make use of use this idea the input signal must be a non-Gaussian i.i.d. signal. Because most real world signals do not have this i.i.d. nature, we will need to preprocess the original signal before the transmission into the channel. Likewise we should assure that the transmitted signal has non-Gaussian statistics in order to achieve the correct function of the algorithm. The strategy used for this preprocessing will be presented in this paper. If the receiver has the inverse of the preprocess, the original signal can be reconstructed without the convolutive distortion.

1 Introduction

Many researches have been done in the identification and/or the inversion of linear and nonlinear systems. These works assume that both the input and the output of the system are available [14]; they are based on higher-order input/output cross-correlation [3], bispectrum estimation [12, 13] or on the application of the Bussgang and Prices theorems [4, 9] for nonlinear systems with Gaussian inputs. However, in real world situations, one often does not have access to the distortion input. In this case, blind identification of the system becomes the only way to solve the problem. In this paper we propose to adapt the method presented in [17, 18] for blind deconvolution by means of source separation techniques, to the case of real world signals which are non i.i.d. This is done by means of a stage of preprocessing before sending the signal and post-processing stage after the reception and deconvolution. The paper is organized as follows. The source separation problem is described in Section 2. The model and assumptions for applying these techniques to blind deconvolution are presented in Section 3. Section 4 contains the proposed preprocessing and

M. Faundez-Zanuy et al. (Eds.): NOLISP 2005, LNAI 3817, pp. 357–367, 2005.

post-processing stages, and the simulation results are presented in Section 5, before concluding in Section 6.

2   Source Separation Review

The problem of source separation may be formulated as the recovering of a set of unknown independent signals from the observations of mixtures of them without any prior information about either the sources or the mixture [10, 6]. The strategy used in this kind of problems is based in obtaining signals as independent as possible at the output of the system. In the bibliography multiple algorithms have been proposed for solving the problem of source separation in instantaneous linear mixtures. These algorithms range from neural networks based methods [1], cumulants or moments methods [7, 5], geometric methods [15] or information theoretic methods [2]. In real world situations, however, the majority of mixtures can not be modeled as instantaneous and/or linear. This is the case of the convolutive mixtures, where the effect of channel from source to sensor is modeled by a filter [11]. Also the case of the post-nonlinear (PNL) mixtures, where the sensor is modeled as a system that performs a linear mixture of sources plus a nonlinear function applied to its output, in order to take into account the possible non-linearity of the sensor (saturation, etc.) [16].

Mathematically, we can write the observed signals in source separation problem of instantaneous and linear mixtures as (see figure 1):

$$e_i(t) = \sum_{j=1}^{n} a_{ij} s_j(t) \tag{1}$$

where $\mathbf{A} = \{a_{ij}\}$ is the mixing matrix. It is well known that such a system is blindly invertible if the source signals are statistically independent and we have no more than one Gaussian signal.

A solution may be found by minimizing the mutual information function between the outputs of the system y_i:

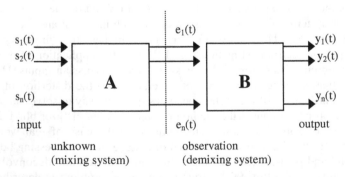

Fig. 1. Block diagram of the mixture system and blind demixing system. Both matrix A and signals $s_i(t)$ on the mixture process are unknown.

$$min_B[I(y)] = min_B\left[\sum_{i=1}^{n} H(y_i) - H(e) - \ln|det(B)|\right]$$

A related problem with blind separation is the case of blind deconvolution, which is presented in figure 2, and can be expressed in the framework of Equation (1). Development of this framework is presented in the following section.

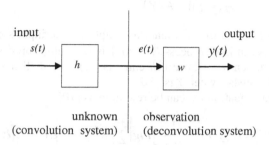

Fig. 2. Block diagram of the convolution system and blind deconvolution system. Both filter h and signal s(t) on the convolution process are unknown.

3 Model and Assumptions

We suppose that the input of the system $S=\{s(t)\}$ is an unknown non-Gaussian independent and identically distributed (i.i.d.) process. We are concerned by the estimation of $s(t)$ from the system's output e. This implies the blind inversion of a filter. From figure 2, we can write the output of filter h in a similar form that obtained in source separation problem, Equation 1, but now with vectors and matrix of infinite dimension:

$$e = Hs$$

where:

$$H = \begin{bmatrix} \cdots & \cdots & & \cdots & \cdots \\ \cdots & h(t+1) & h(t) & h(t-1) & \cdots \\ \cdots & h(t+2) & h(t+1) & h(t) & \cdots \\ \cdots & \cdots & & \cdots & \cdots \end{bmatrix}$$

is a Toeplitz matrix of infinite dimension and represents the action of filter h to the signal s(t). This matrix h is nonsingular provided that the filter h is invertible, i.e. h^{-1} exists and satisfies $h^{-1}*h = h*h^{-1} = \delta_0$, where δ_0 is de Dirac impulse. The solution to invert this systems and the more general nonlinear systems (Wiener systems) are presented and studied in [17, 18] where a Quasi-nonparametric method is presented. In the particular case of windowed signals, h is of finite length, and the product $h^{-1}*h$ yields a Dirac impulse with a delay.

3.1 Summary of the Deconvolution Algorithm

From figure 2 we can write the mutual information of the output of the filter w using the notion of entropy rates of stochastic processes [12] as:

$$I(Y) = \lim_{T \to \infty} \frac{1}{2T+1} \left\{ \sum_{t=-T}^{T} H(y(t)) - H(y_{-T}, ..., y_T) \right\}$$

$$= H(y(\tau)) - H(Y) \tag{2}$$

where τ is arbitrary due to the stationary assumption. The input signal $S=\{s(t)\}$ is an unknown non-Gaussian i.i.d. process, $Y=\{y(t)\}$ is the output process and **y** denotes a vector of infinite dimension whose t-*th* entry is $y(t)$. We shall notice that $I(Y)$ is always positive and vanishes when Y is i.i.d.

After some algebra, Equation (2) can be rewritten as [10]:

$$I(Y) = H(y(\tau)) - \frac{1}{2\pi} \int_0^{2\pi} \log \left| \sum_{t=-\infty}^{+\infty} w(t) e^{-jt\theta} \right| d\theta - E[\mathcal{E}] \tag{3}$$

To derive the optimization algorithm we need the derivative of I(Y) with respect to the coefficients of w filter. For the first term of Equation (3) we have:

$$\frac{\partial H(y(\tau))}{\partial w(t)} = -E\left[\frac{\partial y(\tau)}{\partial w(t)} \psi_y(y(\tau)) \right] = -E\left[e(\tau-t) \psi_y(y(\tau)) \right] \tag{4}$$

where $\psi_y(u) = (\log P_y)'(u)$ is the *score* function. The second term is:

$$\frac{\partial}{\partial w(t)} \frac{1}{2\pi} \int_0^{2\pi} \log \left| \sum_{n=-\infty}^{+\infty} w(n) e^{-jn\theta} \right| d\theta = \frac{1}{2\pi} \int_0^{2\pi} \frac{e^{-jt\theta}}{\sum_{n=-\infty}^{+\infty} w(n) e^{-jn\theta}} d\theta \tag{5}$$

where one recognizes the $\{-t\}$-th coefficient of the inverse of the filter w, which we denote $\overline{w}(-t)$. Combining Equations (5) and (6) leads to:

$$\frac{\partial I(Y)}{\partial w(t)} = -E\left[x(\tau - t) \psi_y(y(\tau)) \right] - \overline{w}(-t)$$

that is the gradient of $I(Y)$ with respect to $w(t)$. Using the concept of natural or relative gradient, the gradient descendent algorithm will be finally as:

$$w \leftarrow w + \mu \left\{ E\left[x(\tau - t) \psi_y(y(\tau)) \right] + \delta \right\} * w$$

It is important to notice that blind deconvolution is a different problem than source separation. In our blind deconvolution scenario the objective is to recover an unknown signal filtered by an unknown filter, using only the received (observed) signal. The main idea, proposed in [17, 18] is to use independence criteria, as in source separation problems, for deal with this problem. The relationship between source separation and deconvolution is shown in figure 3.

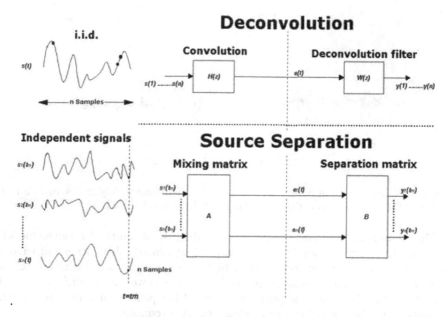

Fig. 3. Relationship between deconvolution and source separation. In source separation we force spatial independence between samples of different signals (at the same time instant). On the other hand, in deconvolution we impose this independence between samples of the unique signal at different times (time independence).

4   Application to Real World Signals

When signals are sent from the transmitter to the receiver, they suffer changes in their waveform form due to the effect of the transmission channel. In real systems, usually the transmission channel is unknown, and therefore the effect over the signals is uncertain. In this case, the use of an equalizer is necessary in order to ensure that the received signal is exactly the sent signal. Our proposed method uses directly the signal of interest to blindly estimate the inverse of the transmission channel, without the need of any other technique for equalizing the channel. In order to apply this method to real world signals we have to deal, as shown in previous Section with the fact that the signals are usually not i.i.d., so we can not use this method directly. In order to use it, we need to preprocess the input signal to ensure its temporal independence between samples and also to ensure its non-Gaussian distribution. In this section we present these preprocessing and post-processing stages in experiments done with speech signals.

4.1   Whitening of the Signal

Speech signals have a clear correlation structure, and therefore are not i.i.d. In order to use source separation techniques for blind deconvolution a preprocessing stage is necessary and it consists on a whitening of the original signal by means of an inverse LPC filter.

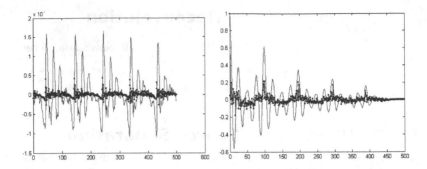

Fig. 4. Left: wave form of Vowel /a/ and the prediction residual. Right: Comparison of the normalized autocorrelations of the vowel signal and the residual of the LPC.

The use of a LPC preprocessing is useful in the case of signals that can be modeled as autoregressive processes. This is the case of speech signal or string and wind instruments, which were our test signals. In figure 4 we can see the autocorrelation sequence of a speech segment corresponding to the vowel /a/, before and after the whitening filter. Note that the prediction residual has peaks at multiples of the pitch frequency, and the residual in between these peaks is reduced.

The resulting residual after the whitening filter is an i.i.d. signal with a Gaussian distribution. Consequently it will be necessary to change de probability density function of the i.i.d. signal in order to acquire the necessary conditions to apply the deconvolution process on the receiver.

4.2 *Des-Gaussianity* of the Signal

In figure 5 we show the quantiles of the signal and the residual vs. the quantiles of a Gaussian distribution. One can see that the whitened signal has a nearly linear relation and is symmetric. For the purposes of the algorithm that we propose the signal can be considered Gaussian.

In order to change its probability density function, we propose a method based on the following observations:

1. Speech signals are periodic signals, with a fundamental period (pitch frequency).
2. A whitening filter removes all the components and keeps only the non-predictable part that corresponds precisely to the excitation of the speech signal (the pitch), and to the excitation of the instrument, with its fundamental period.

Starting from these observations we propose the following method:

1. Normalize the whitened signal provided by a LPC filter in a way that the maximums associated whit the periodic excitation been around ±1.
2. Pass the result signal trough a nonlinear function that maintains the peak values but modify substantially all the other values, therefore the pdf. We propose two functions for this task:

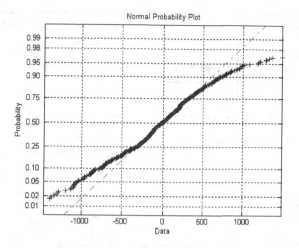

Fig. 5. Plot of the quantiles of the whitened signal versus probability of a Gaussian. One can observe the Gaussianity of the distribution.

    2.1. exponential function $x(\cdot)^n$: attenuate all the values between two consecutive peaks.

    2.2. *tanh*($\cdot$): amplify all the values between two consecutive peaks.

The effect of this process will be that the output signal will maintain important parts of the signal (the peaks of the series) and will change the form of the distribution. In the next figure we can see the proposed method:

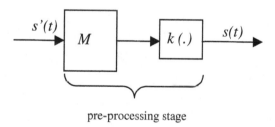

pre-processing stage

Fig. 6. The proposed preprocessing stage, composed by a whitening filter M and a nonlinear function $k(\cdot)$ preserving the most important part of the input signal. This stage is applied before sending the (unknown) signal through the transmitter channel.

5 Experimental Results

The input signal is a fragment of speech signal, preprocessed as shown in the diagram of figure 5. This is the signal that we want to transmit through an unknown channel. Our objective is to recover the original signal $s'(t)$ only from the received observation $e(t)$.

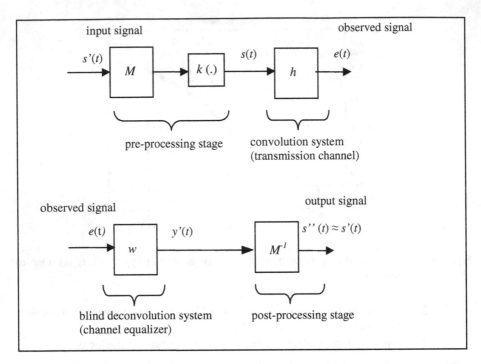

Fig. 7. On the upper part of the figure, the proposed preprocessing stage, and the convolution system. On the lower part, the blind deconvolution system and the post-processing stage.

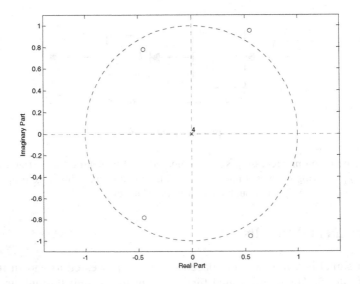

Fig. 8. Poles and zeros of h. We observe two zeros outside the unit circle, so the filter is of non-minimum phase.

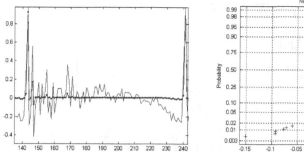
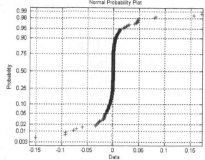

Fig. 9. On the left, the LPC residue of the signal superposed with the residual after the non-linear operation. On the right, the quantile of the processed residual versus the quantiles of a Gaussian distribution.

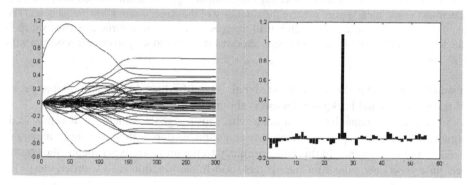

Fig. 10. On the left, filter coefficients evolution of the inverse filter w. The convergence is attained at 150 iterations approx. On the right, convolution between filter h and its estimated inverse w. The result is clearly a delta function.

Consider, now, the filter h as a FIR filter with coefficients: $h=[0.826,-0.165,0.851,0.165,0.810]$. Figure 8 show that h has two zeros outside the unit circle, which indicates that h is non-minimum phase. The system that we propose will always yield a stable equalizer because it computes a FIR approximation of the IIR optimal solution. The algorithm was provided with a signal of size $T = 500$. The size of the impulse response of w was set to 81. In the pre-processing stage the length of LPC was fixed at 12 coefficients and nonlinear function (des-*Gaussianity*) was $k(u) = u^3$.

In figure 9, on the left we show the prediction residual, before and after a cubic non-linearity. On the right we show a quantile plot of the residual vs. a Gaussian distribution. It can be seen that it does not follow a linear relation. The results showed in figure 10 prove the good behavior of the proposed algorithm, i.e. we perform correctly the blind inversion of the filter (channel). The recovered signal at the output of the system has the spectrum showed in figure 11. We can see how, although we have modified in a nonlinear manner the input signal in the preprocess stage, the spectrum matches the original because non-linear function $k(\cdot)$ preserve the structure of the

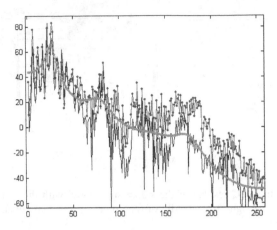

Fig. 11. Input signal spectrum (continuous line), output signal spectrum (dotted line) and LPC spectrum (Thick line). We can observe the similarity of these spectrums at low frequencies. The main difference is related to the high part of the spectrum of the reconstructed signal, which is due to the fact that the LPC reconstruction allocates resources to the parts of the spectrum with high energies.

residual (non-predictable part) of the signal. The difference between harmonic peaks of speech signal and background noise is about 40 dB.

Similar experiments have been done with music signals, which are also non i.i.d. signals and have strong correlations between samples due to the resonance character-istics of the instruments. In this case, similarly to the previous experiments, the results are quite good, using the same parameters (non-linear function k, LPC order, etc.).

6 Summary

In this paper we have presented a method for blind deconvolution of channel applied to real world signals. This method is based on source separation techniques, which can estimate the inverse of a channel if the original signal is non-Gaussian and i.i.d. In order to apply this result to real world signals we will need to preprocess the original signal using a whitening stage by means of an inverse LPC filter and applying after a non-linear function for des-*Gaussianity* the signal. Thereby we modify the density probability function without changing the non predictable part of the signal. In recep-tion, after the deconvolution of the signal, we need a post-processing stage by means of the inverse of the LPC filter in order to reconstruct the original signal.

For a future works we are studying other possibilities in order to apply source sepa-ration techniques to linear or non-linear channel deconvolution problems with real world signals. Our preliminary work indicated that we can effectively invert the nonlin-ear convolution by means of source separation techniques [10, 11] but for a non station-ary signal it is necessary to study how we can preprocess this signals to insure the i.i.d. and non Gaussian distribution necessary conditions to apply these techniques.

Acknowledgment

This work has been in part supported by the University of Vic under de grant R0912 and by the Spanish CICyT project ALIADO.

References

1. Amari, S., Cichocki A., Yang, H.H., "A new learning algorithm for blind signal separation", *NIPS 95*, MIT Press, 8, 1996
2. Bell, A.J., Sejnowski T.J., "An information-maximization approach to blind separation and blind deconvolution", *Neural Computation 7*, 1995.
3. S.A. Bellings, S.Y. Fakhouri. Identification of a class of nonlinear systems using correlation analysis. *Proc. IEEE*, 66 pp. 691-697 (1978)
4. E.D. Boer. Cross-correlation function of a bandpass nonlinear network. *Proc. IEEE*, 64 pp. 1443-1444 (1976)
5. Cardoso, J.P., "Source separation using higher order moments", in *Proc. ICASSP*, 1989.
6. Comon, P., "Independent component analysis, a new concept?", *Signal Processing* 36, 1994.
7. Comon, P., "Separation of sources using higher order cumulants" *Advanced Algorithms and Architectures for Signal Processing*, 1989.
8. Cover, T.M., Thomas, J.A., *Elements of Information Theory*. Wiley Series in Telecommunications, 1991.
9. G. Jacoviti, A. Neri, R. Cusani. Methods for estimating the autocorrelation function of complex stationary process. *IEEE Trans. ASSP*, 35, pp. 1126-1138 (1987)
10. Jutten, C., Hérault, J. ,"Blind separation of sources, Part I: An adaptive algorithm based on neuromimetic architecture", *Signal Processing* 24, 1991.
11. Nguyen Thi, H.L., Jutten, C., "Blind source separation for convolutive mixtures", *IEEE Transactions on Signal Processing*, vol. 45, 2, 1995.
12. C.L. Nikias, A.P. Petropulu. *Higher-Order Spectra Analysis – A Nonlinear Signal processing Framework*. Englewood Cliffs, NJ: Prentice-Hall (1993)
13. C.L. Nikias, M.R.Raghuveer. Bispectrum estimation: A digital signal processing framework. *Proc. IEEE*, 75 pp. 869-890 (1987)
14. S. Prakriya, D. Hatzinakos. Blind identification of LTI-ZMNL-LTI nonlinear channel models. *Biol. Cybern.*, 55 pp. 135-144 (1985)
15. Puntonet, C., Mansour A., Jutten C., "A geometrical algorithm for blind separation of sources", *GRETSI*, 1995
16. Taleb, A., Jutten, C., "Nonlinear source separation: the post-nonlinear mixtures", in *Proc. ESANN*, 1997
17. Taleb, A., Solé, J., Jutten, C., "Blind Inversion of Wiener Systems", in Proc. IWANN, 1999
18. Taleb, A., Solé, J., Jutten, C., "Quasi-Nonparametric Blind Inversion of Wiener Systems", *IEEE Transactions on Signal Processing*, 49, n°5, pp.917-924, 2001

Method for Real-Time Signal Processing Via Wavelet Transform

Pavel Rajmic

Department of Telecommunications,
Faculty of Electrical Engineering and Communication Technologies,
Brno University of Technology,
Purkyňova 118, 612 00 Brno, Czech Republic
rajmic@feec.vutbr.cz

Abstract. The new method of segmented wavelet transform (SegWT) makes it possible to compute the discrete-time wavelet transform of a signal segment-by-segment. This means that the method could be utilized for wavelet-type processing of a signal in "real time", or in case we need to process a long signal (not necessarily in real time), but there is insufficient computational memory capacity for it (for example in the signal processors). Then it is possible to process the signal part-by-part with low memory costs by the new method. The method is suitable also for the speech processing, e.g. denoising the speech signal via thresholding the wavelet coefficients or speech coding. In the paper, the principle of the segmented forward wavelet transform is explained and the algorithm is described in detail.

1 Introduction

There are a number of theoretical papers and practical applications of the wavelet transform. However, all of them approach the problem from such a point of view as if we knew the whole of the signal (no matter how long it is). Due to this assumption, we cannot perform the wavelet-type signal processing in real time in this sense. Of course there are real-time applications of the wavelet type, but, all of them utilize the principle of overlapping segments of the "windowed" signal (see for example [1]). In the reconstruction part of their algorithms they certainly introduce errors into the processing, because the segments are assembled using weighted averages.

Processing a signal in "real time" actually means processing it with minimum delay. A signal, which is not known in advance, usually comes to the input of a system piecewise, by mutually independent segments that have to be processed and, after the modification, sent to the output of the system. This is typically the case of processing audio signals, in particular speech signals in telecommunications.

The new method, the so-called segmented wavelet transform (SegWT[1]), enables this type of processing. It has a great potential application also in cases

---

[1] we introduce abbreviation SegWT (Segmented Wavelet Transform), because SWT is already reserved for stationary wavelet transform.

M. Faundez-Zanuy et al. (Eds.): NOLISP 2005, LNAI 3817, pp. 368–378, 2005.

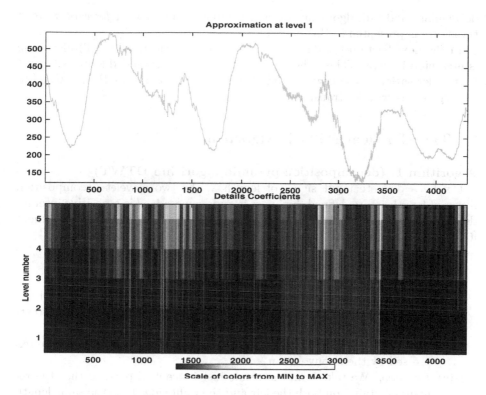

Fig. 1. Signal (top) and its scalogram (bottom). Scalogram is a type of graph representing the frequency contents of a signal in time. It is constructed from the wavelet coefficients.

when it is necessary to process a long signal off-line and no sufficient memory capacity is available. It is then possible to use this new method for equivalent segmentwise processing of the signal and thus save the storage space. In this sense the SegWT algorithm corresponds to overlap-add and overlap-save algorithms in Fourier-type linear filtering.

Another possible application of the SegWT algorithm is the instantaneous visualization of signal using an imaging technique referred to as "scalogram", see Fig. 1. The decomposition depth is $d = 5$ in this Figure. The bigger is the absolute value of the single coefficient, the lighter is the color saturation of the respective cell in the graph. In fact, plotting scalogram is a technique very similar to plotting a spectrogram in real time. In wavelet transformation there is an advantage in that the signal need not be weighted with windows, which results in a distortion of the frequency information, as is the case with the spectrogram. Moreover, there is one more good thing about it: a scalogram created by means of the SegWT is quite independent of the chosen length of segment.

In the available literature, this way of performing the wavelet transform is practically neglected, and this was the reason why our effort was devoted to developing modified algorithm. In fact, a modified method of *forward* wavelet transform is presented in this paper.

In the next Section the discrete-time wavelet transform (DTWT) algorithm is presented in detail. This is because we need to have extended knowledge of it for the derivation of the new method. The subsequent parts are then devoted to the segmented wavelet transform.

2   The Classical DTWT Algorithm

Algorithm 1. (decomposition pyramid algorithm DTWT)
Let $\mathbf{x}$ be a discrete input signal of length s, the two wavelet decomposition filters of length m are defined, highpass $\mathbf{g}$ and lowpass $\mathbf{h}$, d is a positive interger determining the decomposition depth. Also, the type of boundary treatment [6, ch. 8] must be known.

1. We denote the input signal $\mathbf{x}$ as $\mathbf{a}^{(0)}$ and set $j = 0$.
2. One decomposition step:
   (a) *Extending the input vector.* We extend $\mathbf{a}^{(j)}$ from both the left and the right side by $(m-1)$ samples, according to the type of boundary treatment.
   (b) *Filtering.* We filter the extended signal with filter $\mathbf{g}$, which can be expressed by their convolution.
   (c) *Cropping.* We take from the result just its central part, so that the remaining "tails" on both the left and the right sides have the same length $m-1$ samples.
   (d) *Downsampling (decimation).* We downsample this vector, e.g. leave just its even samples (supposing the vector is indexed beginning with 1).
   We denote the resulting vector $\mathbf{d}^{(j+1)}$ and store it.
   We repeat items (b)–(d), now with filter $\mathbf{h}$, denoting and storing the result as $\mathbf{a}^{(j+1)}$.
3. We increase j by one. If it now holds $j < d$, we return to item 2., in the other case the algorithm ends.

Remark. After algorithm 1 has been finished, we hold the wavelet coefficients stored in $d+1$ vectors $\mathbf{a}^{(d)}, \mathbf{d}^{(d)}, \mathbf{d}^{(d-1)}, \ldots, \mathbf{d}^{(1)}$.

One step of the wavelet decomposition principle can be seen in Figure 2.

3   The Method of Segmented Wavelet Transform

3.1   Motivation and Aim of the Method

Regularly used discrete-time wavelet transform (see Section 2) is suitable for processing signals "off-line", i.e. known before processing, even if very long. The task for the segmented wavelet transform, SegWT, is naturally to allow signal processing by its segments, so that in this manner we get the same result (same

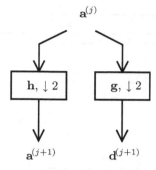

Fig. 2. One step of the wavelet decomposition. Approximate coefficients of the j-th level (vector $\mathbf{a}^{(j)}$) are decomposed into approximate and detail coefficients of level $j + 1$ via filtering with $\mathbf{h}$ and $\mathbf{g}$ and decimation (denoted $\downarrow 2$).

wavelet coefficients) as in the ordinary DTWT case. In this problem, the following parameters play a crucial role.

m wavelet filter length, $m > 0$,
d transform depth, $d > 0$,
s length of segment, $s > 0$.

The derivation of the SegWT algorithm requires a very detailed knowledge of the DTWT algorithm. Thanks to this it is possible to deduce fairly sophisticated rules how to handle the signal segments. We have found that in dependence on m, d, s, it is necessary to extend every segment from the left by an exact number of samples from the preceding segment and from the right by another number of samples from the subsequent segment. However, every segment has to be extended by a different length from the left and the right, and these lengths can also differ from segment to segment! Also the first and the last segments have to be handled in a particular way.

3.2 Important Theorems Derived from the DTWT Algorithm

Before we introduce detailed description of the SegWT algorithm, several theorems must be presented. More of them and their proofs can be found in [5, ch. 8]. We assume that the input signal $\mathbf{x}$ is divided into $S \geq 1$ segments of equal length s. Single segments will be denoted $^1\mathbf{x}, ^2\mathbf{x}, \ldots, ^S\mathbf{x}$. The last one can be of a length lower than s. See Fig. 3.

By the formulation that the *coefficients (or more properly two sets of coefficients) from the k-th decomposition level follow-up on each other* we mean a situation when two consecutive segments are properly extended see Figs. 3, 4, so that applying the DTWT[2] of depth k separately to both the segments and joining the resultant coefficients together lead to the same set of coefficients as computing it via the DTWT applied to the two segments joined first.

---

[2] With step 2(a) omitted.

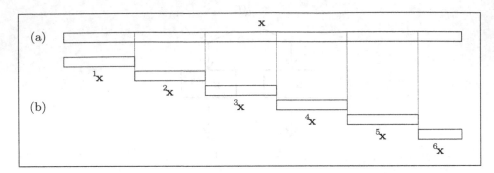

Fig. 3. Scheme of signal segmentation. The input signal $\mathbf{x}$ (a) is divided into segments of equal length, the last one can be shorter than this (b); the n-th segment of $\mathbf{x}$ is denoted by $^n\mathbf{x}$.

Theorem 1. *In case that the consecutive segments have*

$$r(k) = (2^k - 1)(m - 1) \tag{1}$$

common input signal samples, the coefficients from the k-th decomposition level follow-up on each other.

Thus, for a decomposition depth equal to d it is necessary to have $r(d) = (2^d - 1)(m - 1)$ common samples in the two consecutive extended segments.

The aim of the following part of the text is to find the proper extension of every two consecutive signal segments. We will show that the length of such extension must comply with the strict rules derived from the theorems below.

The extension of a pair of consecutive segments, which is of total length $r(d)$, can be divided into the right extension of the first segment (of length P) and the left extension of the following segment (of length L), while $r(d) = P + L$. However, the lengths $L \geq 0, P \geq 0$ cannot be chosen arbitrarily. The lengths L, P are not uniquely determined in general. The formula for the choice of extension $L_{\max}$, which is unique and the most appropriate in case of real-time signal processing, is given in Theorem 2.

Theorem 2. *Let a segment be given whose length including its left extension is l. The maximal possible left extension of the next segment, $L_{\max}$, can be computed by the formula*

$$L_{\max} = l - 2^d \ \mathrm{ceil}\left(\frac{l - r(d)}{2^d}\right). \tag{2}$$

The minimal possible right extension of the given segment is then naturally

$$P_{\min} = r(d) - L_{\max}. \tag{3}$$

For the purposes of the following text, it will be convenient to assign the number of the respective segment to the variables $L_{\max}, P_{\min}, l$, i.e. the left extension of the n-th segment will be of length $L_{\max}(n)$, the right extension

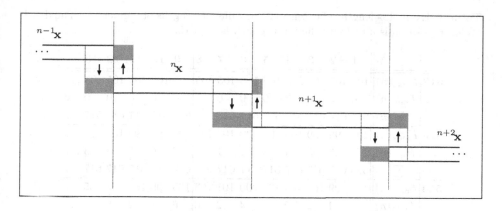

Fig. 4. Illustration of extending of the segments

will be of length $P_{\min}(n)$ and the length of the original n-th segment with the left extension joined will be denoted $l(n)$. Using this notation we can rewrite equation (3) as

$$P_{\min}(n) = r(d) - L_{\max}(n+1). \tag{4}$$

Let us now comment on the special situation of the first or the last segment. These naturally represent the "boundaries" of the signal. The discrete-time wavelet transform uses several modes how to treat the boundaries and we must preserve these modes also in our modified algorithm. Therefore we must treat the first and the last segment separately and a bit differently from the other segments. For details and proofs we again refer to [5]. The appropriate procedure is to extend the first segment from the left by $r(d)$ zero samples, i.e. $L_{\max}(1) = r(d)$, and to process it using Algorithm 4. Similarly the last segment has to be extended by $r(d)$ zeros from the right and processed using Algorithm 5.

Theorem 3. *The length of the right extension of the n-th segment, $n = 1, 2, \ldots,$ $S - 2$, must comply with*

$$P_{\min}(n) = 2^d \operatorname{ceil}\left(\frac{ns}{2^d}\right) - ns, \tag{5}$$

and the length of the left extension of the $(n+1)$-th segment is $L_{\max}(n+1) = r(d) - P_{\min}(n)$.

Remark. From (5) it is clear that $P_{\min}$ is periodic with respect to s with period 2^d, i.e. $P_{\min}(n + 2^d) = P_{\min}(n)$.

This relation and also some more can be seen in Table 1.

Theorem 4. *(on the total length of segment)*
After the extension the n-th segment (of original length s) will be of total length

$$\sum(n) = r(d) + 2^d \left[\operatorname{ceil}\left(\frac{ns}{2^d}\right) - \operatorname{ceil}\left(\frac{(n-1)s}{2^d}\right)\right]. \tag{6}$$

Table 1. Example - lengths of extensions for different lengths of segments s. The depth of decomposition is $d = 3$ and the filter length is $m = 16$.

s	n	1	2	3	4	5	6	7	8	9	10	11	12	13 ...
512	$L_{\max}(n)$	105	105	105	105	105	105	105	105	105	105	105	105	105 ...
	$P_{\min}(n)$	0	0	0	0	0	0	0	0	0	0	0	0	0 ...
	$\sum(n)$	617	617	617	617	617	617	617	617	617	617	617	617	617 ...
513	$L_{\max}(n)$	105	98	99	100	101	102	103	104	105	98	99	100	101 ...
	$P_{\min}(n)$	7	6	5	4	3	2	1	0	7	6	5	4	3 ...
	$\sum(n)$	625	617	617	617	617	617	617	617	625	617	617	617	617 ...
514	$L_{\max}(n)$	105	99	101	103	105	99	101	103	105	99	101	103	105 ...
	$P_{\min}(n)$	6	4	2	0	6	4	2	0	6	4	2	0	6 ...
	$\sum(n)$	625	617	617	617	625	617	617	617	625	617	617	617	625 ...
515	$L_{\max}(n)$	105	100	103	98	101	104	99	102	105	100	103	98	101 ...
	$P_{\min}(n)$	5	2	7	4	1	6	3	0	5	2	7	4	1 ...
	$\sum(n)$	625	617	625	617	617	625	617	617	625	617	625	617	617 ...
516	$L_{\max}(n)$	105	101	105	101	105	101	105	101	105	101	105	101	105 ...
	$P_{\min}(n)$	4	0	4	0	4	0	4	0	4	0	4	0	4 ...
	$\sum(n)$	625	617	625	617	625	617	625	617	625	617	625	617	625 ...
517	$L_{\max}(n)$	105	102	99	104	101	98	103	100	105	102	99	104	101 ...
	$P_{\min}(n)$	3	6	1	4	7	2	5	0	3	6	1	4	7 ...
	$\sum(n)$	625	625	617	625	625	617	625	617	625	625	617	625	625 ...
518	$L_{\max}(n)$	105	103	101	99	105	103	101	99	105	103	101	99	105 ...
	$P_{\min}(n)$	2	4	6	0	2	4	6	0	2	4	6	0	2 ...
	$\sum(n)$	625	625	625	617	625	625	625	617	625	625	625	617	625 ...
519	$L_{\max}(n)$	105	104	103	102	101	100	99	98	105	104	103	102	101 ...
	$P_{\min}(n)$	1	2	3	4	5	6	7	0	1	2	3	4	5 ...
	$\sum(n)$	625	625	625	625	625	625	625	617	625	625	625	625	625 ...
520	$L_{\max}(n)$	105	105	105	105	105	105	105	105	105	105	105	105	105 ...
	$P_{\min}(n)$	0	0	0	0	0	0	0	0	0	0	0	0	0 ...
	$\sum(n)$	625	625	625	625	625	625	625	625	625	625	625	625	625 ...
$\vdots$	$\vdots$	$\vdots$	$\vdots$	$\vdots$	$\vdots$	$\vdots$	$\vdots$	$\vdots$	$\vdots$	$\vdots$	$\vdots$	$\vdots$	$\vdots$	$\ddots$

This expression can acquire only one of two values, either

$$r(d) + 2^d \operatorname{ceil}\left(\frac{s}{2^d}\right) \quad or \quad r(d) + 2^d \operatorname{ceil}\left(\frac{s}{2^d}\right) - 2^d. \tag{7}$$

The relations presented in this Theorem are apparent in Table 1.

3.3 The Algorithm of Segmented Wavelet Transform

The algorithm SegWT (Algorithm 2 below) works such that it reads (receives) single segments of the input signal, then it extends – overlaps them in a proper way, then it computes the wavelet coefficients in a modified way and, in the end, it easily joins the coefficients.

Algorithm 2. Let the wavelet filters $\mathbf{g}, \mathbf{h}$ of length m, the decomposition depth d, and the boundary treatment mode be given. The segments of length $s > 0$ of the input signal $\mathbf{x}$ are denoted $^1\mathbf{x}, ^2\mathbf{x}, ^3\mathbf{x}, \ldots$. The last segment can be shorter than s.

1. Set $N = 1$.
2. Read the first segment, $^1\mathbf{x}$, and label it "current". Extend it from the left by $r(d)$ zero samples.
3. *If* the first segment is at the same time the last one
   (a) It is the case of regular wavelet transform. Compute the DTWT of this single segment using Algorithm 1.
   (b) The Algorithm ends.
4. Read $(N + 1)$-th segment and label it "next".
5. *If* this segment is the last one
   (a) Join the current and next segment together and label it "current". (The current segment becomes the last one now.)
   (b) Extend the current vector from the right by $r(d)$ zero samples.
   (c) Compute the DTWT of depth d from the extended current segment using Algorithm 5.
   Otherwise
   (d) Compute $L_{\max}$ for the next segment and $P_{\min}$ for the current segment (see Theorem 2).
   (e) Extend the current segment from the right by $P_{\min}$ samples taken from the next segment. Extend the next segment from the left by $L_{\max}$ samples taken from the current segment.
   (f) *If* the current segment is the first one, compute the DTWT of depth d from the extended current segment using Algorithm 4. *Otherwise* compute the DTWT of depth d from the extended current segment using Algorithm 3.
6. Modify the vectors containing the wavelet coefficients by trimming off a certain number of redundant coefficients from the left side, specifically: at the k-th level, $k = 1, 2, \ldots, d - 1$, trim off $r(d - k)$ coefficients from the left.
7. *If* the current segment is the last one, then in the same manner as in the last item trim the redundant coefficients, this time from the right.
8. Store the result as $^N\mathbf{a}^{(d)}, ^N\mathbf{d}^{(d)}, ^N\mathbf{d}^{(d-1)}, \ldots, ^N\mathbf{d}^{(1)}$.
9. *If* the current segment is not the last one
   (a) Label the next segment "current".
   (b) Increase N by 1 and go to item 4.

Remark. If the input signal has been divided into $S > 1$ segments, then $(S - 1)(d + 1)$ vectors of wavelet coefficients

$$\{ ^i\mathbf{a}^{(d)}, \ ^i\mathbf{d}^{(d)}, \ ^i\mathbf{d}^{(d-1)}, \ldots, ^i\mathbf{d}^{(1)} \}_{i=1}^{S-1}.$$

are the output of the Algorithm. If we join these vectors together in a simple way, we obtain a set of $d+1$ vectors, which are identical with the wavelet coefficients of signal $\mathbf{x}$.

Next we present the "subalgorithms" of the SegWT method.

Algorithm 3. This algorithm is identical with Algorithm 1 with the exception that we omit step 2(a), i.e. we do not extend the vector.

The next two algorithms serve to process the first and the last segment.

Algorithm 4. This algorithm is identical with Algorithm 1 with the exception that we replace step 2(a) by the step:

Modify the coefficients of vector $\mathbf{a}^{(j)}$ on positions $r(d-j) - m + 2, \ldots,$ $r(d-j)$, as it corresponds to the given boundary treatment mode.

Algorithm 5. This algorithm is identical with Algorithm 1 with the exception that we replace step 2(a) by the step:

Modify the coefficients of vector $\mathbf{a}^{(j)}$ on positions $r(d-j) - m + 2, \ldots,$ $r(d-j)$, as it corresponds to the given boundary treatment mode, however this time taken from the right side of $\mathbf{a}^{(j)}$.

3.4   Corollaries and Limitations of the SegWT Algorithm

In this part of the text we will derive how many coefficients we are able to compute from each segment with SegWT. The minimum length of a segment will also be derived.

Theorem 5. *Let the decomposition depth d be given and let $^{\widetilde{n}}\mathbf{x}$ be the extension of the n-th segment $^n\mathbf{x}$ of original length s. Then we will compute*

$$
q_{\max}(n) = \begin{cases} \operatorname{ceil}\left(\dfrac{s}{2^d}\right) & \text{for} \quad \sum(n) = r(d) + 2^d \operatorname{ceil}\left(\dfrac{s}{2^d}\right) \\ \operatorname{ceil}\left(\dfrac{s}{2^d}\right) - 1 & \text{for} \quad \sum(n) = r(d) + 2^d \operatorname{ceil}\left(\dfrac{s}{2^d}\right) - 2^d \end{cases} \tag{8}
$$

wavelet coefficients at level d from $\mathbf{x}$.

Corollary 1. *(the minimum length of a segment)*
Let the decomposition depth d be given. Assuming $S > 2^d + 1$, the length of the original segment, s, must satisfy the condition $s > 2^d$.

It is clear from the description that the time lag of Algorithm 2 is one segment (i.e. s samples) plus the time needed for the computation of the coefficient from the current segment. In a special case when s is divisible by 2^d it holds even $P_{\min}(n) = 0$ for every $n \in \mathbb{N}$ (see Theorem 3), i.e. the lag is determined only by the computation time!

3.5   A Few Examples

Now we present a few examples of SegWT performance which follow from the above Theorems.

Example 1. For $d = 4$ and $m = 12$, the minimum segment length is just 16 samples. When we set $s = 256$, $P_{\min}$ will always be zero and $L_{\max} = r(4) = 165$. The length of every extended segment will be $256 + 165 = 421$ samples.

Example 2. For $d = 5$ and $m = 8$, the minimum segment length is 32 samples. When we set $s = 256$, $P_{\min}$ will always be zero and $L_{\max} = r(5) = 217$. The length of every extended segment will be $256 + 217 = 473$ samples.

Example 3. For $d = 5$ and $m = 8$ we set $s = 300$, which is not divisible by 2^5. Thus $P_{\min}$ and $L_{\max}$ will alternate such that $0 \leq P_{\min} \leq 31$ and $186 \leq L_{\max} \leq 217$. The length of every extended segment will be $256 + r(5) = 473$ samples.

4   Conclusion

The paper contains a description of the algorithm which allows us to perform the wavelet transform in real time. The algorithm works on the basis of calculating the optimal extension (overlap) of signal segments, and subsequent performance of the modified transform.

In the future it would be convenient to improve the computational effectivity by reducing redundant computations at the borders of the segments, as it follows from the Algorithm 2. There is also a chance to generalize the SegWT method to include biorthogonal wavelets and more general types of decimation [2,4].

Another important part of the future work is the derivation of an efficient counterpart to the introduced method – the segmented *inverse* transform. In fact, we made first experience with such development. It turned out that the algorithm will have to be quite complicated and, above all, that the time lag in the consecutive forward-inverse processing will be, unfortunately, always nonzero.

Acknowledgements. The paper was prepared within the framework of No. 102/04/1097 and No. 102/03/0762 projects of the Grant Agency of the Czech Republic and COST Project No. OC277.

References

1. Darlington, D., Daudet, L., Sandler, M.: Digital Audio Effects in the Wavelet Domain. In *Proc. of the 5th Int. Conf. on Digital Audio Effects (DAFX-02)*, Hamburg (2002)
2. Dutilleux, P.: An implementation of the "algorithme à trous" to compute the wavelet transform. In *Wavelets: Time-Frequency Methods and Phase Space*, Inverse Problems and Theoretical Imaging, editors J.-M. Combes, A. Grossman, P. Tchamitchian. pp. 298–304, Springer-Verlag, Berlin (1989)

3. Mallat, S.: *A Wavelet Tour of Signal Processing.* 2nd edition, Academic Press (1999) ISBN 0-12-466606-X
4. Nason, G.P., Silverman, B.W.: The stationery wavelet transform and some statistical applications. In *Wavelets and Statistics*, volume 103 of *Lecture Notes in Statistics*, editors A. Antoniadis, G. Oppenheim, pp. 281–300, Springer-Verlag, New York (1995)
5. Rajmic, P.: *Využití waveletové transformace a matematické statistiky pro separaci signálu a šumu (Exploitation of the wavelet transform and mathematical statistics for separation signals and noise, in Czech)*, PhD Thesis, Brno University of Technology, Brno (2004)
6. Strang, G., Nguyen, T.: *Wavelets and Filter Banks.* Wellesley Cambridge Press (1996)
7. Wickerhauser, M. V.: *Adapted Wavelet Analysis from Theory to Software.* IEEE Press, A K Peters, Ltd. (1994)

Author Index

Lecture Notes in Artificial Intelligence (LNAI)